T0360399

Differential Geometry of
Curves and Surfaces
with Singularities

Series in Algebraic and Differential Geometry

Series Editor: Phillip A Griffiths *(Institute for Advanced Study, Princeton, USA)*

Published

Vol. 1   *Differential Geometry of Curves and Surfaces with Singularities*
by Masaaki Umehara, Kentaro Saji and Kotaro Yamada
translated by Wayne Rossman

Series in
Algebraic and
Differential
Geometry
Volume 1

Differential Geometry of Curves and Surfaces with Singularities

Masaaki Umehara
Tokyo Institute of Technology, Japan

Kentaro Saji
Kobe University, Japan

Kotaro Yamada
Tokyo Institute of Technology, Japan

Translated by:

Wayne Rossman
Kobe University, Japan

World Scientific

NEW JERSEY · LONDON · SINGAPORE · BEIJING · SHANGHAI · HONG KONG · TAIPEI · CHENNAI · TOKYO

Published by

World Scientific Publishing Co. Pte. Ltd.

5 Toh Tuck Link, Singapore 596224

USA office: 27 Warren Street, Suite 401-402, Hackensack, NJ 07601

UK office: 57 Shelton Street, Covent Garden, London WC2H 9HE

Library of Congress Cataloging-in-Publication Data

Names: Umehara, Masaaki, author. | Saji, Kentaro, author. | Yamada, Kotaro, 1961–   author. |
    Rossman, Wayne, 1965–   translator.
Title: Differential geometry of curves and surfaces with singularities / Masaaki Umehara,
    Tokyo Institute of Technology, Japan; Kentaro Saji, Kobe University, Japan; Kotaro Yamada,
    Tokyo Institute of Technology, Japan ; translated by Wayne Rossman, Kobe University, Japan.
Other titles: Tokuiten wo motsu kyokusen to kyokumen no bibunkikagaku. English.
Description: New Jersey : World Scientific, [2022] | Series: Series in algebraic and differential
    geometry ; volume 1 | Originally published in Japan in 2017 by Maruzen Publishing Co., Ltd,
    as Tokuiten Wo Motsu Kyokusen To Kyokumen No Bibunkikagaku. |
    Includes bibliographical references and index.
Identifiers: LCCN 2021012079 | ISBN 9789811237133 (hardcover) |
    ISBN 9789811237140 (ebook for institutions) | ISBN 9789811237157 (ebook for individuals)
Subjects: LCSH: Geometry, Differential. | Curves on surfaces. | Singularities (Mathematics)
Classification: LCC QA641 .U4413 2022 | DDC 516.3/6--dc23
LC record available at https://lccn.loc.gov/2021012079

British Library Cataloguing-in-Publication Data
A catalogue record for this book is available from the British Library.

For any available supplementary material, please visit
https://www.worldscientific.com/worldscibooks/10.1142/12284#t=suppl

Desk Editors: Jayanthi Muthuswamy/Lai Fun Kwong

Typeset by Stallion Press
Email: enquiries@stallionpress.com

Printed in Singapore

Preface

Addressing the need for a differential geometric treatment of singularity theory, this textbook began as a renewal of the first author's lecture note [82] in 2009 at Keio University, found in the first four chapters and Chapter 10 here, and subsequently other chapters were newly produced.

Several textbooks on singularity theory with applications to differential geometry have already been written. However, those other texts do not treat singularities within the framework of differential geometry, in spite of the fact that when we look at examples of curves and surfaces, singular points are often prominent. To discuss these singular points from within the viewpoint of differential geometry, we need a more careful treatment than is usually given for regular points, and for this reason, they have often been ignored. The purpose of this textbook, rather, is to focus on singular points on curves and surfaces as a part of differential geometry, and allow the reader to appreciate how handling singular points is not as difficult as one might first think.

Singular points do generally appear on differentiable maps between manifolds. In this textbook, we restrict our attention mainly to singularities on curves and surfaces, which allows us to visualize singular points, to help them become familiar objects to the reader. We will introduce several important types of singular points, give useful criteria for determining them, and provide applications to the topology and geometry of curves and surfaces.

We now describe how this textbook is structured, chapter by chapter:

In Chapter 1, curves with singular points are discussed. We first introduce several important types of singular points on curves. We give a criterion for cusp singularities without proof, and introduce cuspidal curvature at cusps using that criterion. After that we consider planar curves admitting cusp singularities as (one-dimensional) wave fronts, and discuss the topology of closed curves with cusps.

In Chapter 2, we give a quick review of elementary surface theory, with referencing, where useful, to the first and third authors' textbook on curves and surfaces [84]. We then give a more general treatment of surfaces, by regarding them as (now two-dimensional) wave fronts, analogous to what we did in Chapter 1 for the case of planar curves. After that, we introduce cuspidal edges and swallowtails as typical types of singular points appearing on wave fronts. Cross caps and cuspidal cross caps are also introduced, and useful criteria for all four of these singularities are given. The reader who completes a study of the first two chapters will have quite a robust understanding of what singularities on curves and surfaces are, from the viewpoint of differential geometry.

In Chapter 3, we prove criteria for the cusps, cross caps and cuspidal edges introduced in the first two chapters. We will wait to prove the criterion for swallowtails until Chapter 8, since it requires that we first introduce the concept of "unfolding".

In Chapter 4, we give variants of criteria for cuspidal edges and swallowtails, which have useful applications. At the end of this chapter, we give a brief introduction to two Gauss–Bonnet type formulas for closed surfaces, generalized to wave fronts, without relying on a knowledge of manifold theory.

To venture beyond Chapter 4, the reader will need a knowledge of manifold theory.

In Chapter 5, singular curvature functions and limiting normal curvature functions along cuspidal edges are introduced, which will play important roles in later chapters. The singular curvatures are used to formulate one of the two Gauss–Bonnet type formulas introduced in Chapter 4. Moreover, we show how the limiting normal curvatures are also related to the behavior of Gaussian curvature near cuspidal edges.

In Chapter 6, we prove the two Gauss–Bonnet type formulas introduced at the end of Chapter 4. These are formulated not only for closed two-dimensional wave fronts in the Euclidean 3-space, but also for their Gauss maps. Because of this, surprisingly, we obtain four genuinely distinct Gauss–Bonnet type formulas for closed surfaces with singularities. If the surfaces have no singular points, then the first two formulas reduce to the classical Gauss–Bonnet formula. But, even then, the other two Gauss–Bonnet formulas for the Gauss maps remain separate from the classical one. As a result, we can give various important applications even for closed surfaces without singular points.

In Chapter 7, we discuss the singular sets of flat surfaces (i.e., surfaces with vanishing Gaussian curvature), and prove several important properties of such surfaces. We also give a global treatment of them in terms of notions of completeness for such surfaces.

In Chapter 8, we prove the criteria for swallowtails given in Chapters 2 and 4 by applying the theory of (uni-)versal unfolding, which is a well-known tool used in singularity theory. The material in Appendix E is applied in the proof of the criterion for swallowtails. Using that material, we can give an alternate proof of the criteria for cusps and cuspidal edges as well.

In Chapter 9, we remove our dependence on the ambient space, and describe our two Gauss–Bonnet type formulas from the viewpoint of manifold theory. For this purpose, we introduce the concept of coherent tangent bundles on manifolds. In this chapter, singular points on a surface are treated as the singular set of a certain vector bundle homomorphism. This abstract treatment of singularities enables us to apply our Gauss–Bonnet type formulas to surfaces in a variety of ambient spaces other than Euclidean 3-space, which is quite useful.

Chapter 10 is a brief review of wave fronts as hypersurfaces in manifolds, which will be of help to the reader who wishes to understand wave fronts from the viewpoint of contact geometry.

This book ends with a number of appendices.

The contents of this textbook are related to the textbook on curves and surfaces written by the first and third authors [84]. In fact, at places where the reader might need to recall concepts and facts from elementary curve and surface theory, that textbook is frequently referred to, and in this sense this book is a continuation of that previous one.

Finally, the authors wish to express their gratitude to Yukio Matsumoto, who encouraged them to write this book, and they heartily thank Go-o Ishikawa, Sumio Yamada, Kosuke Naokawa, Atufumi Honda, Masatoshi Kokubu and Shoichi Fujimori for valuable comments. The authors also give their thanks to the publisher Ichiro Misaki (at Maruzen) for handling the book production procedure.

Preface to the English Edition

This book is an English translation of the Japanese edition of our text on the differential geometry of curves and surfaces with singularities published by Maruzen Co. Ltd. We the authors hope readers will find this text to be a unique introduction to the singularities of curves and surfaces from the viewpoint of differential geometry, and hope it will be useful for students and researchers interested in this subject.

The contents in Chapters 1–4 can be read without knowledge of differential manifolds. To read Chapter 2, fundamental material of surface theory in the Euclidean 3-space (first fundamental form, second fundamental form, Gaussian curvature, mean curvature and principal curvatures) is explained without proofs. However, in all such places, we reference proofs from the first and third authors' book [84], and so non-experts of surface theory can read this text at least up through Chapter 4, using [84] as a supplementary document.

On the other hand, the contents in Chapters 5–10 are written for readers having knowledge of manifolds. In fact, readers who are familiar with manifolds can easily read these chapters, except for Chapter 10, aimed at researchers who intend to investigate in this research area. We remark that cross caps are the most generic singular points appearing in smooth maps from 2-manifolds to $\boldsymbol{R}^3$, but they were not treated in the original Japanese version. In this English translation, we newly add Chapter 10 with the general theory of wave fronts as hypersurfaces in Riemannian manifolds, which was an appendix of the original Japanese version of this text.

There are eight appendices in total. Appendix A presents the division lemmas on smooth functions, and Appendix B explains further properties of cusps, as a continuation of Chapter 1. Appendix C gives a proof of the criterion of 4/3-cusps stated in Chapter 1. Appendix D gives a proof of the criterion of Whitney cusps stated in Chapter 4, and this appendix is not in the original Japanese version, and rather is newly featured in this book. Appendix E is devoted to proving Zakalyukin's lemma, which is applied to prove the criterion for swallowtails in Chapter 8.

Finally, the authors are grateful to Joseph Cho, Atsushi Fujioka, Atsufumi Honda, Shunsuke Ichiki, Keisuke Teramoto, Ryosuke Kinoshita and Masahiro Yamanaka for careful readings and valuable comments.

About the Authors

Masaaki Umehara began his research career as an undergraduate at Keio University and received his doctorate from University of Tsukuba. He initially worked on complex geometry and the differential geometry of submanifolds, later extending into the study of singularities from the perspective of differential geometry. For much of his career, he has been conducting joint research with Kotaro Yamada. He is presently a full professor at Tokyo Institute of Technology.

Kentaro Saji met Masaaki Umehara and Kotaro Yamada while writing his doctoral thesis on singularity theory at Hiroshima University, and the three have been conducting joint research ever since. In addition to singularity theory, his research interests include applications of that theory to the differential geometry of surfaces. He is a full professor at Kobe University, and he shares with his coauthors here a dedication to enhancing accessibility of this research field, as exemplified by the present textbook.

 **Kotaro Yamada**, starting his career as an undergraduate at Keio University, met Masaaki Umehara there and began a research collaboration with him that has spanned three decades. Kotaro continued his studies at Keio University and received his doctorate there. His research interests include differential geometry — of curves and surfaces in particular — and singularity theory. Presently he is at Tokyo Institute of Technology as a full professor.

About the Translator

Wayne Rossman, upon receiving a Ph.D. from the University of Massachusetts (Amherst), began his mathematics career in Japan. Now a full professor at Kobe University, he has been conducting joint research with Masaaki Umehara and Kotaro Yamada for the better part of three decades, and with Kentaro Saji for two decades. His research interests are surface theory and discrete differential geometry.

Contents

Chapter 1

Planar Curves and Singular Points

A regular planar curve is a curve without singular points. We will regard it as a (one-dimensional) wave front, and consider the family of parallel curves as a time evolution of the given curve. The most common type of singularity that one sees in these parallel curves is what is called a *cusp* singularity. More so than with the case of surfaces, the planar curves, whether described via equations or described geometrically, are remarkably accessible objects — and because of this, this chapter in particular is essentially self-contained upon reading.

1.1. Singular Points of Planar Curves

Planar curves. *Planar curves* are C^∞ class maps from an interval[1] I to the plane $\mathbf{R}^2$, i.e.,

$$\gamma(t) = (x(t), y(t)) \quad (t \in I).$$

In this text, we refer to a map of C^∞ class simply as a C^∞-map, or a smooth map. The single independent variable t is called a *parameter* for the curve. We regard vectors as column vectors, although we will write them as horizontal vectors, $\gamma(t) = (x(t), y(t))$, when they appear within text lines. Regarding vectors as column vectors (that is, as 2×1-matrices), we can apply the transposition operator, and thus we can write

$$\gamma(t) = (x(t), y(t))^T = \begin{pmatrix} x(t) \\ y(t) \end{pmatrix}.$$

---

[1] When I is not an open interval, we say that γ is a map of C^∞ class if I is contained in a larger open interval so that γ can be extended to a C^∞ class map on that larger interval.

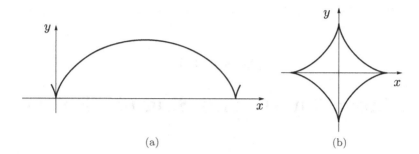

Fig. 1.1.   Cycloid (a) and astroid ((b), see Exercise **1** in this section).

In this text, except in some exceptional cases, we will write the derivative with respect to t with a superscript dash, as in $\gamma' = d\gamma/dt$. Then $\gamma'(t)$ becomes the velocity vector of $\gamma(t)$. At a point $t = c$ where $\gamma'(c)$ is non-zero, we say that the curve has a *regular point*, and a point $t = c$ where $\gamma'(c)$ is zero is called a *singular point*. We write the zero vector as **0**. When $t = c$ is a singular point, we can also refer to the image point $\gamma(c)$ as a singular point as well. Curves without singular points are called *regular curves*.

Example 1.1.1. Taking a circle of radius $a > 0$ and rolling it along a straight horizontal line like a wheel along a road, the path traced out by a given fixed point on the circle is called a *cycloid* (see Fig. 1.1(a)). That path is a curve that can be parametrized as

$$\gamma(t) := a(t - \sin t, 1 - \cos t) \quad (a > 0) \tag{1.1}$$

with derivative $\gamma'(t) = a(1 - \cos t, \sin t)$, and thus one sees that the singular points occur when $(\cos t, \sin t) = (1, 0)$, that is, where $t \in 2\pi\mathbf{Z}$ (here $\mathbf{Z}$ represents the set of all integers). From this we can see how singular points can arise on curves obtained from smooth maps, looking like they do in Fig. 1.1.

Example 1.1.2. For a positive constant a, the solution set of the equation

$$(x^2 + y^2)(x^2 + y^2 - 2ax) - a^2 y^2 = 0$$

in the xy-plane gives a curve called the *cardioid*, which can be parametrized by

$$\gamma(t) := a(1 + \cos t)(\cos t, \sin t) \quad (0 \le t < 2\pi). \tag{1.2}$$

This curve has a singular point at $t = \pi$ (see the central figure in Fig. 1.19). More generally, taking a circle of fixed radius a, and rolling a separate circle

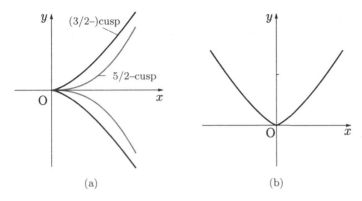

Fig. 1.2. A cusp and a 5/2-cusp ((a), where the inner curve is the 5/2-cusp) and a 4/3-cusp (b).

of the same radius along the outside of it, a fixed point on the rolling circle traces out a curve called an *epicycloid*. The cardioid is a special case of an epicycloid.

The singular points on the curves in Examples 1.1.1 and 1.1.2 are all called *cusp points*, or rather, simply *cusps* (see Section 1.3). However, other types of singularities are possible in general. For example,

$$\gamma(t) := (t^2, t^m) \quad (m = 3, 5, 7, \ldots) \tag{1.3}$$

has a singularity at $t = 0$. Each different value of m gives a different type of singularity. In particular, when $m = 3$, we have the *standard cusp*. When $m = 5$, we have what is called the *standard 5/2-cusp*, which makes a narrower spike (see Fig. 1.2(a)). Also,

$$\gamma(t) := (t^3, t^4) \tag{1.4}$$

has a singular point at $t = 0$, called the *standard 4/3-cusp* (see Fig. 1.2(b)), which is the image of the graph of the C^1-function

$$y = \sqrt[3]{x^4}. \tag{1.5}$$

By the phrase "singular point of a curve", we mean a singular point of a map parametrizing that curve.

Example 1.1.3. The image curve of the parametrization $\gamma_0(t) := (t^3, 0)$ is a straight line with a singular point at the origin in the plane. However, of course a straight line can be parametrized so that it becomes a regular curve, for example by $\gamma_1(t) := (t, 0)$.

Changing parameters for a curve. For a parametrization $\gamma(t)$ ($a \leq t \leq b$) of a curve, we could choose a different parametrization. We now give a proper definition for parameter change: On an interval $[c, d]$ with a C^∞-function $\varphi : [c, d] \to [a, b]$ defined on it so that $\varphi(c) = a$, $\varphi(d) = b$ and

$$\frac{d\varphi(s)}{ds} > 0 \quad (c \leq s \leq d), \tag{1.6}$$

we call φ an *orientation preserving parameter change* or *oriented coordinate transformation* from the interval $[a, b]$ to the interval $[c, d]$. With this, the new planar curve

$$\tilde{\gamma}(s) := \gamma(\varphi(s))$$

has the same image as the original planar curve, and also is parametrized in the same direction. Thus, we can regard both the original curve $\gamma(s)$ and the new curve $\tilde{\gamma}(s)$ as the same, but the new curve has parameter s rather than t, and the new curve is obtained from the original one by applying the coordinate transformation. In this text, the phrase "parameter change for a curve" will refer to this type of oriented coordinate transformation.

Remark 1.1.4. In Example 1.1.3, the maps γ_0 and γ_1 could be imagined as reparametrizations of each other via the relation $\gamma_0(t) = \gamma_1(t^3)$. However, this map $t \mapsto t^3$ does not satisfy the condition (1.6) required of a coordinate transformation, thus these two curves are not equivalent up to reparametrization.

Exercises 1.1

1   For a fixed positive constant a, the implicit function $\sqrt[3]{x^2} + \sqrt[3]{y^2} = \sqrt[3]{a^2}$ in the plane $\mathbf{R}^2$ can be parametrized by

$$\gamma(t) := a(\cos^3 t, \sin^3 t) \quad (0 \leq t < 2\pi) \tag{1.7}$$

and this curve[2] is called an astroid (see Fig. 1.1, (b)). Show that this curve has singular points at $t = 0$, $\pi/2$, π, $3\pi/2$.

2   Imagining the line segment in the y-axis of the plane from $(0, 0)$ to $(0, 1)$ to be a baton, and letting its upper point slide down the y-axis while its lower point moves out along the positive x-axis, show that the baton at all moments

---

[2]This astroid can be constructed by rolling a circle of radius $a/4$ along the inside of a non-moving circle of radius a and taking the path traced out by a fixed point on the rolling circle. Thus, the astroid is one example of a hypocycloid.

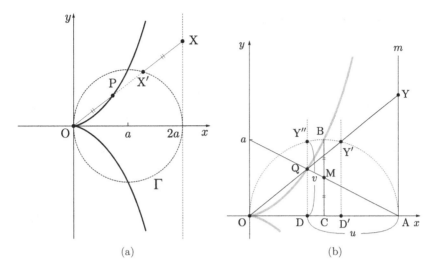

(a) (b)

Fig. 1.3. Cissoid (a) and construction for the doubling of the cube problem (b).

is enveloping[3] (is tangent to) the portion of the previously considered astroid from $t = 0$ to $t = \pi/2$ (in the case that $a = 1$).

3 For a positive number a, the implicit function determining the *cissoid* curve in the plane is $x^3 + xy^2 - 2ay^2 = 0$. The cissoid was introduced to solve the construction problem of doubling the cube, by Diocles (see Fig. 1.3).

(1) Let a line m be the tangent line at $A := (2a, 0)$ of the circle Γ centered at $C := (a, 0)$ with radius a. Let $X = (2a, t)$ be a point on m and let X' be the intersection point of the circle Γ with the line segment OX. Let $P = (x, y)$ be a point on OX such that $|OP| = |X'X|$, where $|OP|$ is the length of the line segment OP. Then show

$$x = \frac{2at^2}{4a^2 + t^2}, \quad y = \frac{t^3}{4a^2 + t^2}, \tag{1.8}$$

and show P lies on the cissoid.

(2) Show $t = 0$ is the only singular point of the parametrization (1.8).

(3) We set $B = (a, a)$ on the cissoid, and let M be the midpoint between B and C. Let Q be the intersection point of the line AM with the cissoid, and let Y'' (respectively, D) be the intersection point between Γ (respectively,

---

[3]In general, for a family of curves $\{C_t\}_{t \in [a,b]}$, when there exists a curve σ that is tangent to all C_t, we call σ an *envelope* of the family $\{C_t\}$. If each C_t satisfies the implicit function equation $F(x, y, t) = 0$, the equation $F = \partial F/\partial t = 0$ gives an envelope. See [84], Appendix B.1.

the x-axis) and the line passing through Q parallel to m. Set

$$u := |DA|, \quad v := |DY''|.$$

Show that $\sqrt[3]{2} = u/v$, according to the following procedure:

(a) Let Y and Y′ be the intersection points of the line OQ with m and the circle Γ, respectively. Verify that Y′ is the point symmetrically opposite to Y″ with respect to the line BC (*Hint*: Use the fact that $|OQ| = |YY'|$).

(b) Let D′ be the foot of the perpendicular line from the point Y′ on the x-axis. Set $w := |OD|$. With this, $|YY'| = |OQ|$ holds, and Q lies on Y″D. Show

$$\frac{|AD|}{|DQ|} = \frac{|AC|}{|CM|} = 2, \qquad \frac{|AD|}{|DY''|} = \frac{|Y''D|}{|DO|} = \frac{|Y'D'|}{|D'A|} = \frac{|OD'|}{|D'Y'|} = \frac{|OD|}{|DQ|}.$$

(c) Using (a), show

$$\frac{u}{v} = \frac{v}{w} = \frac{w}{u/2}$$

and obtain $u^3 = 2v^3$.

1.2. Fundamentals of Regular Curves

Before we rigorously consider singularities appearing on curves, we shall review the basic notions for regular curves.

Definition of the curvature function. For vectors $a, b \in R^2$, we denote the standard inner product by $a \cdot b$. The norm (i.e., length) of a vector $a := (a_1, a_2)$ is defined as

$$|a| := \sqrt{a \cdot a} = \sqrt{(a_1)^2 + (a_2)^2}.$$

For a regular curve $\gamma(t)$ ($a \le t \le b$) in the plane R^2,

$$e(t) := \frac{\gamma'(t)}{|\gamma'(t)|} = (v_1(t), v_2(t))$$

is called the *unit tangent vector*. Furthermore,

$$n_L(t) := (-v_2(t), v_1(t)) \tag{1.9}$$

is called the (leftward) *unit normal vector*, which is oriented to be on the left side of the parameter direction of the curve itself. For a regular curve $\gamma(t)$, we define the *curvature* (or *curvature function*) $\kappa(t)$ by

$$\kappa := \frac{x'y'' - y'x''}{\left((x')^2 + (y')^2\right)^{3/2}} = \frac{\det(\gamma', \gamma'')}{|\gamma'|^3}. \tag{1.10}$$

Here, the vector $\gamma'' := d^2\gamma/dt^2$ is the acceleration vector, and inserting the two vectors γ' and γ'' into the two columns of a 2×2 matrix, we can take the determinant $\det(\gamma', \gamma'')$. The absolute value of the curvature at a point is known to be the inverse of the radius of the best-fitting circle to the curve at that point (see [84], Section 2).

Proposition 1.2.1. *The curvature is invariant under orientation-preserving parameter changes of the curve.*

Proof. Consider the reparametrization $\gamma(s) := \gamma(t(s))$ by the coordinate transformation $t = t(s)$. Writing the derivative with respect to s as $\dot\gamma = d\gamma/ds$ using an overhead dot, and the derivative with respect to t as $\gamma' = d\gamma/dt$ ($s' = ds/dt$) using a superscript dash, we have the relation

$$\gamma' = \frac{ds}{dt}\frac{d\gamma}{ds} = s'\dot\gamma.$$

Taking the derivative with respect to t of both sides of this equation, we have $\gamma'' = (s')^2\ddot\gamma + s''\dot\gamma$. Since $s' = ds/dt > 0$, $|\gamma'|^3 = (s')^3|\dot\gamma|^3$ holds, and, using properties of determinants, we have

$$\frac{\det(\gamma', \gamma'')}{|\gamma'|^3} = \frac{\det(s'\dot\gamma, (s')^2\ddot\gamma + s''\dot\gamma)}{|\gamma'|^3} = \frac{\det(s'\dot\gamma, (s')^2\ddot\gamma)}{(s')^3|\dot\gamma|^3} = \frac{\det(\dot\gamma, \ddot\gamma)}{|\dot\gamma|^3}.$$

This proves the proposition. $\square$

We now write the inverse of the absolute value of the curvature $\kappa(t)$ as

$$\rho(t) := \frac{1}{|\kappa(t)|},$$

and we call this the *curvature radius* at $\gamma(t)$. At points where $\kappa(t) = 0$, the curvature radius diverges to ∞, so let us just start with the case that $\kappa(t) \neq 0$. The circle C_t of radius $\rho(t)$ tangent to the curve at $\gamma(t)$ lying to the left of the curve when $\kappa(t) > 0$ and to the right when $\kappa(t) < 0$ is called the *osculating circle* of the curve. In the case that the curvature is zero ($\kappa(t) = 0$), the osculating circle becomes the tangent line to the curve at that point, which can be regarded as a circle with infinite radius.

The centers of the osculating circles for a regular curve $\gamma(t)$ can be written as

$$\sigma(t) := \gamma(t) + \frac{1}{\kappa(t)}\boldsymbol{n}_L(t), \tag{1.11}$$

and this curve is called the *caustic* or *evolute* or *focal curve* of $\gamma(t)$. We have the following known result (see [84], Theorem B.1.1 in Appendix B.1):

Fact 1.2.2. The envelope of a family of normal lines of a regular curve $\gamma(t)$ is the same as the caustic.

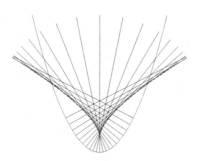

Fig. 1.4.   The family of normal lines of a parabola and their envelope.

For example, the envelope of the normal lines of the parabola $\gamma(t) := (t, t^2)$ as in Fig. 1.4 becomes

$$\sigma(t) := \left(-4t^3, \frac{1}{2} + 3t^2 \right), \tag{1.12}$$

at each point. This curve $\sigma(t)$ has a singular point at $t = 0$, and by Fact 1.2.2, this curve is the same as the caustic of $\gamma(t)$ (see Exercise 1.2 at the end of this section).

As seen in Fig. 1.4, consider laying out a light source along the curve, and emitting laser lights in the normal directions. Then the caustic will appear at the places where the light accumulates the most. The name "caustic" is derived from this observation.

Arc-length parameters. For the sake of ease of computation, we often use what is called an arc-length parametrization. Taking a curve $\gamma(t)$ ($t \in [a, b]$), set

$$s(t) := \int_a^t |\gamma'(u)|\, du,$$

defined on an interval $[a, t]$. This function $s(t)$ represents the length along the curve from the point $\gamma(a)$. When $\gamma(t)$ is a regular curve, we have

$$\frac{ds(t)}{dt} = |\gamma'(t)| > 0,$$

and so $s(t)$ gives a strictly increasing C^∞-function $s : [a, b] \to [0, s(b)]$, and thus has an inverse function

$$t = t(s) \quad (0 \le s \le s(b))$$

so that $s\bigl(t(c)\bigr) = c$ for all $c \in [a, b]$. In this case, we can apply the coordinate transformation

$$\gamma(s) := \gamma\bigl(t(s)\bigr) \quad (0 \le s \le s(b)).$$

We call this an *arc-length parametrization* of the curve.

In this chapter (similarly to what was done in the proof of Proposition 1.2.1), we write the derivative with respect to the arc-length parameter s as $\dot{\gamma} = d\gamma/ds$ using an overhead dot, and the derivative with respect to a general parameter t as $\gamma' = d\gamma/dt$ ($s' = ds/dt$) using a superscript dash. By the definition of arc-length, $|\dot{\gamma}(s)| = 1$ holds. Conversely, a parameter s satisfying this equation coincides with the arc-length up to an additive constant. Thus, a given parameter s is the arc-length parameter if and only if s satisfies $|\dot{\gamma}(s)| = 1$. The arc-length parameter can be regarded as the most natural parametrization for investigating Euclidean geometric properties. However, at singular points, the speed is zero, so one cannot use an arc-length parametrization when including singular points.

Differentiating $\dot{\gamma}(s) \cdot \dot{\gamma}(s)(= |\dot{\gamma}(s)|^2) = 1$, we have $\ddot{\gamma}(s) \cdot \dot{\gamma}(s) = 0$. This means that $\ddot{\gamma}(s)$ is a scalar multiple of the leftward unit normal $\boldsymbol{n}_L(s)$. One can easily see that the proportionality constant coincides with the curvature defined in (1.10). Namely,

$$\ddot{\gamma}(s) = \kappa(s)\boldsymbol{n}_L(s). \tag{1.13}$$

We set $\gamma(s) = (x(s), y(s))$. Then, since s is an arc-length parameter, the unit tangent vector and the leftward unit normal vector are

$$\boldsymbol{e}(s) = \dot{\gamma}(s) = (\dot{x}(s), \dot{y}(s)), \quad \boldsymbol{n}_L(s) = (-\dot{y}(s), \dot{x}(s)),$$

respectively, and (1.13) is equivalent to the two equations $\ddot{x} = -\kappa\dot{y}$ and $\ddot{y} = \kappa\dot{x}$. This is also equivalent to $\dot{\boldsymbol{n}}_L = -\kappa\dot{\gamma}$. Namely, we obtain two equations

$$(\ddot{\gamma}(s) =)\dot{\boldsymbol{e}}(s) = \kappa(s)\boldsymbol{n}_L(s), \quad \dot{\boldsymbol{n}}_L(s) = -\kappa(s)\boldsymbol{e}(s)(= -\kappa(s)\dot{\gamma}(s)). \tag{1.14}$$

These two equations are called the *Frenet formulas*, and by these formulas, we can show that the curvature function $\kappa(s)$ provides complete information about the given curve. In fact, the following theorem is known (cf. [84, Theorem 2.8]).

Theorem 1.2.3 (Fundamental theorem for planar curves). *Let $I = [a, b]$ be an interval, and $\kappa(s)$ ($a \leq s \leq b$) a smooth function on I (namely, a C^∞-function on I). Then, up to rotations and translations in the plane, there exists a unique regular planar curve $\gamma(s)$ ($a \leq s \leq b$) parametrized by the arc-length parameter s so that the curvature is $\kappa(s)$.*

We will not give a proof of this theorem, and one can easily see that one planar curve whose curvature coincides with a given function κ is

$$\gamma(s) := \int_a^s (\cos\theta(t), \sin\theta(t))dt, \quad \left(\theta(t) := \int_a^t \kappa(u)du\right). \tag{1.15}$$

Regular closed planar curves. Let us now turn our attention to the global behavior of planar curves. For a planar curve $\tilde{\gamma} : \boldsymbol{R} \to \boldsymbol{R}^2$ defined on the entire real line $\boldsymbol{R}$, if there exists a positive real number a such that

$$\tilde{\gamma}(t + a) = \tilde{\gamma}(t) \quad (t \in \boldsymbol{R}), \tag{1.16}$$

namely, $\tilde{\gamma}(t)$ is periodic as a vector-valued function, then the restriction

$$\gamma : [0, a] \to \boldsymbol{R}^2 \tag{1.17}$$

of $\tilde{\gamma}$ to the interval $[0, a]$ is called a *closed curve*.

If $\tilde{\gamma}$ has no singular point, γ is called a *regular closed curve*. Regular closed curves can be parametrized by arc-length parameters. Let $\gamma(s)$ ($0 \le s \le a$) be a regular closed curve parametrized by arc-length. Then the total curvature (integration of the curvature $\kappa(s)$ on the domain) divided by 2π is written as

$$i_\gamma := \frac{1}{2\pi} \int_0^a \kappa(s) \, ds, \tag{1.18}$$

which is called the *rotation index* of the regular closed curve. The rotation index takes values in the set of integers. If a closed regular curve has no self-intersections, then it is called a *simple closed curve*. We have the following facts [84, Theorems 3.2 and 3.3].

Fact 1.2.4. The rotation index of a regular simple closed curve is ± 1.

Fact 1.2.5 ([90]). Two regular closed curves which have the same rotation indices can be deformed from one to the other while always remaining a regular curve.

One can take representatives of regular closed curves for each rotation index as in Fig. 1.5.

We now introduce another method to describe maps $\tilde{\gamma} : \boldsymbol{R} \to \boldsymbol{R}^2$ with a period $a(> 0)$ which does not require us to restrict the domain, using a notation appearing only in this chapter. We say that two real numbers $s, t \in \boldsymbol{R}$ satisfy the relation $s \sim t$ if $s - t$ is an integer multiple of a. This

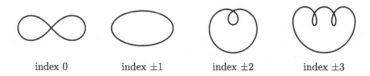

index 0   index ± 1 index ± 2 index ± 3

Fig. 1.5.   Rotation indices of closed curves.

relation $\sim$ is an equivalence relation. The equivalence class of t with respect to this equivalence relation is denoted by $[t]$, and the quotient space $\boldsymbol{R}/\sim$ of $\boldsymbol{R}$ by the equivalence relation $\sim$ is denoted by

$$\boldsymbol{R}/a\boldsymbol{Z} \ (= \{[t] \, ; \, t \in \boldsymbol{R}\}) \, .$$

One can consider $\bar{\gamma} : \boldsymbol{R}/a\boldsymbol{Z} \ni [t] \mapsto \tilde{\gamma}(t) \in \boldsymbol{R}^2$ instead of γ (cf. (1.17)). Then γ gives a simple closed curve if and only if $\bar{\gamma}$ is an injection. In this textbook, we do not wish to distinguish between γ and $\bar{\gamma}$, and we simply write γ instead of $\bar{\gamma}$. Moreover, we do not write $[t] \in \boldsymbol{R}/a\boldsymbol{Z}$, but simply $t \in \boldsymbol{R}/a\boldsymbol{Z}$. This means we can regard $\boldsymbol{R}/a\boldsymbol{Z}$ as the closed interval $[0, a]$ with the endpoints identified.

Exercises 1.2

1. Show the caustic of the parabola $\gamma(t) := (t, t^2)$ is given by (1.12).

2. For the planar curve $\gamma(s)$ given by (1.15), show that s is an arc-length parameter, and $\kappa(s)$ is the curvature. Moreover, consider the reason why the rotation indices of closed curves take values in the set of integers, using (1.15).

1.3. Fundamental Properties of Cusps

The definition of cusps and their criterion. As we have seen in Section 1.1, the planar curves

$$\gamma_0(t) := (t^2, t^3), \quad \gamma_1(t) := (t^2, t^5), \quad \gamma_2(t) := (t^3, t^4) \qquad (1.19)$$

have singular points at $t = 0$. These are in fact called the *standard cusp*, the *standard 5/2-cusp* and the *standard 4/3-cusp*, respectively.

Let $t = c$ be a singular point of a planar curve $\gamma(t)$ ($a < t < b$). If there exist an orientation preserving reparametrization $t = t(s)$ ($c = t(0)$) and a local diffeomorphism[4] Φ from a neighborhood of $\gamma(c)$ into a neighborhood of the origin $(0, 0)$ such that

$$\Phi \circ \gamma\big(t(s)\big) = (s^2, s^3),$$

then the planar curve $\gamma(t)$ is said to have a *cusp* at $t = c$, or rather, $t = c$ is a *cusp* of $\gamma(t)$ (also called a *cusp point*, or to be very exact, a *3/2-cusp*). If it can be expressed as

$$\Phi \circ \gamma\big(t(s)\big) = (s^2, s^5),$$

---

[4]See [84, page 18].

then $\gamma(t)$ is said to have a *5/2-cusp* at $t = c$, or $t = c$ is a *5/2-cusp* of $\gamma(t)$ (or, a *5/2-cusp point*). Also, if it can be expressed as

$$\Phi \circ \gamma\big(t(s)\big) = (s^3, s^4),$$

then the planar curve $\gamma(t)$ is said to have a *4/3-cusp* at $t = c$, or $t = c$ is a *4/3-cusp* of $\gamma(t)$ (or, a *4/3-cusp point*).

Namely, for a given singular point of a curve, if there exist coordinate changes on neighborhoods of the point and the target space so that the singular point can be equated with the standard cusp, the standard *5/2-cusp*, or the standard *4/3-cusp* as a map, respectively, then the point is called a cusp, a *5/2-cusp*, or a *4/3-cusp*, respectively. Since the standard cusp and the standard *5/2-cusp* have "spikes", general cusps and *5/2-cusps* do as well. On the other hand, because the standard *4/3-cusp* can be expressed as the graph of a C^1-differentiable function (cf. (1.5)), general *4/3-cusps* can be as well.

Example 1.3.1. Let n be an integer greater than or equal to 2. In general, algebraic equations of degree n can be written as

$$t^n + a_1 t^{n-1} + \cdots + a_{n-1} t + a_n = 0 \quad (a_1, \ldots, a_n \in \boldsymbol{R}).$$

Substituting $t - a_1/n$ into t, the term whose degree is $n - 1$ is eliminated, and we have

$$t^n + b_2 t^{n-2} + \cdots + b_{n-1} t + b_n = 0 \quad (b_2, \ldots, b_n \in \boldsymbol{R}).$$

We call this form the *normal form of degree n algebraic equations*.

Since the planar curve $\gamma(t) := (-3t^2, 2t^3)$ $(t \in \boldsymbol{R})$ is the composition of the map $\gamma_0(t) = (t^2, t^3)$ with the linear transformation $\boldsymbol{R}^2 \ni (x, y) \mapsto (-3x, 2y) \in \boldsymbol{R}^2$, $t = 0$ is a cusp. The image of this curve γ coincides with the set of the coefficients x and y for which the normal form of the cubic equation $\varphi(t) := t^3 + xt + y$ has a real multiple root. In fact, when $\varphi(t)$ has a real multiple root t_0, then $\varphi'(t_0) = 0$, and we have $3t_0^2 + x = 0$, namely $x = -3t_0^2$. Substituting this into $\varphi(t_0) = 0$,

$$0 = t_0^3 + xt_0 + y = t_0^3 - 3t_0^3 + y = -2t_0^3 + y.$$

Thus, we obtain $y = 2t_0^3$. Namely, coefficients x, y in Chapter 8 for which $\varphi(t)$ has a real multiple root can be obtained by considering t_0 as a parameter. More concretely, setting

$$D_{2,2} := \{(x, y) \in \boldsymbol{R}^2 \, ; \, t^3 + xt + y = 0 \text{ has a real multiple root}\},$$

then (see (8.23) in Chapter 8 for the meaning of the notation $D_{2,2}$)

$$D_{2,2} = \{(-3t^2, 2t^3) \in \boldsymbol{R}^2 \, ; \, t \in \boldsymbol{R}\} = \gamma(\boldsymbol{R}),$$

where $\gamma(\mathbf{R})$ denotes the image of the curve γ. Namely,

for the normal form of the cubic equation, the set of coefficients where the equation has a real multiple root gives a cusp.

For the sake of simplicity, we use $\gamma_0(t) = (t^2, t^3)$ for the standard cusp. The above fact is important for our understanding of cusps later (see Example 8.3.3 in Chapter 8).

Like as in this example, one can see that the singular point of the caustic of the parabola (1.12) is a cusp by directly applying a coordinate transformation of $\mathbf{R}^2$ (see Exercise 1 in this section). In general, it is difficult to show whether a given singular point of a curve is a cusp or not by using only the definition of cusps, and this is why the following criteria are convenient.

Theorem 1.3.2 (A criterion for cusps). *A planar curve* $\gamma(t)$ *has a cusp at* $t = c$ *if and only if* $\gamma'(c) = \mathbf{0}$ *and*

$$\det\bigl(\gamma''(c), \gamma'''(c)\bigr) \neq 0. \tag{1.20}$$

The "only if" part will be shown in Corollary 1.3.6 and the "if" part will be proved in Section 3.2 of Chapter 3, and also in Section 8.3 of Chapter 8 by different methods.

Fact 1.3.3 (A criterion for $5/2$-*cusps* **[65, Theorem 1.23]).** A planar curve $\gamma(t)$ has a $5/2$-*cusp* at $t = c$ if and only if $\gamma'(c) = \mathbf{0}$, $\det\bigl(\gamma''(c), \gamma'''(c)\bigr) = 0$ and

$$3\det(\gamma''(c), \gamma^{(5)}(c))\gamma''(c) - 10\det(\gamma''(c), \gamma^{(4)}(c))\gamma'''(c) \neq \mathbf{0}, \tag{1.21}$$

where $\gamma^{(k)}$ denotes the k-th derivative of γ.

See [65] for a proof of Fact 1.3.3. Next, we introduce the following criterion.

Theorem 1.3.4 (A criterion for $4/3$-*cusps***).** *A planar curve* $\gamma(t)$ *has a* $4/3$-*cusp at* $t = c$ *if and only if* $\gamma'(c) = \gamma''(c) = \mathbf{0}$ *and*

$$\det\bigl(\gamma'''(c), \gamma^{(4)}(c)\bigr) \neq 0. \tag{1.22}$$

We prove this theorem in Appendix C. The claim that the condition (1.20) is a necessary condition for a cusp can be easily shown. In fact, we have the following proposition.

Proposition 1.3.5. *If a planar curve $\gamma(t)$ satisfies the condition* (1.20) *at $t = c$, then the curve $\Phi \circ \gamma \circ \varphi$ obtained by the reparametrization $t = \varphi(s)$ $(c = \varphi(b))$ and diffeomorphism $\Phi : \mathbf{R}^2 \to \mathbf{R}^2$ also satisfies* (1.20) *at $s = b$.*

Proof. Without loss of generality, we may set $c = b = 0$. It is left to the reader to check that $\gamma(s)$ obtained by a reparametrization $t = \varphi(s)$ satisfies the condition (1.20). Then it is enough to show that $\tilde{\gamma}(t) := \Phi \circ \gamma(t)$ satisfies the conditions of the theorem, where $\Phi : \mathbf{R}^2 \to \mathbf{R}^2$ is a diffeomorphism. Let $J(x, y)$ be the Jacobi matrix of the map $\Phi(x, y) = (\Phi_1(x, y), \Phi_2(x, y))$. By the chain rule,

$$\tilde{\gamma}'(t) = J(t)\gamma'(t), \tag{1.23}$$

where we regard γ as a column vector and we set

$$J(t) := J(x(t), y(t)) = \begin{pmatrix} (\Phi_1)_x(x(t), y(t)) & (\Phi_1)_y(x(t), y(t)) \\ (\Phi_2)_x(x(t), y(t)) & (\Phi_2)_y(x(t), y(t)) \end{pmatrix} \tag{1.24}$$

with $\gamma(t) = (x(t), y(t))$. Differentiating (1.23), we have

$$\tilde{\gamma}''(t) = J'(t)\gamma'(t) + J(t)\gamma''(t),$$

and because $\gamma'(0) = \mathbf{0}$,

$$\tilde{\gamma}''(0) = J(0)\gamma''(0). \tag{1.25}$$

On the other hand,

$$\tilde{\gamma}'''(t) = J''(t)\gamma'(t) + 2J'(t)\gamma''(t) + J(t)\gamma'''(t),$$

and by the chain rule, $J'(t) = J_x(t)x'(t) + J_y(t)y'(t)$. Since $t = 0$ is a singular point of $\gamma(t)$, we have $x'(0) = y'(0) = 0$. Thus,

$$J'(0) = O, \tag{1.26}$$

where O is the zero matrix. Hence, using $\tilde{\gamma}'''(0) = J(0)\gamma'''(0)$ together with (1.25), we have

$$\det(\tilde{\gamma}''(0), \tilde{\gamma}'''(0)) = \det(J(0)\gamma''(0), J(0)\gamma'''(0)) \tag{1.27}$$

$$= \det J(0) \det(\gamma''(0), \gamma'''(0)) \neq 0. \qquad \square$$

By Proposition 1.3.5 and the definition of cusps, we have the following corollary.

Corollary 1.3.6. *A cusp satisfies the condition* (1.20).

Proof. The standard cusp $\gamma_0(t) = (t^2, t^3)$ satisfies the condition (1.20) (see Exercise **2** in this section). Thus, general cusps satisfy (1.20) as well. $\quad\square$

When one wants to know whether a singular point of an explicitly described curve is a cusp or not, the criterion for cusps in Theorem 1.3.2 plays a useful role. Let us see this with the following examples.

Example 1.3.7. All singular points appearing on the cycloid (1.1) in Example 1.1.1 are cusps. In fact, since by $(-2n\pi)$-distance parallel translations in the direction of the x-axis, all singular points $t = 2n\pi$ $(n \in \mathbf{Z})$ can be identified with the singular point at $t = 0$, and it is enough to show that the singular point $t = 0$ is a cusp. Differentiating $\gamma'(t)$ from (1.1), we have

$$\gamma''(t) = a(\sin t, \cos t), \quad \gamma'''(t) = a(\cos t, -\sin t). \tag{1.28}$$

In particular, $\gamma'(0) = \mathbf{0}$ and

$$\det(\gamma''(0), \gamma'''(0)) = a^2 \det \begin{pmatrix} 0 & 1 \\ 1 & 0 \end{pmatrix} = -a^2 \neq 0.$$

By Theorem 1.3.2, we see that $t = 0$ is a cusp.

Similarly, all singular points appearing on the astroid (Exercise **1** in Section 1.1), the cardioid (Example 1.1.2) and cissoid (Exercise **3** in Section 1.1) are cusps (Exercise **3** at the end of this section).

Cuspidal curvature: To study the behavior of the curvature function near cusps, we prove the following theorem.

Theorem 1.3.8 ([75]). *Let a planar curve* $\gamma(t)$ *have a cusp at* $t = c$. *Let* $\kappa(t)$ *be the curvature defined everywhere except at* $t = c$. *Then*

$$\lim_{t \to c} \kappa(t) \sqrt{|s(t)|} = \frac{\det(\gamma''(c), \gamma'''(c))}{2\sqrt{2}\, |\gamma''(c)|^{5/2}}, \tag{1.29}$$

where

$$s(t) := \int_c^t |\gamma'(u)|\, du$$

is the signed arc-length measured from $t = c$.

Proof. We may assume that $c = 0$ without loss of generality. By l'Hôpital's rule,

$$\lim_{t \to 0} \left| \frac{s(t)}{t^2} \right| = \lim_{t \to 0} \left| \frac{|\gamma'(t)|}{2t} \right| = \lim_{t \to 0} \left| \frac{\gamma'(t)}{2t} \right| = \frac{|\gamma''(0)|}{2}. \tag{1.30}$$

In particular,

$$\lim_{t \to 0} \frac{\sqrt{|s(t)|}}{|t|} = \frac{\sqrt{|\gamma''(0)|}}{\sqrt{2}}. \tag{1.31}$$

Using (1.10), we have

$$\kappa(t)\sqrt{|s(t)|} = \frac{\det(\gamma'(t), \gamma''(t))}{t^2} \cdot \frac{1}{\left|\dfrac{\gamma'(t)}{t}\right|^3} \cdot \frac{\sqrt{|s(t)|}}{|t|}, \qquad (1.32)$$

and also

$$\lim_{t \to 0} \frac{\det(\gamma'(t), \gamma''(t))}{t^2} = \lim_{t \to 0} \frac{\det(\gamma'(t), \gamma'''(t))}{2t} = \frac{\det(\gamma''(0), \gamma'''(0))}{2}.$$

By this equation,

$$\lim_{t \to 0} \frac{\gamma'(t)}{t} = \gamma''(0),$$

and substituting (1.31) into (1.32) with $t \to 0$, we have (1.29). Here, we see that $\gamma''(0)$ does not vanish, from Corollary 1.3.6. □

Taking this into consideration, we define the cuspidal curvature measuring the "sharpness" of the cusp.

Definition 1.3.9. Let a planar curve $\gamma(t)$ have a cusp at $t = c$. Then we call

$$\mu := \frac{\det(\gamma''(c), \gamma'''(c))}{|\gamma''(c)|^{5/2}} \qquad (1.33)$$

the *cuspidal curvature*.[5] The length and the curvature are geometric invariants of the curve, and the formula (1.33) yields that the cuspidal curvature μ is also a geometric invariant of the curve at $t = c$, that is,

- μ does not depend on the choice of parameter t of the curve, and
- letting $T : \boldsymbol{R}^2 \to \boldsymbol{R}^2$ be an orientation preserving isometry of $\boldsymbol{R}^2$, the cuspidal curvature of $T \circ \gamma(t)$ at $t = c$ coincides with μ. Such a T can be written as $T(\boldsymbol{x}) = A\boldsymbol{x} + \boldsymbol{c}$, where A is an orthogonal matrix satisfying $\det A = 1$ and $\boldsymbol{c} \in \boldsymbol{R}^2$.
- When μ is larger, then the cusp spreads wider.

Remark 1.3.10. In the proof of Theorem 1.3.8, we used only that $\gamma''(c) \neq \boldsymbol{0}$, and the cuspidal curvature can be defined on all planar curves having just the properties $\gamma'(c) = \boldsymbol{0}$ and $\gamma''(c) \neq \boldsymbol{0}$. By Theorem 1.3.2, the cuspidal curvature is non-zero if and only if $t = c$ is a cusp.

---

[5]The cuspidal curvature was introduced by the first author [81].

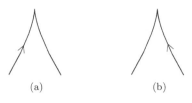

(a)   (b)

Fig. 1.6.   Zig (a) and zag (b).

In particular, we have the following corollary to Theorem 1.3.8.

Corollary 1.3.11. *The curvature function on both sides of a cusp has the same sign, and diverges.*

If the cuspidal curvature μ is positive (namely, the curvature function on a neighborhood is positive), then the cusp is called a *zig*, and if μ is negative (namely, the curvature function on a neighborhood is negative), then the cusp is called a *zag*.

A zig changes to a zag, and vice-versa, if we reverse the direction of the curve. We have the following proposition (see Fig. 1.6).

Proposition 1.3.12. *Let a planar curve $\gamma(t)$ have a cusp singularity at $t = c$. If $t = c$ is a zig, then the curve turns to the right, and if $t = c$ is a zag, then it turns to the left.*

Proof. We may assume $c = 0$ without loss of generality. By the definition of cusps, there exists a local diffeomorphism Φ of $\mathbf{R}^2$ satisfying

$$\Phi \circ \gamma \circ \varphi(t) = (t^2, t^3)(=: \gamma_0(t))$$

and an (orientation preserving) reparametrization φ satisfying $\varphi(0) = 0$. If Φ is an orientation preserving (reversing) diffeomorphism, namely, the determinant of the Jacobi matrix of Φ is positive (negative), then (cf. (1.27))

$$\det(\gamma''(0), \gamma'''(0)) \text{ and } \det(\gamma_0''(0), \gamma_0'''(0))$$

have the same sign (opposite sign). In other words, the sign of the cuspidal curvature of γ is positive (negative). The standard cusp γ_0 turns to the right at $t = 0$. If Φ is orientation preserving (reversing), then γ turns to the right (to the left) at $t = 0$. This proves the assertion. $\qquad\square$

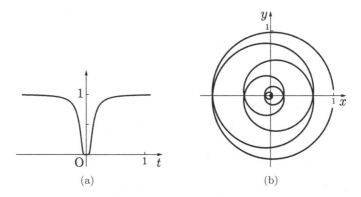

Fig. 1.7.   The function $e^{-1/(at^2)}$ and the curve $\sigma(t)$ for $a = 100$.

By a direct calculation (see Example 1.3.7), one can show the following.

Example 1.3.13. The cuspidal curvature of the singular points of the cycloid (1.1) is $-1/\sqrt{a}$. Cycloids are obtained by taking a circle of radius $a > 0$ and rolling it along a straight line, and taking the path traced out by a given fixed point. The absolute value of the cuspidal curvature is equal to the square root of the inverse of that radius. The cycloid obtained by rolling the circle of radius $1/\mu^2$ gives the best approximation of a given cusp amongst the cusps of cycloids (cf. Corollary B.1.5 in Appendix B).

Exercises 1.3

1 Show the singular point appearing on the caustic of the parabola $y = x^2$ is a cusp, directly from the definition (without using the criterion of Theorem 1.3.2).

2 Confirm that the standard cusp $\gamma_0(t)$ and the standard $5/2$-*cusp* $\gamma_1(t)$ given in (1.19) satisfy the conditions of Theorem 1.3.2 and Fact 1.3.3 at $t = 0$.

3 Using Theorem 1.3.2, show that all singular points on the astroid given in Section 1.1 (Exercise **1** in Section 1.1), on the cardioid (Example 1.1.2) and on the cissoid (Exercise **3** in Section 1.1) are cusps.

1.4.   The Unit Normal Vector for Curves with Singular Points

The unit normal vector and singular points. For a regular planar curve $\gamma(t)$, the 90°-rotation of $\gamma'(t)/|\gamma'(t)|$ gives a smooth parametrization

Fig. 1.8.   Behavior of the unit normal vector at a cusp.

of a unit normal vector. On the other hand, if there is an isolated singular point on a planar curve, the unit normal vector cannot be extended to the singular point in general.

For $a > 0$, we set

$$\sigma(t) := \begin{cases} e^{-1/(at^2)}(\cos(1/t), \sin(1/t)) & (\text{if } t \in \mathbf{R} \setminus \{0\}), \\ (0,0) & (\text{if } t = 0). \end{cases}$$

Since the function $e^{-1/(at^2)}$ rapidly approaches 0 (respectively, 1) at $t = 0$ (respectively, $t = \infty$), $\sigma(t)$ defines a smooth closed curve on $\mathbf{R} \cup \{\infty\}$, and $t = 0$ (respectively, $t = \infty$) is a singular point (respectively, a regular point). As we see in Fig. 1.7, the curve has infinitely many self-intersections and its velocity vector rotates infinitely many times. In fact, calculating the derivative, we have

$$\sigma'(t) = \frac{e^{-1/(at^2)}}{t^3} \left(t \sin\frac{1}{t} + \frac{2}{a}\cos\frac{1}{t}, \frac{2}{a}\sin\frac{1}{t} - t\cos\frac{1}{t} \right)$$

when $t \neq 0$. Thus when $n \to \infty$, the direction tends to $(1,0)$ at $t = 1/(2n\pi)$ $(n \in \mathbf{Z})$, and it tends to $(0,1)$ at $t = 1/((2n+1)\pi)$. Therefore, the unit tangent vector (and thus automatically the unit normal vector as well) cannot be extended to the origin.

In spite of the existence of this wild example, ordinarily the unit normal vector of planar curves can be extended smoothly to the set of singular points.[6] We now demonstrate this with some examples.

Example 1.4.1. For the standard cusp $\gamma_0(t) = (t^2, t^3)$ at $t = 0$,

$$\boldsymbol{n}(t) = \frac{1}{\sqrt{4 + 9t^2}}(-3t, 2) \tag{1.34}$$

---

[6] For example, noting the existence of a first non-vanishing term of the Taylor expansion, any real analytic plane curve defined on an open interval has such a property.

gives a smooth unit normal vector for $t \in \mathbf{R}$. The vector $\mathbf{n}(t)$ points to the right hand side with respect to the parameter direction of γ_0 when $t < 0$, and it points to the left hand side with respect to the parameter direction when $t > 0$ (cf. Fig. 1.8). Thus, different from the case of regular curves, considering only the leftward unit normal vector results in a loss of continuity. A valuable property of $\mathbf{n}(t)$ is that it does not lose regularity at $t = 0$, namely, $\mathbf{n}'(0) \neq \mathbf{0}$.

Example 1.4.2. For the standard $5/2$-*cusp* $\gamma_1(t) = (t^2, t^5)$ at $t = 0$,

$$\mathbf{n}(t) = \frac{1}{\sqrt{4 + 25t^6}}(-5t^3, 2) \tag{1.35}$$

gives a smooth unit normal vector for $t \in \mathbf{R}$. Like as in the case of γ_0, the vector $\mathbf{n}(t)$ of γ_1 changes direction from the right to the left. A difference from the case of a cusp is that $\mathbf{n}(t)$, although smooth, has a singular point at $t = 0$, namely, $\mathbf{n}'(0) = \mathbf{0}$.

Example 1.4.3. For the standard $4/3$-*cusp* $\gamma_2(t) = (t^3, t^4)$ at $t = 0$,

$$\mathbf{n}(t) = \frac{1}{\sqrt{9 + 16t^2}}(-4t, 3) \tag{1.36}$$

gives a smooth unit normal vector for $t \in \mathbf{R}$. In this case, the unit normal vector does not change direction, and it always points to the left with respect to the parameter direction of γ_2. It also shares the same property $\mathbf{n}'(0) \neq \mathbf{0}$ with the standard cusp.

Example 1.4.4. For the parametrization $\gamma_3(t) = (t^3, 0)$ of a line, which has a singular point at the origin, the constant vector $\mathbf{n} = (0, 1)$ can be taken as a unit normal vector, and it points to the left-hand side with respect to the parameter direction of γ_3 even when it crosses the singular point.

Parallel curves of regular curves: As an application of such extension of normal vectors across singular points, we now study the parallel curves of regular curves.

For a given regular curve $\gamma(t)$ and a real number δ, we call the curve defined by

$$\gamma_\delta(t) := \gamma(t) + \delta \mathbf{n}_L(t) \tag{1.37}$$

the *parallel curve* of $\gamma(t)$ (at distance δ), where $\mathbf{n}_L(t)$ is the leftward unit normal vector of $\gamma(t)$. Imagining a wave front traveling in the plane, when one moment is expressed by a curve γ, the wave front moving as time δ elapses, is, by the Huygens principle, an envelope of a family of circles centered at each point with radius δ on the curve γ. This coincides with the image of γ_δ of the parallel curves in the formula (1.37). For example, the parallel curves as wave fronts moving over elapsing time δ in

- the upper direction of the line $\gamma(t) := (t, 0)$, and
- the inner direction of the unit circle $\tilde{\gamma}(t) := (\cos t, \sin t)$

are

$$\gamma_\delta(t) = (t, \delta), \quad \tilde{\gamma}_\delta(t) = (1 - \delta)(\cos t, \sin t),$$

respectively. We have the following proposition.

Proposition 1.4.5. *For the parallel curves* $\{\gamma_\delta(t)\}_{\delta \in \mathbf{R}}$ *of the regular curve* $\gamma(t)$, *we have:*

(1) *The leftward unit normal vector* $\boldsymbol{n}_L(t)$ *of a regular curve* $\gamma(t)$ *gives the (not always leftward) unit normal vector of* $\gamma_\delta(t)$.

(2) *Let* $\kappa(t)$ *be the curvature of the regular curve* $\gamma(t)$. *Then* $\gamma_\delta(t)$ *has a singular point at* $t = c$ *if and only if* $\kappa(c) = 1/\delta$.

(3) *Moreover, that singular point is a cusp if and only if* $\kappa'(c) \neq 0$.

Proof. Since the curvature does not depend on the parametrization of the curve, we may assume t is an arc-length parameter of γ, without loss of generality. Differentiating (1.37), we have

$$\gamma_\delta'(t) = \gamma'(t) + \delta \boldsymbol{n}_L'(t) = \gamma'(t) - \delta\kappa(t)\gamma'(t) = (1 - \delta\kappa(t))\gamma'(t), \quad (1.38)$$

and since $\boldsymbol{n}_L(t)$ is perpendicular to $\gamma'(t)$, (1) is proven. Noting that $\gamma'(t) \neq \boldsymbol{0}$, we see that $\gamma_\delta'(t) = \boldsymbol{0}$ is equivalent to $\delta\kappa(t) = 1$. This proves (2). Differentiating again, we have

$$\gamma_\delta''(t) = -\delta\kappa'(t)\gamma'(t) + (1 - \delta\kappa(t))\gamma''(t), \quad (1.39)$$

and differentiating once again and noting that $1 - \delta\kappa(c) = 0$, we have

$$\gamma_\delta'''(c) = -\delta\kappa''(c)\gamma'(c) - 2\kappa'(c)\boldsymbol{n}_L(c).$$

Then we have

$$\det\left(\gamma_\delta''(c), \gamma_\delta'''(c)\right) = 2\delta\kappa'(c)^2 \det(\gamma'(c), \boldsymbol{n}_L(c)) = 2\delta\kappa'(c)^2.$$

Thus $\kappa'(c) \neq 0$ is equivalent to $t = c$ being a cusp, because of Theorem 1.3.2. $\square$

In particular, by Proposition 1.4.5 (2), the caustic, which is the trajectory of the center of the curvature circle (see (1.11)), has the following relation with the parallel curves.

Corollary 1.4.6. *When the real number δ varies, the trajectory of the singular points of the family of the parallel curves γ_δ coincides with the caustic of the image of γ. In particular, singular points on the parallel curves γ_δ occur at their intersection points with the caustic.*

Remark 1.4.7. We call a point $t = c$ on a regular curve $\gamma(t)$ where the curvature vanishes an *inflection point*. At inflection points, the curvature radius tends to infinity, the denominator of the right-hand side of (1.11) is zero, and the caustic $\sigma(c)$ cannot be defined. In fact, $|\sigma(t)|$ diverges when t approaches c.

Remark 1.4.8. Denoting the leftward unit normal vector as $n_L(t)$ for a given curve $\gamma(t)$, then for all $\delta \in \mathbf{R}$, $n_L(t)$ gives a unit normal vector of the parallel curve $\gamma_\delta(t)$. Since when passing through a singular point, the leftward and rightward unit normal vectors usually interchange, the smooth unit normal vector $n(t)$ along $\gamma_\delta(t)$ is not a leftward unit normal vector for γ_δ in general. Thus, when we deal with curves with singular points, the property that the normal is leftward or rightward loses its meaning.

Example 1.4.9 (Parallel curves of an ellipse). Drawing a time development of parallel curves of an ellipse $\gamma(t) = (\cos t, 2\sin t)$, we have Fig. 1.9. This figure is obtained by considering the ellipse as the initial wave front and flowing to parallel curves in the inner direction. If the wave is moving to the outer direction, there are no singular points. However, when the wave moves to the inner direction, as we see in the figure, in the beginning, the waves are only shrinking and staying regular, but after a short time, four cusps occur, and continue to change shape, and eventually, the singular

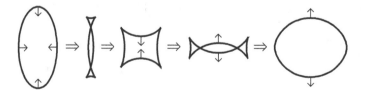

Fig. 1.9.   Parallel curves of an ellipse.

Fig. 1.10.   Parallel curves of a cycloid.

points disappear. Then we end up with a convex curve,[7] and the wave spreads out toward infinity. By Corollary 1.4.6, singular points on the parallel curves of an ellipse are the intersection points of the parallel curves with the caustic.

Example 1.4.10 (Parallel curves of a cycloid). Since the velocity vector of the cycloid (1.1) is

$$\gamma'(t) = a(1 - \cos t, \sin t) = a\left(2\sin^2\frac{t}{2}, 2\sin\frac{t}{2}\cos\frac{t}{2}\right)$$

$$= 2a\sin\frac{t}{2}\left(\sin\frac{t}{2}, \cos\frac{t}{2}\right), \tag{1.40}$$

we can take a smooth unit normal vector

$$\boldsymbol{n}(t) := \left(-\cos\frac{t}{2}, \sin\frac{t}{2}\right) \tag{1.41}$$

along $\gamma(t)$. On a period of the cycloid, $\boldsymbol{n}(t)$ reverses direction, and the cycloid together with unit normal vector return to the original state after two periods (namely, the period 4π). The vector $\boldsymbol{n}(t)$ is a leftward unit normal vector on the open interval $(0, 2\pi)$, and is a rightward unit normal vector on the open interval $(2\pi, 4\pi)$. Considering parallel curves (1.37) under this representation, we can draw the time development of the parallel curves in the past ($\delta < 0$), the present ($\delta = 0$) and the future ($\delta > 0$) as in Fig. 1.10. We can observe a cycloid that is the trajectory of singular points of the parallel curves, and it is congruent to the original cycloid. (This corresponds to the property that the caustic of a cycloid is congruent to the original cycloid. See Exercise **3** at the end of this section. This fact was discovered by Huygens, and applied to the pendulum clock. See [84, Appendix B.1] for details).

---

[7]In this text, a *convex curve* is a simple closed regular curve whose curvature is always positive. See Section 1.7 for details.

Exercises 1.4

1 Orienting the parabola $y = x^2$ with the leftward unit normal vector pointing upward, consider the parallel curves (1.37) for positive δ. Gradually increasing δ from $\delta = 0$, find the first value δ for which γ_δ has a singular point, and find a parametrization of γ_δ. Moreover, show that this singular point is a $4/3$-*cusp*.

2 Recall that we call a point where the curvature $\kappa(t)$ of a curve γ vanishes an inflection point for a given curve. Show that if a curve $\gamma(t)$ has a inflection point at $t = c$, then each parallel curve also has an inflection point at c. Moreover, show that if $\gamma(t)$ has no inflection points, then the caustic $\sigma(t)$ given in (1.11) has a singular point exactly at a critical point of the curvature, namely a point where $\kappa'(t) = 0$. Furthermore, show that this singular point is a cusp if and only if $\kappa''(t) \neq 0$.

3 Show that the caustic of a cycloid is again a cycloid congruent to the original cycloid.

1.5. Abstract Definition of Wave Fronts

As described in the previous section, the parallel curve γ_δ of a regular curve γ can be considered as the wave front of the original curve after a time δ has elapsed. In this section, we call a planar curve obtained as a parallel curve of a regular curve a *classical wave front*. Here, we give a definition of *abstract wave fronts* using properties of such classical wave fronts.

Definition 1.5.1. A C^∞-map $\gamma : I \to \mathbf{R}^2$ defined on an interval I is called a *frontal* if

(a) there exists an $\mathbf{R}^2$-valued C^∞-function $\mathbf{n}(t)$ $(t \in I)$ such that $|\mathbf{n}(t)| = 1$ and $\mathbf{n}(t)$ is perpendicular to $\gamma'(t)$ (that is, $\mathbf{n}(t) \cdot \gamma'(t) = 0$) for all t.

Moreover, γ is called a *wave front*, or a *front* for short, if

(b) $\mathbf{n}'(t) \neq \mathbf{0}$ when $\gamma'(t) = \mathbf{0}$ (that is, t is a singular point of γ).

Remark 1.5.2. This is a remark for readers with knowledge of manifolds. A differentiable map is called an *immersion* if the rank of the Jacobi matrix coincides with the dimension of the domain. The condition (b) is equivalent to the map $(\gamma, \mathbf{n}) : I \to \mathbf{R}^2 \times S^1$ being an immersion, where $S^1 := \{(x, y) \in \mathbf{R}^2 \,;\, x^2 + y^2 = 1\}$.

Under the situation as in Definition 1.5.1, each vector $\mathbf{n}(t)$ is considered as a unit vector which points in the normal direction of the curve at $\gamma(t)$. So, we call $\mathbf{n}(t)$ the *unit normal vector field* of γ. The condition (a) in

Definition 1.5.1 means that the vector field $n(t)$ can be extended smoothly across isolated singular points. One can easily verify that the conditions (a) and (b) for wave fronts do not depend on the choice of parameters.

By definition, a regular planar curve is a wave front. Historically, the definition of wave fronts was extracted from properties of parallel curves. In fact, as we will show later in Proposition 1.5.4, the parallel curves of a plane curve satisfy (a) and (b) in Definition 1.5.1.

Typical examples of wave fronts are cycloids. In fact, although a cycloid is not a regular curve, the unit normal vector field $n(t)$ as in (1.41) of a cycloid is smooth, and $n'(t)$ is non-vanishing everywhere. Hence a cycloid is a wave front. We will show in Proposition 1.5.7 that $3/2$-*cusps* are wave fronts.

As already pointed out (cf. Example 1.4.10), if a wave front has singular points, the direction of unit normal vector fields (leftward or rightward) may change at those singular points.

Example 1.5.3. As seen in Examples 1.4.1 and 1.4.3, the standard cusp and the standard $4/3$-*cusp* are wave fronts. The standard $5/2$-*cusp* is a frontal but not a wave front, as seen in Example 1.4.2.

Let $\gamma(t)$ be a wave front with smooth unit normal vector field $n(t)$. Then we can define its parallel curves

$$\gamma_\delta(t) := \gamma(t) + \delta n(t) \tag{1.42}$$

in the same way as for regular curves. We have the following proposition.

Proposition 1.5.4. *A parallel curve γ_δ of a wave front γ is a wave front for any $\delta \in \mathbf{R}$. The unit normal vector field $n(t)$ of $\gamma(t)$ is also the unit normal vector field of $\gamma_\delta(t)$ for any δ. If $t = c$ is a singular point of $\gamma(t)$, $t = c$ is a regular point of $\gamma_\delta(t)$ for any $\delta \neq 0$. Namely, a singular point moves according to wave propagation.*[8]

Proof. We may assume $\delta \neq 0$. Differentiating $n(t) \cdot n(t) = 1$ in t, we have $2n(t) \cdot n'(t) = 0$, that is, $n'(t)$ is perpendicular to $n(t)$. Hence the tangent vector $\gamma'_\delta(t) := \gamma'(t) + \delta n'(t)$ of a parallel curve is perpendicular to $n(t)$, and so $n(t)$ satisfies (a) of Definition 1.5.1. Assume $\gamma'_\delta(c) = \mathbf{0}$. Then it holds that

$$\gamma'(c) = -\delta n'(c). \tag{1.43}$$

---

[8]In general, the position of a singular point moves to a different point (in the domain of definition). In special cases, singular points can vanish or appear, see (1.62).

Moreover, assume $\gamma'(c) = \mathbf{0}$. Then by condition (b) of Definition 1.5.1 for $\gamma(t)$, $\mathbf{n}'(c) \neq \mathbf{0}$ holds, contradicting (1.43), and thus we have $\gamma'(c) \neq \mathbf{0}$. Since $\delta \neq 0$, we have $\mathbf{n}'(c) \neq \mathbf{0}$ by (1.43), proving the condition (b) for $\gamma_\delta(t)$. By the argument above, $\gamma'(c) = \gamma_\delta'(c) = \mathbf{0}$ never holds for $\delta \neq 0$. Hence the last conclusion is obtained. □

Proposition 1.5.4 implies that if $t = c$ is a singular point of $\gamma(t)$, it is a regular point of $\gamma_\delta(t)$ for $\delta(\neq 0)$. Hence, letting $\sigma := \gamma_\delta$ be an initial curve, its parallel curve

$$\sigma_{-\delta} := \sigma - \delta\mathbf{n}$$

coincides with γ. This yields the following:

Corollary 1.5.5. *Let $\gamma(t)$ be a wave front and $t = c$ a singular point. Then, restricting the domain to a sufficiently small neighborhood of c, γ is a parallel curve of some regular curve.*

Namely, the newly defined notion of "wave fronts" in Definition 1.5.1 is locally obtained as a "classical wave front" as parallel curves associated with a regular curve.

A composition of a regular curve and a diffeomorphism of $\mathbf{R}^2$ is also a regular curve. The same statement holds for wave fronts:

Theorem 1.5.6. *Let $\gamma(t)$ be a wave front (respectively, frontal). Then a curve obtained by changing the parameter of γ is also a wave front (respectively, frontal). On the other hand, let Φ be a local diffeomorphism of a domain in $\mathbf{R}^2$ containing the image of γ. Then $\Phi \circ \gamma(t)$ is a wave front (respectively, frontal) if $\gamma(t)$ is a wave front (respectively, frontal).*

Proof. The first part of the statement is evident. So, we shall prove the second claim. Let $\mathbf{n}(t) = \big(n_1(t), n_2(t)\big)$ be a (smooth) unit normal vector field of $\gamma(t) = \big(x(t), y(t)\big)$. The vector

$$e(t) = \begin{pmatrix} n_2(t) \\ -n_1(t) \end{pmatrix} = R\mathbf{n}(t), \quad R := \begin{pmatrix} 0 & 1 \\ -1 & 0 \end{pmatrix}$$

obtained by a 90° counterclockwise rotation of $\mathbf{n}(t)$ is linearly dependent with $\gamma'(t)$, where we consider $\mathbf{n}(t)$ as a column vector. We remark that $e(t)$ does not necessarily point in the direction of travel. We write $\Phi(x, y) := \big(u(x, y), v(x, y)\big)$ and let $J(\mathrm{P})$ be the Jacobi matrix of Φ at each point $\mathrm{P} = (x, y)$, that is,

$$J(\mathrm{P}) := \begin{pmatrix} u_x(\mathrm{P}) & u_y(\mathrm{P}) \\ v_x(\mathrm{P}) & v_y(\mathrm{P}) \end{pmatrix}.$$

Then $E(t) := J(\gamma(t)) e(t)$ gives a tangent vector of the curve $\Phi \circ \gamma(t)$. In fact, setting

$$\hat{\gamma}(t) := \Phi \circ \gamma(t) = \begin{pmatrix} u(x(t), y(t)) \\ v(x(t), y(t)) \end{pmatrix},$$

we have

$$\hat{\gamma}' = \begin{pmatrix} u_x x' + u_y y' \\ v_x x' + v_y y' \end{pmatrix} = \begin{pmatrix} u_x & u_y \\ v_x & v_y \end{pmatrix} \begin{pmatrix} x' \\ y' \end{pmatrix} = J\gamma'.$$

Let

$$N(t) := R^{-1} E(t),$$

that is, $N(t)$ is a vector obtained by $90°$ rotation of $E(t)$ clockwise. This gives a normal vector field of $\Phi \circ \gamma(t)$. Hence $\Phi \circ \gamma$ is a frontal. Next, we assume $\gamma(t)$ is a wave front. Let $t = c$ be a singular point of $\gamma(t)$. Then $x'(c) = y'(c) = 0$ yields that $\Phi \circ \gamma(t)$ has a singularity at $t = c$, by the chain rule. Since $\gamma(t)$ is a wave front, it holds that $n'(c) \neq 0$. Since

$$\frac{d}{dt}\bigg|_{t=c} J(\gamma(t)) = \begin{pmatrix} u_{xx} x'(c) + u_{xy} y'(c) & u_{yx} x'(c) + u_{yy} y'(c) \\ v_{xx} x'(c) + v_{xy} y'(c) & v_{yx} x'(c) + v_{yy} y'(c) \end{pmatrix} = O,$$

where O is the zero matrix, it holds that

$$N'(c) = R^{-1} \left(\frac{d}{dt}\bigg|_{t=c} J(\gamma(t)) \right) Rn(c) + R^{-1} J(\gamma(c)) \cdot Rn'(c)$$
$$= R^{-1} J(\gamma(c)) Rn'(c) \neq 0,$$

because $J(\gamma(t))$ is a non-singular matrix. The vector field $\hat{n}(t) := N(t)/|N(t)|$ is a unit normal vector field for $\hat{\gamma}(t)$, and it holds that

$$\hat{n}'(t) = \left(\frac{1}{|N(t)|} \right)' N(t) + \left(\frac{1}{|N(t)|} \right) N'(t).$$

Here, $N(c)$ and $N'(c)$ are obtained by multiplying $n(c)$ and $n'(c)$, respectively, by the regular matrix $R^{-1} JR$, and $N(c)$ and $N'(c)$ are linearly independent. In particular, $N'(c) \neq 0$ implies $\hat{n}'(c) \neq 0$. $\quad\square$

Cusps and $4/3$-*cusps* are singularities of wave fronts as follows.

Proposition 1.5.7. *If a planar curve has a cusp or a $4/3$-cusp, the curve is a wave front in a neighborhood of such a singular point.*

Proof. The property of being a wave front is invariant under coordinate changes of R^2 and parameter changes. Since the standard cusp and the standard 4/3-*cusp* are wave fronts, as seen in Example 1.5.3, general cusps and 4/3-*cusps* are also wave fronts.   □

Since a regular curve is a wave front, we have:

Corollary 1.5.8. *A plane curve defined on an interval* $I(\subset \mathbf{R})$ *whose singular points are all cusps is a wave front.*

Proof. Since a unit normal vector field is smoothly defined on a neighborhood of each cusp, one can obtain a unit normal vector field globally defined on I. The property $\mathbf{n}'(t) \neq \mathbf{0}$ is a local property, and we have the conclusion.   □

On the other hand, the standard 5/2-*cusp* is a frontal but not a wave front, so one has an analogous results for general 5/2-*cusps*, that is, a planar curve whose singular points are all cusps or 5/2-*cusps* are frontal.

Remark 1.5.9 (A remark for readers with knowledge of manifolds). Let S^1 be a unit circle in $\mathbf{R}^2$ centered at the origin. Then $\mathbf{R}^2 \times S^1$ can be considered as a three-dimensional submanifold of $\mathbf{R}^4$:

$$\mathbf{R}^2 \times S^1 = \{(x, y, \cos s, \sin s) \in \mathbf{R}^4 \,;\, x, y, s \in \mathbf{R}\},$$

which is identified with the unit tangent bundle $S(T\mathbf{R}^2)$, the set of unit vectors in the tangent bundle $T\mathbf{R}^2$. Using a coordinate system (x, y, s) of $\mathbf{R}^2 \times S^1$, we define the differential 1-form

$$\theta := (\cos s)dx + (\sin s)dy$$

on $\mathbf{R}^2 \times S^1$, and the 3-form

$$\theta \wedge d\theta = (\cos s\, dx + \sin s\, dy) \wedge (-\sin s\, ds \wedge dx + \cos s\, ds \wedge dy)$$
$$= -dx \wedge dy \wedge ds$$

does not vanish on $\mathbf{R}^2 \times S^1$. Thus, θ is a contact differential form on $\mathbf{R}^2 \times S^1$ (see Chapter 10). A map

$$L : I \ni t \longmapsto (\gamma(t), \mathbf{n}(t)) \in \mathbf{R}^2 \times S^1$$

defined on an interval I is called a *Legendrian immersion* if L is an immersion and the pull-back $L^*\theta$ of θ by L vanishes identically. The condition (a) in Definition 1.5.1 is equivalent to $L^*\theta = 0$. In fact, if we set $\gamma(t) = (x(t), y(t))$ and $\mathbf{n}(t) = (\cos s(t), \sin s(t))$, then we have

$$L^*\theta = (x'(t)\sin s(t) + y'(t)\cos s(t))dt = (\gamma'(t) \cdot \mathbf{n}(t))dt.$$

Moreover, the condition (b) is equivalent to L being an immersion. Therefore, *a wave front can be interpreted as the projection of a Legendrian immersion* (cf. Chapter 10).

Exercises 1.5

1 Find the unit normal vector field of the astroid in Exercise **1** in Section 1.1, and directly prove that the astroid satisfies the definition (Definition 1.5.1) of wave fronts, that is, without applying Proposition 1.5.7.

2 Prove from the definition that the cissoid in Exercise **3** in Section 1.1 is a wave front.

3 Determine the family of parallel curves $\{\gamma_\delta\}$ of the astroid γ in Exercise **1** in Section 1.1. Show that the curve γ_δ is a convex closed curve for $|\delta| > 3/2$.

1.6. Curvature Maps and Closed Wave Fronts

Redefinition of curvature for frontals. Let $\gamma(t) = \big(x(t), y(t)\big)$ be a regular curve defined on an interval I, and let $\boldsymbol{n}_L(t)$ be a leftward unit normal vector field. Since

$$\gamma' \cdot \boldsymbol{n}_L' = (\gamma' \cdot \boldsymbol{n}_L)' - \gamma'' \cdot \boldsymbol{n}_L = -\gamma'' \cdot \boldsymbol{n}_L, \tag{1.44}$$

the curvature κ of γ satisfies

$$\kappa = \frac{x'y'' - y'x''}{\big((x')^2 + (y')^2\big)^{3/2}} = \frac{\gamma'' \cdot \boldsymbol{n}_L}{\gamma' \cdot \gamma'} = -\frac{\gamma' \cdot \boldsymbol{n}_L'}{\gamma' \cdot \gamma'}. \tag{1.45}$$

As pointed out at the beginning of Section 1.4, a smooth unit normal vector field for frontals may change from leftward (respectively, rightward) to rightward (respectively, leftward) at a singular point. Thus, if one defines the curvature function using a leftward unit normal vector field via (1.48) below, it may be discontinuous at singular points, which seems unnatural. So, for a frontal (a wave front), fixing a smooth unit normal vector $\boldsymbol{n}(t)$, we define

$$\tilde{\kappa}(t) := \frac{\gamma''(t) \cdot \boldsymbol{n}(t)}{\gamma'(t) \cdot \gamma'(t)}. \tag{1.46}$$

This $\tilde{\kappa}(t)$ is called the *curvature function* with respect to the unit normal vector $\boldsymbol{n}(t)$. Since $\boldsymbol{n}(t)$ might not be leftward, $\tilde{\kappa}$ and κ may have different signs, although their absolute values will coincide. Since $\boldsymbol{n}(t)$ is also the unit normal vector of each parallel curve $\gamma_\delta(t)$ for each δ, the curvature functions of $\gamma_\delta(t)$ satisfy the following proposition.

Proposition 1.6.1. Let $\gamma(t)$ $(a \leq t \leq b)$ be a regular curve, $\mathbf{n}_L(t)$ its leftward unit normal vector field, and $\kappa(t)$ the curvature function with no zeros. Then the curvature function $\tilde{\kappa}_\delta(t)$ of the parallel curve $\gamma_\delta(t)$ with respect to $\mathbf{n}_L(t)$ at a regular point of γ_δ satisfies

$$\frac{1}{\tilde{\kappa}_\delta(t)} = \frac{1}{\kappa(t)} - \delta. \tag{1.47}$$

Remark 1.6.2. Let $\gamma(t)$ be a frontal. Then the curvature function $\kappa_\delta(t)$ of $\gamma_\delta(t)$ with respect to the leftward unit normal of $\gamma_\delta(t)$ satisfies

$$\kappa_\delta(t) = \frac{\kappa(t)}{|1 - \delta\kappa(t)|}. \tag{1.48}$$

Proof of Proposition 1.6.1. Let t be an arc-length parameter of γ. By (1.38), it holds that

$$\gamma_\delta'(t) \cdot \gamma_\delta'(t) = (1 - \delta\kappa(t))^2. \tag{1.49}$$

On the other hand, since $\gamma''(t) = \kappa(t)\mathbf{n}_L(t)$, (1.39) yields

$$\gamma_\delta''(t) = (1 - \delta\kappa(t))'\gamma'(t) + (1 - \delta\kappa(t))\kappa(t)\mathbf{n}_L(t).$$

Substituting this and (1.49) into (1.46), we have

$$\tilde{\kappa}_\delta(t) = \frac{(1 - \delta\kappa(t))\kappa(t)}{(1 - \delta\kappa(t))^2}\mathbf{n}_L(t) \cdot \mathbf{n}_L(t) = \frac{\kappa(t)}{1 - \delta\kappa(t)}, \tag{1.50}$$

which yields the conclusion. □

The curvature map: Since a wave front $\gamma(t)$ has a smooth unit normal vector field $\mathbf{n}(t)$ as in Definition 1.5.1 (even at singular points), the curvature function $\tilde{\kappa}(t)$ with respect to $\mathbf{n}(t)$ is well defined as in (1.46) at regular points. On the other hand, at a singular point, as the denominator of (1.46) vanishes, the curvature function may diverge. For example, a calculation shows that $\tilde{\kappa}(t)$ diverges at a cusp singularity. We now discuss how to extend curvature functions even at singular points.

Considering a fraction a/b as a ratio $a : b$, we can interpret a "fraction" $a/0$ as a ratio $a : 0$ as follows: Consider the set $\mathbf{R}_*^2 := \mathbf{R}^2 \setminus \{(0,0)\}$ of pairs of two real numbers which are not simultaneously 0. We define two pairs (x_1, y_1) and $(x_2, y_2) \in \mathbf{R}_*^2$ to be *equivalent* if the two ratios $x_1 : y_1$ and $x_2 : y_2$ coincide. The quotient space of $\mathbf{R}_*^2$ by such an equivalence relation is called the *real projective line*, denoted by P^1. In other words, P^1 is the set of "ratios of two real numbers". We denote by $[x : y]$ the equivalence class of $(x, y) \in \mathbf{R}^2 \setminus \{(0,0)\}$. Then we can write

$$P^1 := \{[x : y] ; (x, y) \in \mathbf{R}_*^2\}$$

and the map

$$\iota_0 : \boldsymbol{R} \ni t \mapsto [t : 1] \in P^1 \tag{1.51}$$

is an identification of fractions with ratios. Through this map, we can consider $\boldsymbol{R} \subset P^1$ and $1/0 = [1 : 0]$ can be considered as ∞, so P^1 can be identified with $\boldsymbol{R} \cup \{\infty\}$. A map $\varphi : I \to P^1$ is said to be *of class* C^∞ if there exist two C^∞-functions $x(t)$, $y(t)$ such that $\varphi(t) = [x(t) : y(t)]$. Though the denominator of the definition (1.46) of the curvature function $\tilde{\kappa}(t)$ may vanish at singular points of wave fronts, we can extend to that case by using the notion of a projective line.

Let $\gamma : I \to \boldsymbol{R}^2$ be a wave front with unit normal vector field $\boldsymbol{n}$. Then γ' and $\boldsymbol{n}'$ are linearly dependent, and at least one of them is not zero. Thus, the proportionality constant of these two vectors can be considered as a ratio

$$[\gamma'(t) : \boldsymbol{n}'(t)] \in P^1.$$

For example, if there exists a function $a(t)$ (respectively, $b(t)$) such that $\gamma'(t) = a(t)\boldsymbol{n}'(t)$ (respectively, $\boldsymbol{n}'(t) = b(t)\gamma'(t)$), then $[\gamma' : \boldsymbol{n}'] = [a : 1]$ (respectively, $[\gamma' : \boldsymbol{n}'] = [1 : b]$). In particular, the map

$$\omega : I \ni t \mapsto [-\boldsymbol{n}'(t) : \gamma'(t)] \in P^1 \tag{1.52}$$

is called the *curvature map*. By definition, $\omega(c) = [1 : 0]$ if and only if $t = c$ is a singular point of γ. Moreover, we call $t = c$ an *inflection point* of γ if $\omega(c) = [0 : 1]$. Summarizing these arguments, we obtain the following:

Theorem 1.6.3. *The curvature map of a wave front* $\gamma : I \to \boldsymbol{R}^2$ *is of class* C^∞. *If the set of regular points* I_R *of* γ *is dense in* I, *it holds that*

$$\omega(t) = [-\gamma'(t) \cdot \boldsymbol{n}'(t) : \gamma'(t) \cdot \gamma'(t)] = [\tilde{\kappa}(t) : 1] \quad (t \in I_R).$$

Here, $\tilde{\kappa}(t)$ *is the curvature function of* γ *on* I_R *with respect to the unit normal vector field* $\boldsymbol{n}(t)$. *Moreover, any inflection points of* γ *are regular points.*

Let $\gamma(t)$ be a wave front without inflection points defined on an interval I. There exists a unique C^∞-function $\rho : I \to \boldsymbol{R}$ such that

$$[1 : \rho(t)] = \omega(t). \tag{1.53}$$

By definition,

$$\gamma'(t) + \rho(t)\boldsymbol{n}'(t) = \boldsymbol{0} \tag{1.54}$$

holds. In particular, a singular point is a point where $\rho(t) = 0$. We define the *caustic* (or the *evolute*) σ of the wave front γ, which has no inflection points, as follows:

$$\sigma(t) := \gamma(t) + \rho(t)\boldsymbol{n}(t). \tag{1.55}$$

This is a generalization of caustics of regular curves (cf. (1.11)). Note that the caustic cannot be defined at inflection points, because $\rho(t) = \infty$ at an inflection point t.

Proposition 1.6.4. *The caustic of a wave front γ without inflection points is also a wave front.*

Proof. Differentiating (1.55), we have

$$\sigma'(t) = \gamma'(t) + \rho(t)\boldsymbol{n}'(t) + \rho'(t)\boldsymbol{n}(t).$$

Then by (1.54), it holds that

$$\sigma'(t) = \rho'(t)\boldsymbol{n}(t).$$

Thus, the unit tangent vector (that is, the unit vector perpendicular to $\boldsymbol{n}(t)$) is a unit normal vector of $\sigma(t)$. Hence σ is a frontal.

The unit normal vector is obtained by $90°$-rotation of the unit tangent vector. Hence to show that σ is a wave front, it is sufficient to show that $\boldsymbol{n}'(t) \neq \boldsymbol{0}$ at a singular point of $\sigma(t)$. If t is regular point of γ, $\boldsymbol{n}'(t) \neq \boldsymbol{0}$ holds, because t is not an inflection point of γ. On the other hand, if t is a singular point of γ, $\boldsymbol{n}'(t) \neq \boldsymbol{0}$ holds, because γ is a wave front. $\square$

Remark 1.6.5 (Invariance of vertices under wave propagation). A point on a regular curve is called a *vertex* if it is a critical point (that is, a point where the derivative vanishes) of the curvature function. By the proof of Proposition 1.6.4, a vertex of a curve corresponds to a singular point of its caustic. It is well-known that there exist at least four vertices on a simple closed curve (this is by the four-vertex theorem, see [84], Section 4). In particular, the caustic of a simple closed curve with positive curvature function (that is, a convex curve) has at least four singular points. See the upcoming Section 1.7.

Differentiating (1.50), we have

$$\tilde{\kappa}_\delta'(t) = \frac{\kappa'(t)}{(\delta\kappa(t) - 1)^2}.$$

The denominator of the right-hand side vanishes only at singular points of $\gamma_\delta(t)$. Hence the property of being a vertex is common to all parallel curves whenever they are regular. Generalizing the definition of vertices, we also call a critical point of the curvature map ω a *vertex*. In fact, if the curvature map ω to P^1 is represented as $\omega(t) = [\kappa(t) : 1]$ (respectively, $\omega(t) = [1 : \rho(t)]$), a point t where $\kappa'(t) = 0$ (respectively, $\rho'(t) = 0$) is the "critical point" of $\Phi(t)$. Indeed, $\kappa(t)\rho(t) = 1$ if $\kappa(t), \rho(t) \neq 0$. In this case, $\kappa'(t) = 0$ if and only if $\rho'(t) = 0$. One can then easily prove that the property of being a vertex is a property common to the family of wave fronts.

Closed fronts and curvature maps

Definition 1.6.6. Assume a C^∞-frontal $\tilde{\gamma} \colon \mathbf{R} \to \mathbf{R}^2$ has a period a for some positive number a, that is, it satisfies (1.16), and $\tilde{\gamma}$ induces a map $\gamma \colon \mathbf{R}/a\mathbf{Z} \to \mathbf{R}^2$, called a *closed frontal*. Since γ is a frontal, there exists a C^∞-map $\mathbf{n} \colon \mathbf{R} \to \mathbf{R}^2$ such that $\mathbf{n}(t)$ is a unit normal vector of γ at t for each $t \in \mathbf{R}$. Since $\gamma(t)$ is a-periodic, either

$$\mathbf{n}(t) = \mathbf{n}(t+a) \tag{1.56}$$

or

$$\mathbf{n}(t) = -\mathbf{n}(t+a). \tag{1.57}$$

In the former case (1.56), γ is said to be *co-orientable*, and in the latter case (1.57), γ is said to be *non-co-orientable*.

When all singularities of a closed frontal are cusps, it is co-orientable if and only if the number of cusps is even, since the normal vector field changes from leftward (respectively, rightward) to rightward (respectively, leftward) at each cusp. As seen in Fig. 1.19(c) and Fig. 1.11(b), there exists a closed curve with an odd number of cusps. These are considered wave fronts locally, and can be considered wave fronts globally if one allows the curves to be traversed twice: This corresponds to regarding γ to be $2a$-periodic. Then its unit normal vector field $\mathbf{n}$ satisfies

$$\mathbf{n}(t + 2a) = -\mathbf{n}(t+a) = \mathbf{n}(t).$$

Thus $\tilde{\gamma}|_{[0,2a]}$ gives a co-orientable front as a double covering of the non-co-orientable front γ.

For this reason, we now consider only co-orientable wave fronts $\gamma \colon \mathbf{R}/a\mathbf{Z} \to \mathbf{R}^2$, and the phrase "closed fronts" then means closed co-orientable wave fronts. If, for each t, the pair of vectors $(\gamma'(t), \mathbf{n}'(t))$ does not vanish simultaneously, γ is called a *closed front* (see Fig. 1.11).

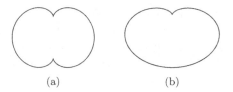

(a)   (b)

Fig. 1.11. Closed curves with even and odd numbers of cusps.

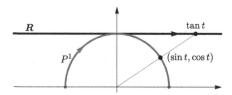

Fig. 1.12. The identification of P^1 with $\boldsymbol{R} \cup \{\infty\}$ by the central projection.

We normalize the period a of a closed front to be $a = 2\pi$, by rescaling $\boldsymbol{R}$ if necessary. Since the parametrization

$$\boldsymbol{R}/2\pi\boldsymbol{Z} \ni t \longmapsto (\sin t, \cos t) \in S^1 := \{(x, y) \in \boldsymbol{R}^2 \; ; \; x^2 + y^2 = 1\} \quad (1.58)$$

of the unit circle S^1 is bijective, we identify S^1 with $\boldsymbol{R}/2\pi\boldsymbol{Z}$. Then γ is a map from S^1 to $\boldsymbol{R}^2$.

On the other hand, a C^∞-map

$$\iota_1 : \boldsymbol{R}/\pi\boldsymbol{Z} \ni t \longmapsto [\sin t : \cos t] \in P^1 \quad (1.59)$$

is also a bijection, and (cf. (1.51))

$$\iota_0 \circ \iota_1(t) = \sin t / \cos t = \tan t$$

is monotone increasing. So this map $\iota_0 \circ \iota$ preserves the canonical orientations of $\boldsymbol{R}/\pi\boldsymbol{Z}$ and $\boldsymbol{R}$, and so the orientation of P^1 is uniquely determined so that ι_0 and ι_1 are both orientation preserving maps. Since, we can identify $\boldsymbol{R}/\pi\boldsymbol{Z}$ with S^1 by the map

$$\boldsymbol{R}/\pi\boldsymbol{Z} \ni t \mapsto (\sin 2t, \cos 2t) \in S^1, \quad (1.60)$$

P^1 can be canonically identified with S^1 by the maps (1.59) and (1.60). Identifying P^1 with S^1 in this way, $[0 : 1](= \iota_1(0))$, $[1 : 1](= \iota_1(\pi/4))$ and $[1 : 0](= \iota(\pi/2))$ on P^1 (i.e., 0, 1 and ∞ on $\boldsymbol{R} \cup \{\infty\}$) correspond to $(0, 1)$, $(1, 0)$ and $(0, -1)$ on S^1, respectively. Thus, clockwise rotation of S^1 corresponds to the natural orientation of P^1 (see Fig. 1.12).

Remark 1.6.7. The ratio $[a : b]$ is usually identified with a/b. However, if we identify $[a : b]$ with b/a, then the map (1.59) should be changed to $\iota(t) := [\cos t : \sin t] \in P^1$.

Let $\gamma : \boldsymbol{R}/2\pi\boldsymbol{Z} \to \boldsymbol{R}^2$ be a frontal. Then in the same way as for the regular curves, the unit normal vector field $\boldsymbol{n}$ of γ can be considered as a map $\boldsymbol{n} : \boldsymbol{R}/2\pi\boldsymbol{Z} \to S^1$, which is called the *Gauss map* of γ. Then the winding number i_γ of this map is called the *rotation index* of the frontal γ. In fact, there exists a smooth function $\alpha : [0, 2\pi] \to \boldsymbol{R}$ such that

$$\boldsymbol{n}(t) = (\cos\alpha(t), \ \sin\alpha(t)).$$

Then the rotation index of γ can be written as

$$i_\gamma = \frac{\alpha(2\pi) - \alpha(0)}{2\pi}.$$

When $\gamma(t)$ is a regular closed curve, i_γ coincides with the rotation index defined in Section 1.2 for regular closed curves.

We now assume that γ is a wave front. Then the winding number of the curvature map $\omega : \boldsymbol{R}/2\pi\boldsymbol{Z} \longrightarrow P^1$ is defined as follows: As the parameter runs over $\boldsymbol{R}/2\pi\boldsymbol{Z}$, the image of the curvature map ω turns about P^1. Then the winding number w_γ of ω is the number of turns of ω, taking the direction into account. In fact, since ω is a smooth map, there exists a smooth function $\beta : [0, 2\pi] \to \boldsymbol{R}$ such that

$$\omega(t) = [\sin\beta(t), \ \cos\beta(t)].$$

Then w_γ can be written as

$$w_\gamma = \frac{\beta(2\pi) - \beta(0)}{\pi}.$$

Replacing the (continuous) unit normal vector field $\boldsymbol{n}(t)$ with $-\boldsymbol{n}(t)$, the winding number of ω changes to $-w_\gamma$. Regarding this, we set

$$z_\gamma := \frac{|w_\gamma|}{2}, \tag{1.61}$$

which we call the *zig-zag number* or the *Maslov index* of the closed front.[9] If γ has no singularities, the curvature map $\omega(t)$ cannot pass through the point $[1 : 0]$, by Theorem 1.6.3. Hence the zig-zag number of a regular closed curve is zero.

Remark 1.6.8. The rotation index i_γ does not change when replacing $\boldsymbol{n}$ with $-\boldsymbol{n}$, and does change sign when the direction of γ is inverted. On the other hand, the winding number of ω does not change when the direction

---

[9]By definition, it is a half-integer. But we will see later that its range is the set of non-negative integers (see Corollary 1.6.12).

of γ is inverted, and changes sign when $\boldsymbol{n}$ is replaced with $-\boldsymbol{n}$. So, in the definition of zig-zag number (cf. (1.61)) we use the "absolute value" of the winding number of ω, which makes the zig-zag number independent of choice of unit normal vector field. Since $\boldsymbol{n}$ is globally well defined on the curve, the zig-zag number is a non-negative integer.

We introduce the equivalence relation of closed fronts:

Definition 1.6.9. Two closed wave fronts $\sigma_i(t) : \boldsymbol{R}/2\pi\boldsymbol{Z} \to \boldsymbol{R}^2$ $(i = 1, 2)$ are *equivalent* (more precisely, *Legendrian homotopic*) if there exists a C^∞ map $\Gamma : \boldsymbol{R}/2\pi\boldsymbol{Z} \times [0, 1] \ni (t, s) \mapsto \gamma_s(t) \in \boldsymbol{R}^2$ and

$$\boldsymbol{N} : \boldsymbol{R}/2\pi\boldsymbol{Z} \times [0, 1] \ni (t, s) \mapsto \boldsymbol{n}_s(t) \in S^1$$

such that

(1) $\Gamma(t, 0) = \sigma_1(t)$ and $\Gamma(t, 1) = \sigma_2(t)$ hold for each t,
(2) for each $s \in [0, 1]$, $\gamma_s : \boldsymbol{R}/2\pi\boldsymbol{Z} \to \boldsymbol{R}^2$ is a wave front, and
(3) for each $s \in [0, 1]$, $\boldsymbol{n}_s : \boldsymbol{R}/2\pi\boldsymbol{Z} \to S^1$ is a normal vector field of γ_s.

Then the following fact holds:

Fact 1.6.10 ([2]). Two closed fronts are equivalent in the sense of Definition 1.6.9 if and only if both the rotation index and the zig-zag number coincide.

In other words, any closed wave front can be deformed to a front as in Fig. 1.13, preserving the properties for being a wave front. The curvature function $\tilde{\kappa}(t)$ changes its sign across a cusp. Thinking of P^1 as $\boldsymbol{R} \cup \{\infty\}$, if a point t in P^1 travels from a positive value to a negative one passing through ∞, this movement is compatible with respect to the natural orientation of P^1.

At a neighborhood of a cusp, one can choose the unit normal vector $\boldsymbol{n}(t)$ such that the curvature $\tilde{\kappa}(t)$ is positive before the point passes through the cusp. When the cusp is zig (respectively, zag), this unit normal vector $\boldsymbol{n}(t)$

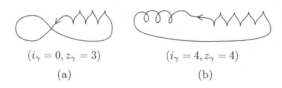

$(i_\gamma = 0, z_\gamma = 3)$ $(i_\gamma = 4, z_\gamma = 4)$

(a)   (b)

Fig. 1.13.   Normal form of wave fronts with rotation index 0 and positive rotation index.

points to the left side (respectively, right side) of the curve before the cusp. The following proposition holds:

Proposition 1.6.11. *The image of the curvature map between two consecutive cusps travels around P^1 once (resp. does not produce any rotation effect) if the two cusps are both zigs or both zags (resp. the two cusps consist of one zig and one zag).*

Proof. If the two cusps are one zig and one zag (resp. are of the same type), then the image of the curvature map approaches the point $[1 : 0] \in P^1$ from the same direction (resp. opposite directions) at the two cusps. So we obtain the assertion. □

We have the following corollary.

Corollary 1.6.12. *Assume we have a closed wave front whose singular points are all cusps, such that zigs and zags appear alternatingly $2l$ times. Then the zig-zag number of the wave front is l. In particular, the zig-zag number is a non-negative integer, because any closed wave front is equivalent to the normal form as in Fig. 1.12.*

Proof. Focusing on a subarc from a zig to the following zag, if the unit normal vector field points to the left side (respectively, right side) before the zig, then it points to the right side (respectively, left side) before the zag. Hence the curvature map turns in the same direction at the two cusps, and then it turns around from $[1 : 0]$ and comes back to the same point. Such an arc continues to appear $2l$ times, so the winding number of ω is $\pm 2l$. □

The rotation index of a regular curve is invariant under deformations as in Fig. 1.14.

A regular closed curve is said to be *generic* if all self-intersections are double points where two pieces of curves meet transversely. Any regular closed curve can be deformed to a generic closed curve by a small perturbation. The same statement is true for closed wave fronts. In fact, if we define generic wave front as follows, then it is known that any closed wave front can be deformed to a generic wave front by a small perturbation.

Definition 1.6.13. A closed wave front γ is said to be *generic* if

(1) all singularities of γ are cusps, and such points are not self-intersections of γ, and

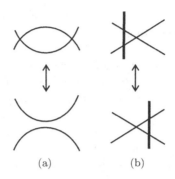

(a)   (b)

Fig. 1.14.   Deformations of wave fronts preserving the rotation index.

(2) the number of self-intersections of γ is finite, and each self-intersection is a double point and is transversal (that is, the two pieces of the curve at the double point have linearly independent tangential directions).

Let $\gamma : \mathbf{R}/2\pi\mathbf{Z} \to \mathbf{R}^2$ be a generic closed wave front and $\mathbf{n}$ its unit normal vector field, and consider the map

$$L := (\gamma, \mathbf{n}) : \mathbf{R}/2\pi\mathbf{Z} \longrightarrow \mathbf{R}^2 \times S^1.$$

At each self-intersection of γ, the unit normal vectors of the two pieces of the curve are linearly independent. This means that L has no self-intersections. In other words, L is a knot in the 3-manifold $\mathbf{R}^2 \times S^1$. This is called a *Legendrian knot* in (the contact manifold) $\mathbf{R}^2 \times S^1$. For the definition of contact manifolds, see Chapter 10, and see [2] for Legendrian knots.

Deformations of regular curves as in Fig. 1.14 are considered as deformations of wave fronts. We consider another deformation of wave fronts as follows. Fix a real number c and set

$$\gamma_c(t) := (3t^4 - ct^2, 4t^3 - 2ct) \quad (t \in \mathbf{R}). \tag{1.62}$$

Then γ_c is a wave front for each real number c. In fact, γ_c is a regular curve when $c < 0$, and when $c = 0$, the corresponding curve γ_0 has a 4/3-cusp at $t = 0$, and one can check that γ_c $(c > 0)$ has cusp singularities at $t = \pm\sqrt{c/6}$, by Theorem 1.3.2. Hence, by Proposition 1.5.7, each γ_c is a wave front for each $c \in \mathbf{R}$. Moreover, the unit normal vector field of γ_c does not depend on the parameter c, so this gives a Legendrian homotopic deformation of wave fronts. Figure 1.15 shows images of γ_c, and this operation for wave fronts generates a pair of cusps and one self-intersection.

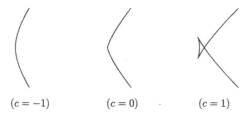

$$(c = -1) \qquad (c = 0) \qquad (c = 1)$$

Fig. 1.15. The images of γ_c.

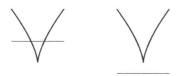

Fig. 1.16. A fundamental operation for cusps.

In addition, we consider another operation as in Fig. 1.16. By this operation, one can remove (or generate) a pair of crossings on the curve.

Summing up, we have introduced four operations on generic closed wave fronts, two as in Fig. 1.14, and two as in Figs. 1.15 and 1.16. We call these the four *fundamental operations of wave fronts*. The following holds, which is a generalization of Fact 1.2.5 for regular closed curves.

Fact 1.6.14. Two generic closed wave fronts can be deformed from one to another using finite numbers of fundamental deformations if and only if both the rotation index and the zig-zag number coincide.

Recall that a cusp is called a zig (respectively, zag) if it is right-turning (respectively, left-turning) (cf. Proposition 1.3.12). Suppose that there are two cusps on the curve, and there are no other cusps between them. Then, by deformations as in Fig. 1.14 and Fig. 1.16, we can move the two cusps to consecutive cusps. In fact, by deformations as in Fig. 1.14 and Fig. 1.16, a piece of a curve can move across a cusp as shown in Fig. 1.17. As an application, we show that the order of zigs and zags is not essential for closed wave fronts. In fact, using the deformation above, if a zig is followed by a zag, one can move a zig around the curve, and put it before the zag. If we deform two zigs to consecutive zigs, then the deformation as in Fig. 1.15 can eliminate these zigs. Applying these operations to a given generic closed wave front γ, we can eliminate consecutive zigs and zags. After these deformations, zigs and zags are located alternately on the curve

Fig. 1.17.   Deformations of cusps.

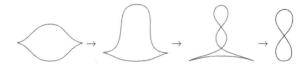

Fig. 1.18.   The "eye shape" to the "figure eight".

and half of the number of remaining cusps is nothing but the zig-zag number of the closed wave front γ (see Fig. 1.18).

Remark 1.6.15. Fact 1.6.14 can be proved with respect to induction on the number of crossings. In fact, one can decrease the number of crossings by the operation as in Fig. 1.15. Using the fact that any closed (co-orientable) wave front can be deformed to a generic one by a small perturbation, Fact 1.6.14 is a consequence of Fact 1.6.10.

Example 1.6.16. For each non-negative number a, set

$$\gamma_a(t) := (1 + a\cos t)(\cos t, \sin t) \quad (0 \le t \le 2\pi).$$

These curves are called *limaçon*, which is a circle when $a = 0$, and the cardioid (1.2) when $a = 1$ (Fig. 1.19(b)). When $a = 2$, it is a locally convex regular closed curve having one self-intersection (see Fig. 1.19(c)).

The curve is a simple closed curve of rotation index 1 when $a < 1$ (Fig. 1.19(a)). On the other hand, it is a closed curve of rotation index 2 when $a > 1$. This means that γ_a is not a continuous deformation as wave fronts, although each γ_a itself is a wave front. In fact, the cusp appears along the way in the deformation (at $a = 1$ and $t = \pi$), and the unit normal vector field $\boldsymbol{n}_a$ of γ_a is not continuous as a vector-valued function of two variables. More precisely, setting $a = -\cos t$, one can check that

$$\lim_{t \to \pi+0} \boldsymbol{n}_{-\cos t}(t) = -\lim_{t \to \pi-0} \boldsymbol{n}_{-\cos t}(t) = (1, 0).$$

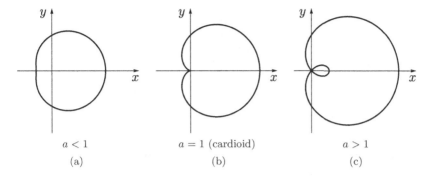

Fig. 1.19. Limaçon (Example 1.6.16).

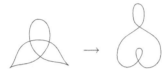

Fig. 1.20. Cancellation of cusps by the four fundamental deformations.

Exercises 1.6

1 Show that two closed wave fronts as in Fig. 1.20 below can be deformed from one to the other by the four fundamental deformations (like as in Fig. 1.18).

2 Regarding the deformation in Example 1.6.16, show that the unit normal vectors of $\gamma_c(t)$ are discontinuous with respect to t when $c = 1$.

1.7. Caustics of Convex Curves

The locus of the centers of osculating circles of a plane regular curve is called the caustic or evolute, as in the previous section, which is a wave front in general (cf. Proposition 1.6.4). The caustic is the envelope of the normals of a given curve, by Fact 1.2.2. For example, Fig. 1.21 shows the family of normals of an ellipse.

As shown in the figure, four cusps appear in the caustic of an ellipse, which correspond to the vertices (that is, critical points of the curvature function of the ellipse). Using the curvature function $\kappa(t)$ of a given regular curve $\gamma(t)$, the caustic is given by

$$\sigma(t) := \gamma(t) + \frac{1}{\kappa(t)}\boldsymbol{n}_L(t). \tag{1.63}$$

Fig. 1.21.   Normals of an ellipse and its envelope (that is, the caustic).

Here, $n_L(t)$ is a leftward unit normal vector field of the regular curve $\gamma(t)$. If t is the arc-length parameter, $n_L'(t) = -\kappa(t)\gamma'(t)$ holds. Hence we have

$$\sigma'(t) = -\frac{\kappa'(t)}{\kappa(t)^2}n_L(t).$$

This yields that

(i) a singular point of the caustic, that is, a point where $\sigma'(t) = \mathbf{0}$, is a critical point of the curvature function $\kappa(t)$, namely a point where $\kappa'(t) = 0$, of the original curve $\gamma(t)$,

(ii) the tangent direction of the caustic σ is the normal direction of the original curve γ, and

(iii) an inflection point of the original curve γ, that is, a point where $\kappa(t_0)$ vanishes, is a point where $\lim_{t \to t_0} |\sigma(t)| = \infty$.

Hence the caustic of $\gamma(t)$ is bounded if and only if the curvature function $\kappa(t)$ of $\gamma(t)$ has no zeros. By the property (ii), $n_L(t)$ of $\gamma(t)$ gives a unit tangent vector of the caustic $\sigma(t)$, and so σ is a frontal. So one can define the rotation index of σ. It is known that the line segment joining two points on a convex curve lies in the interior of the convex curve except at its endpoints (see [84, Appendix B2]). Here, the interior of the simple closed curve is a bounded domain whose boundary is the image of the curve. We have the following:

Proposition 1.7.1. *The caustic of a convex curve is a closed wave front whose rotation index is* 1.

Proof. The caustic is a wave front by Proposition 1.6.4. Let $\gamma(t)$ $(0 \le t \le a)$ be a convex curve and n_L its leftward unit normal vector field. Then n_L is a unit tangent vector field of the caustic $\sigma := \gamma + (1/\kappa)n_L$. Since the rotation index of a convex curve is 1, n_L turns around once about the unit circle from $t = 0$ to $t = a$, and so does the tangent vector of σ. $\qquad\square$

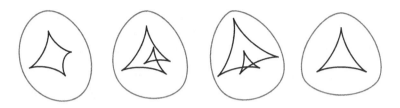

Fig. 1.22. Convex curves (gray) and their caustics (black).

From now on, convex curves will be parametrized by the arc-length parameter with positive curvature function. Then the caustic σ of a convex curve γ satisfies

$$\sigma''(t) = -\left(\frac{\kappa'(t)}{\kappa(t)^2}\right)' \boldsymbol{n}_L(t) + \frac{\kappa'(t)}{\kappa(t)}\gamma'(t). \tag{1.64}$$

Hence

$$\det\big(\sigma'(t), \sigma''(t)\big) = \det\left(-\frac{\kappa'(t)}{\kappa(t)}^2 \boldsymbol{n}_L(t), \frac{\kappa'(t)}{\kappa(t)}\gamma'(t)\right) = \frac{\kappa'(t)^2}{\kappa(t)^3} \tag{1.65}$$

holds. As the right-hand side is positive whenever $\kappa'(t) \neq 0$, we have the following proposition:

Proposition 1.7.2. *The caustic of a convex curve has no inflection points.*

In addition, we have (cf. Exercise **2** in Section 1.4) the following propositions:

Proposition 1.7.3. *The caustic $\sigma(t)$ of a given convex curve $\gamma(t)$ has a cusp at $t = c$ if and only if $\kappa'(c) = 0$ and $\kappa''(c) \neq 0$ hold.*

Cusps in caustics have the following property:

Proposition 1.7.4. *Cusps of the caustics σ of a given convex curve γ are all zigs or all zags. In particular, the zig-zag number of the caustic is 0.*

Proof. By Corollary 1.3.11, the curvature function κ_σ of the caustic σ (with respect to the leftward normal vector field) does not change sign at each cusp point. By Proposition 1.7.2, σ does not have inflection points. So the sign of κ_σ does not change anywhere (cf. (1.65)). Thus, the cusps of the caustic are all zigs or all zags. $\qquad\square$

Figure 1.22 shows some convex curves and their caustics. In the rightmost figure of Fig. 1.22, it seems that the number of cusps of the caustic is odd. However, in this case, the caustic travels twice over its image as the original curve is traversed only once. This case occurs only when γ is a *curve of constant width*.[10] More precisely, the following fact holds:

Fact 1.7.5 ([24, Proposition 2.6]). Let $\sigma(t)$ be the caustic of a convex curve $\gamma(t)$, and assume all singularities of σ are cusps. Then the following three assertions are equivalent:

(1) The caustic σ travels twice over the image.
(2) The number of cusps on the image of the caustic is odd.
(3) γ is a curve of constant width.

For two given convex curves, one curve can be deformed to the other while preserving the property of convexity (see [62, Proposition 1.1]). The following "zero sum law of the signed lengths" [20] is an important property for characterizing caustics.

Fact 1.7.6. Let $\sigma(t)$ be the caustic of a convex curve $\gamma(t)$, and assume all singularities of σ are cusps. If subarcs between two consecutive cusps are numbered in order, then the total sum of the lengths of odd subarcs equals the total sum of the lengths of even subarcs.

For a precise proof, see [20, 24]. Conversely, the following holds:

Fact 1.7.7 ([24]). Suppose a simple closed curve σ satisfies that

(1) all singular points are cusps,
(2) σ has no inflection points,
(3) the rotation number is 1, and
(4) the sum of signed lengths is zero.

Then σ is a caustic of some convex curve.

For example, the astroid is a caustic of a certain convex curve. When the number of cusps is exactly four, the topological type of the image of the caustic is determined in [24]. For more advanced topics on cusps, see Appendix B.

Finally, it should be remarked that an excellent survey of the geometry of spherical curves as wave fronts is given by Arnol'd [3].

---

[10]A (closed) convex curve whose width for every direction is the same constant is called a *curve of constant width*.

Chapter 2

Singularities of Surfaces

In this chapter, we first review the classical surface theory. Then, based on properties of parallel surfaces for regular surfaces, the definition and properties of (abstract) "wave fronts" are given. We introduce typical singularities of surfaces from the viewpoint of wave fronts.

2.1. Regular Surfaces

Regular surfaces. In this book, we denote the *inner product* of a and $b \in \mathbf{R}^3$ by $a \cdot b$, and the *vector product* by $a \times b$. The *norm* of a is denoted by $|a| = \sqrt{a \cdot a}$.

Let $U \subset \mathbf{R}^2$ be a domain (e.g., a connected open subset) of the uv-plane $\mathbf{R}^3$ and consider a C^∞-map $f \colon U \to \mathbf{R}^2$. A point $p \in U$ where the two partial derivatives f_u and f_v are linearly independent at p, namely,

$$f_u(p) \times f_v(p) \neq \mathbf{0} \quad \left(f_u := \frac{\partial f}{\partial u}, \quad f_v := \frac{\partial f}{\partial v} \right)$$

holds, is called a *regular point* of f, and p is said to be a *singular point* if p is not a regular point. We call a C^∞-map $f \colon U \to \mathbf{R}^3$ a *surface*, and call it a *regular surface* if f has no singular points.[1] To represent a surface in this way (as a map from a domain of $\mathbf{R}^2$ into $\mathbf{R}^3$) is called a *parametrization*, and we call such f a *parametrized surface*, with the coordinates (u, v) of $\mathbf{R}^2$ being called *parameters* or a *local coordinate system*.

Let (u_0, v_0) be a regular point of f. Then the tangent plane of the surface at $f(u_0, v_0)$ is spanned by the vectors $f_u(u_0, v_0)$ and $f_v(u_0, v_0)$,

---

[1] In the terminology of manifold theory, that "f has no singular points" means that "f is an immersion".

since they are the tangent vectors of space curves $u \mapsto f(u, v_0)$ (called the *u-curve*) and $v \mapsto f(u_0, v)$ (called the *v-curve*) passing through $f(u_0, v_0)$, respectively. Thus, the non-zero vector product $f_u \times f_v$ is a normal vector of f, and then

$$\pm \frac{f_u(u, v) \times f_v(u, v)}{|f_u(u, v) \times f_v(u, v)|} \qquad (2.1)$$

gives a *unit normal vector* of the surface at $f(u, v)$. Here, we fix the $\pm$-ambiguity and denote this unit normal vector by ν.

A typical example of a regular surface is the graph of a function φ of two variables, which is parametrized as

$$f(x, y) := (x, y, \varphi(x, y)).$$

Its unit normal can be chosen as

$$\nu := \frac{1}{\sqrt{1 + (\varphi_x)^2 + (\varphi_y)^2}} (-\varphi_x, -\varphi_y, 1). \qquad (2.2)$$

The first and second fundamental forms: Here, we define quantities to express fundamental properties, called the first and second fundamental forms. Let $f : U \to \mathbf{R}^3$ be a regular surface.

The first fundamental form: Define three C^∞-functions

$$E := f_u(u, v) \cdot f_u(u, v), \quad F := f_u(u, v) \cdot f_v(u, v), \quad G := f_v(u, v) \cdot f_v(u, v)$$

on U, and consider the formal sum

$$ds^2 := df \cdot df = (f_u \, du + f_v \, dv) \cdot (f_u \, du + f_v \, dv)$$

$$= (f_u \cdot f_u) du^2 + 2(f_u \cdot f_v) du dv + (f_v \cdot f_v)^2 dv^2$$

$$= E \, du^2 + 2F \, du \, dv + G \, dv^2 = (du \ dv) \begin{pmatrix} E & F \\ F & G \end{pmatrix} \begin{pmatrix} du \\ dv \end{pmatrix}. \qquad (2.3)$$

We call ds^2 the *first fundamental form*, and E, F, G its coefficients. The matrix $\left(\begin{smallmatrix} E & F \\ F & G \end{smallmatrix} \right)$ is called the *first fundamental matrix*.

Regarding (f_u, f_v) as a 3×2-matrix consisting of two column vectors f_u and f_v, we can write

$$\begin{pmatrix} E & F \\ F & G \end{pmatrix} = (f_u, f_v)^T (f_u, f_v), \qquad (2.4)$$

where $(f_u, f_v)^T$ is the transpose of (f_u, f_v). By Lagrange's identity [84, (A.3.15)],

$$EG - F^2 = (f_u \cdot f_u)(f_v \cdot f_v) - (f_u \cdot f_v)^2 = |f_u \times f_v|^2 > 0$$

holds, and the first fundamental matrix is a symmetric matrix whose eigenvalues are positive.

Remark 2.1.1. With a knowledge of manifold theory, one can interpret the formal sum (2.3) as follows: Let "$\otimes$" be the tensor product, and consider du^2, $du\,dv$, dv^2 in (2.3) as the *symmetric tensor products* of du and dv, that is,

$$du^2 = du \otimes du, \quad du\,dv = (du \otimes dv + dv \otimes du)/2, \quad dv^2 = dv \otimes dv.$$

Then ds^2 is a symmetric covariant tensor on the surface. Under these terminologies, the first fundamental form is the pull-back of the canonical Riemannian metric (e.g., the canonical inner product) of $\boldsymbol{R}^3$ by the map f.

Take a coordinate change[2] $u = u(a,b)$, $v = v(a,b)$. Then we obtain a new parametrization $\tilde{f}(a,b) = f\big(u(a,b), v(a,b)\big)$. For simplicity, we may write $\tilde{f}(a,b)$ by $f(a,b)$. Then we have (see [84, Section 7])

$$\begin{pmatrix} du \\ dv \end{pmatrix} = J \begin{pmatrix} da \\ db \end{pmatrix}, \quad J := \begin{pmatrix} u_a & u_b \\ v_a & v_b \end{pmatrix}, \tag{2.5}$$

where J is the Jacobi matrix of the coordinate change, which is regular at each point. We denote by $\tilde{E}$, $\tilde{F}$, $\tilde{G}$ the coefficients of the first fundamental form with respect to new coordinate system (a,b). By (2.4) and (2.5), we have

$$(da, db) \begin{pmatrix} \tilde{E} & \tilde{F} \\ \tilde{F} & \tilde{G} \end{pmatrix} \begin{pmatrix} da \\ db \end{pmatrix} = ds^2 = (du, dv) \begin{pmatrix} E & F \\ F & G \end{pmatrix} \begin{pmatrix} du \\ dv \end{pmatrix}$$

$$= (da, db) J^T \begin{pmatrix} E & F \\ F & G \end{pmatrix} J \begin{pmatrix} da \\ db \end{pmatrix},$$

and then

$$\begin{pmatrix} \tilde{E} & \tilde{F} \\ \tilde{F} & \tilde{G} \end{pmatrix} = J^T \begin{pmatrix} E & F \\ F & G \end{pmatrix} J \tag{2.6}$$

holds.

The second fundamental form: Let $\nu(u,v)$ be the unit normal vector field of a regular surface f, and set

$$L := -f_u(u,v) \cdot \nu_u(u,v) = f_{uu}(u,v) \cdot \nu(u,v),$$

$$M := -f_u(u,v) \cdot \nu_v(u,v) = -f_v(u,v) \cdot \nu_u(u,v) = f_{uv}(u,v) \cdot \nu(u,v),$$

$$N := -f_v(u,v) \cdot \nu_v(u,v) = f_{vv}(u,v) \cdot \nu(u,v).$$

---

[2]Here, a *coordinate change* is a diffeomorphism between domains in $\boldsymbol{R}^2$.

Then the formal sum

$$II = -df \cdot d\nu = -(f_u du + f_v dv) \cdot (\nu_u du + \nu_v dv)$$

$$= -(f_u \cdot \nu_u)du^2 + (-(f_u \cdot \nu_v) - (f_v \cdot \nu_u))dudv - (f_v \cdot \nu_v)dv^2$$

$$= L\, du^2 + 2M\, du\, dv + N\, dv^2 = (du\ dv)\begin{pmatrix} L & M \\ M & N \end{pmatrix}\begin{pmatrix} du \\ dv \end{pmatrix}$$

is called the *second fundamental form*. We call the symmetric matrix in the equation above the *second fundamental matrix*, which can be written as

$$\begin{pmatrix} L & M \\ M & N \end{pmatrix} = -(f_u, f_v)^T (\nu_u, \nu_v). \tag{2.7}$$

Remark 2.1.2. The second fundamental form II can be considered as a symmetric covariant tensor on the surface, as in Remark 2.1.1.

Similar to the case for the first fundamental form, the coefficients $\tilde{L}$, $\tilde{M}$, $\tilde{N}$ of the second fundamental form with respect to the new coordinate system (a, b) induced by a coordinate change $u = u(a, b)$, $v = v(a, b)$ satisfy

$$\begin{pmatrix} \tilde{L} & \tilde{M} \\ \tilde{M} & \tilde{N} \end{pmatrix} = J^T \begin{pmatrix} L & M \\ M & N \end{pmatrix} J, \tag{2.8}$$

where J is the Jacobi matrix of the coordinate change $(a, b) \mapsto (u, v)$ (cf. (2.5)).

Using the coefficients of the first and second fundamental forms, we define

$$W := \begin{pmatrix} E & F \\ F & G \end{pmatrix}^{-1} \begin{pmatrix} L & M \\ M & N \end{pmatrix}$$

$$= \frac{1}{EG - F^2}\begin{pmatrix} GL - FM & GM - FN \\ -FL + EM & -FM + EN \end{pmatrix} \tag{2.9}$$

which is called the *Weingarten matrix*. Under the coordinate change $(a, b) \mapsto (u, v)$ the Weingarten matrix changes to

$$\tilde{W} = J^{-1}WJ, \tag{2.10}$$

because of (2.6) and (2.8). Therefore, the eigenvalues, determinant, and trace of W are invariant under coordinate changes, and so in particular are the following two functions:

$$K := \det W = \frac{LN - M^2}{EG - F^2}, \qquad H := \frac{1}{2}\operatorname{tr} W = \frac{EN - 2FM + GL}{2(EG - F^2)},$$

where $\operatorname{tr} W$ denotes the trace of the square matrix W. We call K the *Gaussian curvature* and H the *mean curvature* of the surface. One can easily check that these functions are invariant under rotations and translations of $\boldsymbol{R}^3$ (cf. [84, Section 8]).

Example 2.1.3 (The graph of a function). Consider a surface given by the graph $z = \varphi(x, y)$ of a function $\varphi(x, y)$. Since the unit normal vector is given by (2.2), direct calculations yield

$$E = 1 + \varphi_x^2, \quad F = \varphi_x \varphi_y, \quad G = 1 + \varphi_y^2,$$

$$L = \frac{\varphi_{xx}}{\alpha}, \quad M = \frac{\varphi_{xy}}{\alpha}, \quad N = \frac{\varphi_{yy}}{\alpha},$$

where $\alpha := \sqrt{1 + \varphi_x^2 + \varphi_y^2}$. In particular, the Gaussian curvature and mean curvature are represented as

$$K = \frac{\varphi_{xx}\varphi_{yy} - \varphi_{xy}^2}{\alpha^4}, \quad H = \frac{\varphi_{xx}(1 + \varphi_y^2) - 2\varphi_{xy}\varphi_x\varphi_y + \varphi_{yy}(1 + \varphi_x^2)}{2\alpha^3}.$$

Fact 2.1.4 ([84, Theorem 8.7]). Near a point where the Gaussian curvature is positive, the surface is convex, as shown in Fig. 2.1(a). On the other hand, near a point where the Gaussian curvature is negative, the surface takes the shape of a saddle, as shown in Fig. 2.1(b).

Let ν be a unit normal vector of a regular surface $f \colon U \to \boldsymbol{R}^3$, and consider $\nu(u, v)$ as a unit vector emanating from the origin $(0, 0, 0)$ by a suitable translation of $\boldsymbol{R}^3$, then $\nu(u, v)$ can be identified with a point on the unit sphere

$$S^2 := \{(x, y, z) \in \boldsymbol{R}^3 \,;\, x^2 + y^2 + z^2 = 1\}.$$

Thus, we obtain a map $\nu \colon U \to S^2 \subset \boldsymbol{R}^3$, called the *Gauss map* of the surface.

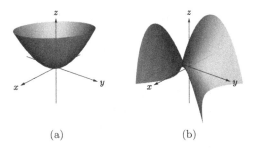

(a)   (b)

Fig. 2.1.   An elliptic paraboloid (a) and a hyperbolic paraboloid (b).

Proposition 2.1.5 (The Weingarten formula [84, Proposition 8.5]). *Let $f\colon U \to \mathbf{R}^3$ be a regular surface and ν its Gauss map. Then*

$$(\nu_u, \nu_v) = -(f_u, f_v)W, \qquad (2.11)$$

where W is the Weingarten matrix defined in (2.9) and the right-hand side has matrix multiplication of (f_u, f_v) with W. In particular, a singular point of the Gauss map $\nu\colon U \to S^2 \subset \mathbf{R}^3$ is a point where the Gaussian curvature K of f vanishes.

Proof. Define two 3×3-matrices $M_1 := (f_u, f_v, \nu)$, $M_2 := (\nu_u, \nu_v, \nu)$. Then by (2.4) and (2.7), we have

$$\begin{pmatrix} E & F & 0 \\ F & G & 0 \\ 0 & 0 & 1 \end{pmatrix} = M_1^T M_1, \qquad \begin{pmatrix} L & M & 0 \\ M & N & 0 \\ 0 & 0 & -1 \end{pmatrix} = -M_1^T M_2,$$

and then

$$-M_1^{-1} M_2 = -(M_1^T M_1)^{-1} M_1^T M_2$$

$$= \begin{pmatrix} E & F & 0 \\ F & G & 0 \\ 0 & 0 & 1 \end{pmatrix}^{-1} \begin{pmatrix} L & M & 0 \\ M & N & 0 \\ 0 & 0 & -1 \end{pmatrix} = \begin{pmatrix} W & \mathbf{0} \\ \mathbf{0} & -1 \end{pmatrix}.$$

Hence we have

$$(\nu_u, \nu_v, \nu) = -(f_u, f_v, \nu) \begin{pmatrix} W & \mathbf{0} \\ \mathbf{0} & -1 \end{pmatrix},$$

and in particular (2.11) holds. □

The geodesic curvature. Let $f\colon U \to \mathbf{R}^3$ be a regular surface defined on a domain U in the uv-plane, and let $\gamma\colon (a, b) \to U$ be a regular curve in the uv-plane. We then have the space curve

$$\hat{\gamma}(t) := f \circ \gamma(t) \quad (a < t < b).$$

We call

$$\hat{e}(t) := \frac{\hat{\gamma}'(t)}{|\hat{\gamma}'(t)|}$$

the *unit tangent vector* of the curve $\hat{\gamma}$. In addition, letting ν be a unit normal vector field of f defined on U, and setting $\hat{\nu}(t) := \nu(\gamma(t))$ to be the restriction of ν to the curve γ, we call $\hat{\nu}$ the *unit normal vector of f along* γ. Moreover, we set

$$\hat{n}(t) := \hat{\nu}(t) \times \hat{e}(t) \qquad (2.12)$$

and call this the *conormal vector* of the curve $\hat{\gamma}(t)$. The vector $\hat{\boldsymbol{n}}(t)$ is a tangent vector of the surface perpendicular to the unit tangent vector $\hat{\boldsymbol{e}}(t)$, and $\{\hat{\boldsymbol{e}}(t), \hat{\boldsymbol{n}}(t), \hat{\nu}(t)\}$ gives an orthonormal basis[3] of $\boldsymbol{R}^3$ for each choice of t, that is, it always gives an orthonormal basis of $\boldsymbol{R}^3$, which generally changes as one moves along γ.

Assume t is an arc-length parameter of $\hat{\gamma}$. Then $\hat{\gamma}'(t) \cdot \hat{\gamma}'(t) = 1$ holds. Differentiating this, we have $\hat{\gamma}''(t) \cdot \hat{\gamma}'(t) = 0$. Hence we can write

$$\hat{\gamma}''(t) = (\hat{\gamma}''(t) \cdot \hat{\boldsymbol{n}}(t))\hat{\boldsymbol{n}}(t) + (\hat{\gamma}''(t) \cdot \hat{\nu}(t))\hat{\nu}(t).$$

Using this decomposition, we define

$$\kappa_g(t) := \hat{\gamma}''(t) \cdot \hat{\boldsymbol{n}}(t), \quad \kappa_n(t) := \hat{\gamma}''(t) \cdot \hat{\nu}(t), \tag{2.13}$$

which are called the *geodesic curvature* (function) and the *normal curvature* (function) of $\gamma(t)$, respectively (see [84, pages 89 and 111]). The curvature function $\kappa(t) := |\hat{\gamma}''(t)|$ of the space curve $\hat{\gamma}(t)$ satisfies

$$\kappa(t) = \sqrt{\kappa_g(t)^2 + \kappa_n(t)^2}, \tag{2.14}$$

by definition, and this identity (2.14) will hold even for a non-arc-length parameter t of the curve. With respect to a general parameter t that might not be arc-length, the geodesic curvature is given by (see Exercise **1** in this section)

$$\kappa_g(t) = \frac{\det(\hat{\gamma}'(t), \hat{\gamma}''(t), \hat{\nu}(t))}{|\hat{\gamma}'(t)|^3}. \tag{2.15}$$

In particular, if the image of the surface is the xy-plane, that is, when $f(u, v) = (u, v, 0)$, the definition of the geodesic curvature coincides with that for plane curves if one takes the unit normal vector as $\nu = (0, 0, 1)$.

On the other hand, the normal curvature of the curve $\gamma(t)$ can be represented as

$$\kappa_n(t) = \frac{\hat{\gamma}''(t) \cdot \hat{\nu}(t)}{\hat{\gamma}'(t) \cdot \hat{\gamma}'(t)} \tag{2.16}$$

when t is not necessarily arc-length (cf. Exercise **2** in this section).

Definition 2.1.6. A curve $\hat{\gamma}(t)$ on the surface is called a *geodesic* if the parameter t is proportional to arc-length (that is, the norm $|\hat{\gamma}'(t)|$ of the velocity vector is constant) and the geodesic curvature is identically zero.

---

[3]These vectors should be considered as the basis of the tangent space at $\hat{\gamma}(t)$. In this sense, $\{\hat{\boldsymbol{e}}(t), \hat{\boldsymbol{n}}(t), \hat{\nu}(t)\}$ is often called an *orthonormal frame* at $\hat{\gamma}(t)$.

A geodesic in the plane is a straight line. In general, geodesics on a surface play the role of "straight curves" in the Gauss–Bonnet formula in Chapter 6. We note that the definition of geodesics does depend on the choice of parameter (that is, only parameters which are proportional to the arc-length parameter can be a parametrization of geodesics). It is known that for two points on a given geodesic, when the two points are sufficiently close, the shortest of the geodesic arcs joining the points is the shortest path on the surface between those points. See [84, Section 10] for details.

Geometric properties of the Gaussian curvature

Proposition 2.1.7 (Gauss' theorem [84, page 118]). *The Gaussian curvature of a regular surface, which has been defined above using both the first and second fundamental forms, can be expressed using only the coefficients of the first fundamental form and their first and second derivatives, as follows*:

$$K = \frac{E\big(E_v G_v - 2F_u G_v + (G_u)^2\big)}{4(EG - F^2)^2}$$

$$+ \frac{F(E_u G_v - E_v G_u - 2E_v F_v - 2F_u G_u + 4F_u F_v)}{4(EG - F^2)^2}$$

$$+ \frac{G\big(E_u G_u - 2E_u F_v + (E_v)^2\big)}{4(EG - F^2)^2}$$

$$- \frac{E_{vv} - 2F_{uv} + G_{uu}}{2(EG - F^2)}. \tag{2.17}$$

When the parameters (u, v) are chosen so that F is identically zero, then K is expressed as

$$K = -\frac{1}{eg}\left[\left(\frac{g_u}{e}\right)_u + \left(\frac{e_v}{g}\right)_v\right], \tag{2.18}$$

where $e := \sqrt{E} = |f_u|$, $g := \sqrt{G} = |f_v|$ (see Exercise **6** in this section).

In the context of Riemannian geometry, the *sectional curvature* for the special case of a surface can be defined by the right-hand side of (2.17). That is, the sectional curvature in Riemannian geometry can be regarded as a generalization of the Gaussian curvature for surfaces.

Fact 2.1.8 (Perimeter of geodesic circles [84, Section 11]). Let $\mathcal{L}_P(r)$ be the length (perimeter) of the geodesic circle[4] with center point P on the surface and with radius r. Then

$$\lim_{r \to 0} \frac{3}{\pi} \left(\frac{2\pi r - \mathcal{L}_P(r)}{r^3} \right)$$

coincides with the Gaussian curvature K at P. In particular, the ratio of the perimeter and the diameter is less than (respectively, greater than) π when $K > 0$ (respectively, $K < 0$).

Principal curvatures and principal directions. The Gaussian curvature K and mean curvature H are the product and the average, respectively, of the two eigenvalues of the Weingarten matrix W. We define the normal curvature for the purpose of explaining a geometric meaning for these eigenvalues:

Definition 2.1.9. Let $f : U \to \mathbf{R}^3$ be a regular surface and fix $p \in U$. Then, for each $(\alpha, \beta) \in \mathbf{R}^2 \setminus \{(0,0)\}$, we define

$$\kappa_n := \frac{L\alpha^2 + 2M\alpha\beta + N\beta^2}{E\alpha^2 + 2F\alpha\beta + G\beta^2}, \tag{2.19}$$

and call this the *normal curvature* of the surface f with respect to the direction (α, β) at p. The normal curvature κ_n will change sign when the unit normal vector ν is replaced by $-\nu$.

The normal curvature depends on the direction at the point p, that is, if we set

$$\alpha = r \cos \theta, \quad \beta = r \sin \theta \quad (r > 0),$$

the normal curvature with respect to the direction (α, β) can be expressed as

$$\kappa_n := \frac{L \cos^2 \theta + 2M \cos \theta \sin \theta + N \sin^2 \theta}{E \cos^2 \theta + 2F \cos \theta \sin \theta + G \sin^2 \theta}, \tag{2.20}$$

in terms of only θ. The normal curvature defined here is related to the normal curvature function along curves given in (2.16), as follows:

---

[4]The *geodesic circle* centered at P with radius r is the locus of terminal points of the geodesics emanating from P with length r.

Proposition 2.1.10 ([84, Proposition 9.2]). *Let* $f \colon U \to \mathbf{R}^3$ *be a regular surface defined on a domain in the uv-plane, and fix* $p \in U$. *For a regular curve* $\gamma \colon [0, \varepsilon) \to U$ ($\varepsilon > 0$) *in the uv-plane starting at* $p \in U$, *e.g.,* $\gamma(0) = p$, *the normal curvature of the curve* $\hat{\gamma}(t) = f \circ \gamma(t)$ *at* $t = 0$ *coincides with* κ_n *as in* (2.19) *when* $(\alpha, \beta) := \gamma'(0)$.

Proof. Let $\gamma(t) = (u(t), v(t))$. Differentiating $\hat{\gamma}(t) = f(u(t), v(t))$, we have

$$\hat{\gamma}' = u' f_u + v' f_v, \quad \hat{\gamma}'' = u'' f_u + v'' f_v + (u')^2 f_{uu} + 2u'v' f_{uv} + (v')^2 f_{vv}.$$

By the definitions of the first and second fundamental forms, we have

$$\hat{\gamma}' \cdot \hat{\gamma}' = (u')^2 E + 2u'v' F + (v')^2 G,$$
$$\hat{\gamma}'' \cdot \hat{\nu} = (u')^2 L + 2u'v' M + (v')^2 N.$$

Noticing that $(\alpha, \beta) = (u'(0), v'(0))$, we have (2.19) by substituting $t = 0$ into (2.16).   □

Example 2.1.11. The coefficients of the first and second fundamental forms of the surface given by the graph $z = (x^2 + ay^2)/2$ ($0 < a < 1$) satisfy

$$E = G = 1, \quad F = 0, \quad L = 1, \quad M = 0, \quad N = a$$

at the origin. Then $\kappa_n = \cos^2 \theta + a \sin^2 \theta$ for $(\alpha, \beta) = (\cos \theta, \sin \theta)$.

Fact 2.1.12 ([84, Theorem 9.1]). Let $f \colon U \to \mathbf{R}^3$ be a regular surface and fix $p \in U$, and let Π be the plane passing through the point $f(p)$ on the surface containing the two vectors

$$X := \alpha f_u(p) + \beta f_v(p), \quad Y := \nu(p)$$

with initial point $f(p)$, where $\nu(p)$ is a unit normal vector of the surface at $f(p)$. Then the intersection of Π and $f(U)$ can be parametrized by a regular curve on the surface through $f(p)$. Taking a coordinate system (x, y) on Π such that the x-axis (respectively, y-axis) corresponds to X (respectively, Y), the curvature of that plane curve in Π, at $f(p)$, is equal to the normal curvature κ_n of the surface at $f(p)$ with respect to the direction (α, β).

The eigenvalues of the Weingarten matrix W satisfy the following:

Fact 2.1.13 ([84, Section 9]). Let $f \colon U \to \mathbf{R}^3$ be a regular surface. The eigenvalues $\lambda_1(p)$, $\lambda_2(p)$ of the Weingarten matrix W at $p \in U$ are both real numbers. Moreover,

(1) the maximum and minimum of the function $[0, 2\pi) \ni \theta \mapsto \kappa_n(\theta) \in \mathbf{R}$ coincide with $\lambda_1(p)$ and $\lambda_2(p)$, where $\kappa_n(\theta)$ is the normal curvature at $f(p)$ with respect to the direction $(\alpha, \beta) = (\cos\theta, \sin\theta)$ (we call these eigenvalues the *principal curvatures*),

(2) the normal curvature of f at p with respect to the eigenvector (α_i, β_i) of W corresponding to the eigenvalue $\lambda_i(p)$ $(i = 1, 2)$ coincides with $\lambda_i(p)$ itself, and

(3) at a point p where $\lambda_1(p) \neq \lambda_2(p)$, the two tangent vectors

$$V_1 := \alpha_1 f_u(p) + \beta_1 f_v(p), \quad V_2 := \alpha_2 f_u(p) + \beta_2 f_v(p)$$

of the surface at $f(p)$ are orthogonal.

(4) The one-dimensional vector space L_j in $\mathbf{R}^3$ spanned by V_j $(j = 1, 2)$ is independent of the choice of parametrization of the surface. These vector spaces L_j $(j = 1, 2)$ are called the *principal directions*.

Summing up, the principal curvatures are the maximum and minimum of the normal curvature at the given point on the surface, and the corresponding directions are the principal directions. We remark that the principal curvatures change sign when ν is replaced by $-\nu$, and that the directions of the maximum and minimum are interchanged.

As an example, recall the quadric in Example 2.1.11. The maximum and minimum of the normal curvature are 1 and a, which are the principal curvatures at the origin. Moreover, $(1, 0, 0)$ and $(0, 1, 0)$ are the principal directions corresponding 1 and $a(\neq 1)$, respectively.

A point on a surface where the two principal curvatures coincide is called an *umbilic point*. Having an umbilic point is equivalent to having

$$H^2 - K = \frac{1}{4}(\lambda_1 - \lambda_2)^2 = 0$$

at the point. At the umbilic, any direction in the surface is a principal direction. On the other hand, as noted in Fact 2.1.13 above, the two principal directions are fixed, and perpendicular to each other, at each non-umbilic point.

A curve $\gamma(t)$ is said to be a *line of curvature* if the velocity vector $\gamma'(t)$ gives a principal direction at each t. At each non-umbilic point, there exist exactly two lines of curvature, which are orthogonal, cf. [84, Theorem B.5.1]. Moreover, the following assertion holds.

Theorem 2.1.14 (Monge's form). Let $f : U \to \mathbf{R}^3$ be a regular surface and $p \in U$ an arbitrarily fixed point. Then, by a suitable translation and *rotation, the surface f can be written as*

$$f(x,y) = \left(x, y, \frac{\lambda_1 x^2 + \lambda_2 y^2}{2}\right) + (0, 0, o(x^2 + y^2)), \qquad (2.21)$$

such that p corresponds to the origin $(0,0)$, where λ_i $(i = 1, 2)$ are principal curvatures at p and $o(x^2 + y^2)$ denotes a function $h(x, y)$ satisfying

$$\lim_{(x,y)\to(0,0)} \frac{h(x,y)}{x^2 + y^2} = 0.$$

Moreover, if p is a non-umbilical point, then the x-axis is parallel to the principal direction L_1.

Proof. If we take the canonical coordinates (x, y, z) of $\mathbf{R}^3$ so that $f(p)$ will be the origin, and the tangent plane at $f(p)$ is the xy-plane, then we can write (cf. [84, page 83])

$$f(x,y) = \left(x, y, \frac{ax^2 + 2bxy + cy^2}{2}\right) + (0, 0, o(x^2 + y^2)).$$

Then the coefficient of the second fundamental form at $(x, y) = (0, 0)$ satisfies $L = a$, $M = b$ and $N = c$. So, if p is an umbilical point, then $(\lambda :=)a = c$ and $b = 0$ and λ is the principal curvature of f at p. Thus, we may assume that p is a non-umbilical point. Since the two principal directions L_1, L_2 are orthogonal, we can rotate the xy-plane so that the x-axis and the y-axis are parallel to V_1 and V_2, respectively. Then $b = 0$ holds, and we get $a = \lambda_1(p)$ and $c = \lambda_2(p)$. $\qquad\square$

Corollary 2.1.15. *Let $f : U \to \mathbf{R}^3$ be a regular surface and $p \in U$ an arbitrarily fixed point. Then, there exists a local coordinate system (u, v) centered at p such that $F = M = 0$ at p.*

Proof. Let (x, y) be the coordinates given in (2.21). Then the desired coordinates (u, v) are obtained by setting $u = x$ and $v = y$. $\qquad\square$

Curvature line coordinates: We introduce the notion of *curvature line coordinates*, a notion that is common to the full family of parallel surfaces.

Definition 2.1.16 (Curvature line coordinate system). Let $f : U \to \mathbf{R}^3$ be a regular surface defined on a domain U in the uv-plane $\mathbf{R}^2$. The parameters (u, v) are a *curvature line coordinate system* if

(1) there are no umbilic points of f on U,
(2) the u-curves $u \mapsto f(u, v)$ and the v-curves $v \mapsto f(u, v)$ are both curvature lines.

One can examine whether a given coordinate system is a curvature line coordinate system by utilizing the first and second fundamental forms, as follows:

Proposition 2.1.17 ([84, Theorem B.5.2]). *Let $f\colon U \to \mathbf{R}^3$ be a regular surface with no umbilic points. The coordinate system (u, v) is a curvature line coordinate system if and only if the coefficients of the first and second fundamental forms with respect to the parameters (u, v) satisfy $F = M = 0$.*

Proof. Since the two principal directions are orthogonal, by Fact 2.1.13, f_u and f_v are also orthogonal if (u, v) are curvature line coordinates, which implies $F = 0$. Then the first fundamental matrix is diagonal. On the other hand, the principal directions are the directions of the eigenvectors of the Weingarten matrix W, and so W is diagonal if (u, v) is a curvature line coordinate system. It follows that the second fundamental matrix is diagonal as well, and in particular, $M = 0$.

Conversely, if $F = M = 0$, the first and second fundamental matrices are both diagonal, and hence W is as well, which implies that (u, v) is a curvature line coordinate system. □

On a coordinate system (u, v) such that the coefficient F of the first fundamental form is zero, the u-curves and v-curves are orthogonal. Such a coordinate system is called an *orthogonal net*. A curvature line coordinate system is a typical example of an orthogonal net.

We provide some examples.

Example 2.1.18 (Cylinders). Consider a regular plane curve $\gamma(u)$,

$$\gamma(u) := (x(u), y(u), 0) \quad (a \le u \le b)$$

in the xy-plane in $\mathbf{R}^3$, and let

$$f(u, v) := \gamma(u) + v e_3 \quad (e_3 = (0, 0, 1)).$$

This surface $f(u, v)$ is a ruled surface generated by vertical lines passing through $\gamma(u)$ for each $u \in [a, b]$. We call such a surface $f : [a, b] \times \mathbf{R} \to \mathbf{R}^3$ the *cylinder* over the base curve $\gamma(u)$. The coefficients of the first fundamental form satisfy $F = 0$, since $f_u = \gamma'(u)$ is orthogonal to $f_v = e_3$. Moreover, since $f_{vu} = (e_3)_u = \mathbf{0}$, the coefficients of the second fundamental form satisfy $M = 0$, and we conclude that (u, v) is a curvature line coordinate system of the cylinder. Since the v-curve is a line, the principal curvature of f with respect to v-direction vanishes. In particular, the Gaussian curvature of f is identically 0.

The existence of curvature line coordinate systems is stated in the next fact:

Fact 2.1.19 ([84, Appendix B.5]). For each non-umbilic point on a regular surface, there exists a curvature line coordinate system in a neighborhood of that point.

On the other hand, for surfaces having open sets consisting of umbilic points, the following fact holds.

Fact 2.1.20 ([84, Proposition 9.7]). Let $f : U \to \mathbf{R}^3$ be a regular surface defined on a domain of $\mathbf{R}^2$. If all points in U are umbilic points of f, the image $f(U)$ is a part of a plane or a sphere.

Fact 2.1.19 tells us that there exists a local curvature line coordinate system, then necessarily satisfying $F = M = 0$, in a neighborhood of a non-umbilic point. In particular, for the tori of revolution (more generally, surfaces of revolution), the canonical parametrizations give curvature line coordinate systems (cf. Exercise **5** in this section). An umbilic point is a point where the flow of curvature lines cannot be defined. For the behavior of curvature lines near umbilic points, see [84].

Exercises 2.1

1 Derive the expression for the geodesic curvature in terms of a general parameter, by using the definition of geodesic curvature. Verify that the geodesic curvature of curves in the plane $f(u, v) := (u, v, 0)$ with unit normal vector $\nu := (0, 0, 1)$ coincides with the curvature for planar curves.

2 Derive the identity (2.16) for the normal curvature with respect to general parametrizations, using the definition of normal curvature.

3 Verify the expressions for the Gaussian curvature and mean curvature in Example 2.1.3.

4 Compute the first and second fundamental forms for the parametrization of the sphere of radius a

$$f(u, v) := \frac{a}{1 + u^2 + v^2}(2u, 2v, u^2 + v^2 - 1).$$

Also, compute the Gaussian curvature and mean curvature.

5 Let a, b be positive constants satisfying $a > b$, and consider the torus

$$f(u, v) := ((a + b\cos u)\cos v, (a + b\cos u)\sin v, b\sin u) \quad (0 \le u, v < 2\pi).$$
$$(2.22)$$

By computing the first and second fundamental forms of f, verify that (u, v) is a curvature line coordinate system. Using this coordinate system, determine the set where the Gaussian curvature vanishes, and discuss geometric meanings of this.

6 Prove (2.18) from (2.17).

2.2. Principal Curvatures of Parallel Surfaces

Let $f(u, v)$ be a surface and ν its unit normal vector field. A surface

$$f^t(u, v) := f(u, v) + t\,\nu(u, v) \qquad (2.23)$$

obtained from parallel translation of f in the direction ν with (signed) distance t is called a *parallel surface* of f. (We use the notation f^t with superscript t, to avoid confusion with the partial derivative $f_t = \partial f / \partial t$).

Parallel surfaces may have singular points in general. At regular points of parallel surfaces, the following result holds, which corresponds to the formula for the curvatures of parallel plane curves (in Proposition 1.6.1):

Theorem 2.2.1. *Let $f : U \to \mathbf{R}^3$ be a regular surface and f^t its parallel surface as in (2.23). Then the unit normal vector field of f is also the unit normal vector field of f^t, and the principal curvatures λ_1^t, λ_2^t at each regular point of f^t are given by*

$$\lambda_j^t = \frac{\lambda_j}{1 - t\lambda_j} \qquad (j = 1, 2), \qquad (2.24)$$

where λ_1 and λ_2 are the principal curvatures of f. Equation (2.24) can be rewritten as

$$\frac{1}{\lambda_j^t} = \frac{1}{\lambda_j} - t \qquad (j = 1, 2).$$

Note that this equation looks similar to (1.47) for plane curves. In fact, (2.24) can be obtained from (1.47), see [84, Appendix B.7] for details.

By (2.24), the Gaussian curvature K^t and mean curvature H^t of f^t can be expressed as

$$K^t = \frac{K}{1 - 2tH + t^2 K}, \qquad H^t = \frac{H - tK}{1 - 2tH + t^2 K}, \qquad (2.25)$$

where K and H are the Gaussian curvature and mean curvature of f, respectively.

Proof. We fix a point $p \in U$ arbitrarily. It is sufficient to show the theorem holds at p. Without loss of generality, we may assume that the coordinate (u, v) of U satisfies $F = M = 0$ at p (see Corollary 2.1.15). On such a coordinate system, the Weingarten matrix of f at p is written as

$$W = \begin{pmatrix} \lambda_1(p) & 0 \\ 0 & \lambda_2(p) \end{pmatrix},$$

where $\lambda_1(p)$ and $\lambda_2(p)$ are the principal curvatures at p. Then, by Weingarten's formula (Proposition 2.1.5), we have

$$\nu_u(p) = -\lambda_1(p) f_u(p), \quad \nu_v(p) = -\lambda_2(p) f_v(p), \tag{2.26}$$

and hence

$$(f^t)_u(p) = (1 - t\lambda_1(p)) f_u(p), \quad (f^t)_v(p) = (1 - t\lambda_2(p)) f_v(p). \tag{2.27}$$

Thus, $\nu(p)$ is also the unit normal vector of f^t. Since

$$(f^t)_u(p) \cdot \nu_v(p) = (1 - t\lambda_1(p)) f_u(p) \cdot \nu_v(p) = 0,$$

the coefficients of the first and second fundamental forms of f^t with respect to (u, v) satisfy $F^t(p) = M^t(p) = 0$.

By (2.27) and the fact that $f_u(p)$ and $f_v(p)$ are orthogonal, $(f^t)_u(p)$ and $(f^t)_v(p)$ are orthogonal as well. Hence, the point p is a singular point of f^t if and only if $(f^t)_u(p)$ or $(f^t)_v(p)$ vanishes. Here, since both f_u, f_v do not vanish, p is a singular point of f^t if and only if

$$(1 - t\lambda_1(p))(1 - t\lambda_2(p)) = 0 \tag{2.28}$$

holds at the point.

We now assume p is a regular point of f_t, and compute the principal curvatures of f^t at p. By definition,

$$\nu_u(p) = -\lambda_1(p) f_u(p) = -\frac{\lambda_1(p)}{1 - t\lambda_1(p)} (f^t)_u(p),$$

$$\nu_v(p) = -\lambda_2(p) f_v(p) = -\frac{\lambda_2(p)}{1 - t\lambda_2(p)} (f^t)_v(p) \tag{2.29}$$

hold. Applying Weingarten's formula, we have that $1/\lambda_i(p) - t$ $(i = 1, 2)$ are the inverses of the principal curvatures of f^t at p. $\qquad \square$

By (2.27), we have the following corollary:

Corollary 2.2.2. *A point $p \in U$ is a singular point of f^t if and only if $1/t = \lambda_1(p)$ or $1/t = \lambda_2(p)$ holds.*

As seen in the proof of Theorem 2.2.1, we have shown the property that $F = M = 0$ is common to the entire parallel family f^t. By Proposition 2.1.17, we get the following:

Corollary 2.2.3. *Assume the parameters (u, v) of a regular surface $f(u, v)$ are a curvature line coordinate system at a neighborhood of a non-umbilical point. Then (u, v) is also a curvature line coordinate system of any parallel surface f^t ($t \in \mathbf{R}$) on neighborhoods of regular points. Furthermore, the property that $p \in U$ is an umbilic point is a property common to all the parallel surfaces f^t.*

The identity (2.25) yields the following:

Corollary 2.2.4. *If the Gaussian curvature of a surface f is zero at a point $p \in U$, then the Gaussian curvature of its parallel surfaces is also zero at p.*

In this way, the property that the Gaussian curvature be zero is common to all parallel surfaces. We call a surface whose Gaussian curvature vanishes identically a *flat surface*. Like as in the case of curves, parallel surfaces of a front can be regarded as wave propagation from the original surface. Then Corollary 2.2.4 can be restated as "flatness is invariant under wave propagation".

The following important fact can be proved easily (see Exercise 1).

Corollary 2.2.5. *If the Gaussian curvature K of a surface f is the positive constant $1/a^2$, then the mean curvature of f^t in (2.23) is constantly $\pm 1/|2a|$ when $t = \pm a$.*

Thus, there is a (local) 1 to 2 correspondence between surfaces of positive Gaussian curvature and surfaces of non-zero constant mean curvature. The two constant mean curvature surfaces corresponding to a positive constant Gaussian curvature surface are not congruent in general, and one may have singular points even when the other is regular. We introduce typical examples: The first and third figures in Fig. 2.2 are the examples of surfaces of revolution with constant mean curvature $1/2$. It is known that non-zero constant mean curvature surfaces of revolution are in these two families called *nodoids* and *unduloids*, with the exception of only the circular cylinders and spheres. The third figure is an unduloid, whose profile curve the locus of a focal point of an ellipse rolling without slippage along a fixed line. The first one is a nodoid, whose profile curve is the locus of a focal

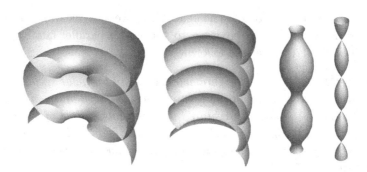

Fig. 2.2. Delaunay surfaces and their parallel surfaces: a *nodoid* (left) and its parallel surface (second from left), and an *unduloid* (second from right) and its parallel surface (right).

point of a similarly rolling hyperbola. For details, see [84, page 266]. These surfaces are called *Delaunay surfaces*. Consider the parallel surface $f + \nu$ of distance 1 in the direction of the mean curvature vector for such a surface. Then the second and fourth (rightmost) surfaces in Fig. 2.2 are obtained. These surfaces are of constant Gaussian curvature 1, having singularities (cf. Corollary 2.2.5). In this way, one can obtain different kinds of surfaces by taking parallel surfaces. In general, singularities will appear.

Exercises 2.2

1   Prove Corollary 2.2.5 using (2.25).

2.3.   Definition of Wave Fronts

Recall that for a regular surface $f : U \to \mathbf{R}^3$ and its Gauss map $\nu : U \to S^2$, the parallel surfaces of f are given by

$$f^t := f + t\nu, \qquad (2.30)$$

as defined in (2.23) of Section 2.2.

Similar to the case of curves, we consider the image of f as the initial wave front. Then by the Huygens principle, *the wave front after time t elapses is the envelope of spheres of radius $|t|$ centered at points in the image of f*, and is the image of f^t. As seen in Section 2.2, f^t may have singularities even if f is a regular surface.

In this section, we define (abstract) wave fronts for the case of surface in the same way as we did for the case of curves.

Definition 2.3.1. Let $f : U \to \boldsymbol{R}^3$ be a C^∞-map defined on a domain U in the uv-plane $\boldsymbol{R}^2$. Then f is said to be a (co-orientable) *frontal* if there exists a C^∞-map $\nu : U \to S^2$ such that

(a) $\nu(p) \cdot f_u(p) = \nu(p) \cdot f_v(p) = 0$ holds for each point $p \in U$,

and we call ν the *unit normal vector field* of the frontal f. Moreover, if

(b) the C^∞-map $L := (f, \nu)$ from U into $\boldsymbol{R}^3 \times S^2$ is an immersion, that is, the rank of the 6×2-matrix

$$\boldsymbol{M} := \begin{pmatrix} f_u & f_v \\ \nu_u & \nu_v \end{pmatrix} \tag{2.31}$$

is 2 at each point $p \in U$,

then the map f is called a *wave front*, or simply *front*.

When a surface is obtained by gluing together several parametrizations, the unit normal vector ν might not be defined on the entire surface, though it will be well defined on each sufficiently small neighborhood of each point on the surface. For regular surfaces, orientability of a surface is equivalent to having ν well defined on the entirety of the surface.

On the other hand, for a frontal with singularities, the unit normal vector field might not be well defined on the entire surface (including singularities) even if the surface is orientable. A frontal having globally defined unit normal vector field is said to be *co-orientable*, and a *co-orientation* is a choice of ν. Orientable surfaces may not be co-orientable in general. In fact, the cylinder over the cardioid

$$f(t, v) = ((1 + \cos t) \cos t, (1 + \cos t) \sin t, v) \tag{2.32}$$

defined on $\boldsymbol{R}/2\pi\boldsymbol{Z} \times \boldsymbol{R}$ has no globally defined unit normal vector field. A non-co-orientable frontal induces a co-orientable frontal by taking the double cover (see Section 10.3 in Chapter 10). In this book, fronts and frontals are assumed to be orientable and co-orientable, if not specified.

Remark 2.3.2 (For readers having a knowledge of manifolds). We regard $\boldsymbol{R}^3 \times S^2$ as a submanifold of $\boldsymbol{R}^6$, via $\boldsymbol{R}^3 \times S^2 = \{(x, y, z, u, v, w) \in \boldsymbol{R}^6 ; u^2 + v^2 + w^2 = 1\}$ in $\boldsymbol{R}^6$, and consider a differential form $\theta := u\, dx + v\, dy + w\, dz$ on $\boldsymbol{R}^6$. Restricting θ to $\boldsymbol{R}^3 \times S^2$, we obtain a differential 5-form $\theta \wedge d\theta \wedge d\theta$ on $\boldsymbol{R}^3 \times S^2$ that is nowhere vanishing. Namely, $\boldsymbol{R}^3 \times S^2$ has the structure of a contact manifold (see Definition 10.2.4 in Chapter 10). A C^∞-map

$$L : U \ni (u, v) \mapsto (f(u, v), \nu(u, v)) \in \boldsymbol{R}^3 \times S^2$$

is called a *Legendrian immersion* if the partial derivatives L_u and L_v are linearly independent at each point U, and the pull back $L^*\theta$ vanishes identically on U if and only if

$$\nu(p) \cdot f_u(p) = \nu(p) \cdot f_v(p) = 0 \quad (p \in U).$$

This means that f is a wave front if and only if L is a Legendrian immersion in the sense of Section 10.3 in Chapter 10 (this is an analogue of Remark 1.5.9 in Chapter 1).

We define parallel surfaces f^t of a frontal by (2.30). Then we have the following:

Proposition 2.3.3. *The unit normal vector field ν of a frontal f is also a unit normal vector field of the parallel surface f^t for each t. In particular, f^t is also a frontal. Moreover, if f is a front, then so is f^t.*

Proof. The first claim can be obtained in the same way as for regular surfaces. So it is sufficient to show that f^t is a wave front if f is as well. A frontal f is a wave front if the 6×2-matrix $\mathbf{M}$ in (2.31) is of rank 2. On the other hand, the parallel surface $f^t = f + t\nu$ is a wave front if and only if the matrix

$$\mathbf{M}_t := \begin{pmatrix} f_u + t\nu_u & f_v + t\nu_v \\ \nu_u & \nu_v \end{pmatrix}$$

is of rank 2. Here, $\mathbf{M}_0(=\mathbf{M})$ and $\mathbf{M}_t$ have the same rank. Hence f^t is a wave front. □

We refer to a wave front obtained as a parallel surface of a regular surface as a *classical wave front*. Proposition 2.3.3 says that a classical wave front is a wave front in the sense of Definition 2.3.1. The inverse assertion holds locally as follows:

Proposition 2.3.4. *Let $p \in U$ be a singular point of a wave front $f : U \to \mathbf{R}^3$. Then p is a regular point of f^t for sufficient small $t \neq 0$. Furthermore, since $f = (f^t)^{-t}$, singular points of wave fronts can be considered as singular points of parallel surfaces of regular surfaces.*

Before proving the proposition, we prepare the following lemma:

Lemma 2.3.5. *Let $p \in U$ be a singular point of a C^∞-map $f : U \to \mathbf{R}^3$, where U is a domain of the uv-plane $\mathbf{R}^2$. Suppose that $f_u(p) \neq \mathbf{0}$, then there exists a constant $k \in \mathbf{R}$ such that $f_{\tilde{u}}(p) = \mathbf{0}$ holds with respect to the new coordinates given by*

$$u = k\tilde{u} + \tilde{v}, \quad v = -\tilde{u}.$$

Proof. Since $f_u(p) \neq \mathbf{0}$, there exists a constant $k \in \mathbf{R}$ such that $f_v(p) = kf_u(p)$. By the chain rule, we have

$$f_{\tilde{u}}(p) = f_u(p)u_{\tilde{u}} + f_v(p)v_{\tilde{u}} = kf_u(p) - f_v(p) = \mathbf{0},$$

proving the assertion. $\qquad\qquad\square$

Proof of Proposition 2.3.4. If $f_u(p) = f_v(p) = \mathbf{0}$, ν_u and ν_v are linearly independent at the point p. Then the parallel surfaces $f^t := f + t\nu$ are regular at p for all $t \neq 0$. In fact, one can easily show that $(f^t)_u$ and $(f^t)_v$ are linearly independent. Next we assume that p is a singular point which does not satisfy $f_u(p) = f_v(p) = \mathbf{0}$. In this case, by Lemma 2.3.5, we can change coordinates so that $f_u(p) = \mathbf{0}$ and $f_v(p) \neq \mathbf{0}$ without loss of generality. Then we have

$$0 = f_u(p) \cdot \nu_v(p) = -f_{uv}(p) \cdot \nu(p) = f_v(p) \cdot \nu_u(p). \qquad (2.33)$$

By the condition (b) in the definition (Definition 2.3.1) of wave fronts, both $\nu_u(p)$ and $f_v(p)$ are non-zero vectors. Hence (2.33) implies that ν_u and f_v are linearly independent at the point p. In particular, since

$$(f^t)_u(p) = t\nu_u(p), \quad (f^t)_v(p) = f_v(p) + t\nu_v(p),$$

we have

$$(f^t)_u(p) \times (f^t)_v(p) = t\nu_u(p) \times (f_v(p) + t\nu_v(p)).$$

The right-hand side of this equality does not vanish at $t = 0$. Hence, for sufficiently small t, we have $(f^t)_u(p) \times (f^t)_v(p) \neq \mathbf{0}$, which implies that f^t is a regular surface on a neighborhood of p. $\qquad\qquad\square$

Before introducing two important examples of wave fronts, we define flatness of wave fronts.

Definition 2.3.6. A wave front $f : U \to \mathbf{R}^3$ is said to be *flat* if the Gauss map $\nu : U \to S^2$ has a singularity at every point in U.

When f is a regular surface, the condition that the Gaussian curvature vanishes is equivalent to the Gauss map having a singularity, because of the Weingarten formula (Proposition 2.1.5). So this flatness is an extension of that defined in Section 2.2. Parallel surfaces of a flat wave front are flat wave fronts because of Corollary 2.2.4.

Typical examples of flat wave fronts are cylinders, as in Example 2.1.18. We introduce here other examples: cones and tangential developables.

Example 2.3.7 (Cones). Let a, b ($a < b$) be real numbers and $\gamma(u)$ ($a < u < b$) a regular curve in $\mathbf{R}^3$ lying on a plane $\Pi \subset \mathbf{R}^3$. Assume Π does not pass through the origin $(0,0,0)$ and let

$$f(u,v) := v\gamma(u).$$

Then the surface $f \colon (a,b) \times \mathbf{R} \to \mathbf{R}^3$ is called the *cone* associated to the curve γ. Since the velocity vector $\gamma'(u)$ is parallel to the plane Π, and $(0,0,0) \notin \Pi$, $\gamma(u)$ and $\gamma'(u)$ are linearly independent. Then, noticing that

$$f_u(u,v) = v\gamma'(u) \quad \text{and} \quad f_v(u,v) = \gamma(u),$$

the singular points of f correspond to $v = 0$, that is, the u-axis of the uv-plane. Moreover,

$$\nu(u) := \frac{\gamma'(u) \times \gamma(u)}{|\gamma'(u) \times \gamma(u)|}$$

is a unit normal vector field to f at (u,v). As this is independent of v, the Gauss map ν has (as a map into the unit sphere S^2) singularities everywhere. Since

$$\begin{pmatrix} f_u \; f_v \\ \nu_u \; \nu_v \end{pmatrix} = \begin{pmatrix} \mathbf{0} \; \gamma \\ \nu_u \; \mathbf{0} \end{pmatrix}$$

on $v = 0$, f is a flat front if and only if $\nu_u(u,0) \neq \mathbf{0}$. By the formula for scalar triple products, it holds that

$$\nu_u(u,0) \cdot \gamma'(u) = \frac{(\gamma''(u) \times \gamma(u)) \cdot \gamma'(u)}{|\gamma'(u) \times \gamma(u)|} = \frac{\det(\gamma''(u), \gamma(u), \gamma'(u))}{|\gamma'(u) \times \gamma(u)|}.$$

$$(2.34)$$

If γ has no inflection points (as a plane curve), γ'' and γ' are linearly independent as vectors parallel to Π. Then, noticing that Π does not pass through the origin, γ'', γ' and γ are linearly independent on $\mathbf{R}^3$. Hence $\nu_u \cdot \gamma' \neq 0$ by (2.34), and then f is a wave front (see Fig. 2.3(a)).

Example 2.3.8 (Tangential developables). Let $\gamma(u)$ ($a < u < b$) be a space curve satisfying $\gamma'(u) \times \gamma''(u) \neq \mathbf{0}$, which is equivalent to the curvature function never vanishing. The *tangential developable* of the curve γ is the map $f \colon (a,b) \times \mathbf{R} \to \mathbf{R}^3$ defined by (see Fig. 2.3(b))

$$f(u,v) := \gamma(u) + v\gamma'(u).$$

Since

$$f_u(u,v) = \gamma'(u) + v\gamma''(u), \quad f_v(u,v) = \gamma'(u),$$

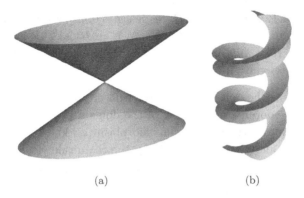

(a)   (b)

Fig. 2.3.   The cone associated to an ellipse, and the tangential developable of a helix.

it holds that $f_u(u, v) \times f_v(u, v) = v\gamma''(u) \times \gamma'(u)$. Hence the singular points of f correspond to the u-axis of the uv-plane. Moreover, the unit binormal vector $\boldsymbol{b}(u)$ (cf. [84, Section 5]) is the unit normal vector field of $f(u, v)$. Since $\boldsymbol{b}(u)$ does not depend on v, the Gauss map of f is singular everywhere. The singular set of f is $\{v = 0\}$, and the image of the singular set under f coincides with that of the curve $\gamma(u)$. Since

$$\begin{pmatrix} f_u & f_v \\ \nu_u & \nu_v \end{pmatrix} = \begin{pmatrix} \gamma' & \gamma' \\ \boldsymbol{b}' & \boldsymbol{0} \end{pmatrix}$$

when $v = 0$, the necessary and sufficient condition for the rank of the right-hand side to be 2 is $\boldsymbol{b}'(u) \neq \boldsymbol{0}$. Hence $f(u, v)$ is a wave front if and only if the torsion function of γ [84, Section 5] does not vanish (see Fig. 2.3(b)).

Exercises 2.3

1   Let $\gamma(u) = \big(x(u), y(u), z(u)\big)$ be a regular space curve satisfying $\gamma'(u) \times \gamma''(u) \neq \boldsymbol{0}$, and consider the tangent developable of this curve (cf. Example 2.3.8)

$$f(u, v) := \gamma(u) + v\gamma'(u).$$

Prove that the mean curvature of f at its regular points is $H = \tau/(2\kappa v)$. Here, κ and τ are the curvature and the torsion of γ, respectively.

2.4.   Parametrization-invariant Notions for Wave Fronts

We prove the following theorem, which is the surface version of Theorem 1.5.6 for curves.

Theorem 2.4.1. *Let $f : U \to \mathbf{R}^3$ be a wave front (respectively, frontal), $\varphi : V \to U$ a diffeomorphism from domain V in $\mathbf{R}^2$ to a domain U in $\mathbf{R}^2$, and $\Psi : \mathbf{R}^3 \to \mathbf{R}^3$ a diffeomorphism. Then the composition $\Psi \circ f \circ \varphi$ is also a wave front (respectively, frontal).*

Proof. Let ν be the unit normal vector field of f defined on U. First, we prove that the coordinate change φ of the domain preserves the property of being a frontal. Writing the diffeomorphism $\varphi \colon V \to U$ as $\varphi(a,b) = (u(a,b), v(a,b))$ and $f \circ \varphi(a,b)$ as $f(a,b)$, the chain rule yields

$$\begin{pmatrix} f_u & f_v \\ \nu_u & \nu_v \end{pmatrix} P = \begin{pmatrix} f_a & f_b \\ \nu_a & \nu_b \end{pmatrix}, \quad P := \begin{pmatrix} u_a & u_b \\ v_a & v_b \end{pmatrix}. \tag{2.35}$$

Since f_a, f_b are expressed by linear combinations of f_u and f_v, they are perpendicular to ν if f is a frontal. Then $f(a,b) := f(u(a,b), u(a,b))$ is a frontal. Moreover, $f(a,b)$ is a wave front if $f(u,v)$ is as well, because P is a regular matrix.

Next, we prove that $\hat{f} := \Psi \circ f$ is a frontal. Writing Ψ as

$$\Psi(x,y,z) = (\xi(x,y,z), \eta(x,y,z), \zeta(x,y,z)),$$

consider the Jacobian matrix J of Ψ composed with the map $f(u,v)$:

$$\begin{aligned} \hat{J}(u,v) &:= \begin{pmatrix} \xi_x(f(u,v)) & \xi_y(f(u,v)) & \xi_z(f(u,v)) \\ \eta_x(f(u,v)) & \eta_y(f(u,v)) & \eta_z(f(u,v)) \\ \zeta_x(f(u,v)) & \zeta_y(f(u,v)) & \zeta_z(f(u,v)) \end{pmatrix} \\ &= (\Psi_x(u,v), \Psi_y(u,v), \Psi_z(u,v)). \end{aligned} \tag{2.36}$$

Since Ψ is a diffeomorphism, $\hat{J}(u,v)$ is a regular matrix. Let

$$\hat{\nu}(u,v) := A(u,v)\nu(u,v), \quad A(u,v) := (\hat{J}(u,v)^T)^{-1},$$

where ν is the unit normal vector of f considered as a column vector, and A is the inverse of the transpose of the matrix in (2.36). Since $\nu \neq \mathbf{0}$ and J is a regular matrix, $A\nu \neq \mathbf{0}$ holds. Now, we show that the unit vector $\mathbf{N}(u,v) := \hat{\nu}(u,v)/|\hat{\nu}(u,v)|$ is the unit normal vector field of $\hat{f}$, that is, $\hat{f}_u$ and $\hat{f}_v$ are perpendicular to $\mathbf{N}$. Write the components of f as $f = (f_1, f_2, f_3)$. Then, by the chain rule,

$$\begin{aligned} \hat{f}_u(u,v) &= \Psi(f(u,v))_u = (\xi(f(u,v)), \eta(f(u,v)), \zeta(f(u,v)))_u \\ &= (\xi_x(f_1)_u + \xi_y(f_2)_u + \xi_z(f_3)_u, \eta_x(f_1)_u + \eta_y(f_2)_u + \eta_z(f_3)_u, \\ &\quad \zeta_x(f_1)_u + \zeta_y(f_2)_u + \zeta_z(f_3)_u) \\ &= \hat{J}(u,v) \begin{pmatrix} (f_1)_u(u,v) \\ (f_2)_u(u,v) \\ (f_3)_u(u,v) \end{pmatrix} = \hat{J}(u,v) f_u(u,v) \end{aligned} \tag{2.37}$$

holds. One can compute $\hat{f}_v(u, v)$ in the same way, and we have

$$\hat{f}_u = \hat{J} f_u, \quad \hat{f}_v = \hat{J} f_v. \tag{2.38}$$

By the formula $(C\boldsymbol{a}) \cdot \boldsymbol{b} = \boldsymbol{a} \cdot (C^T \boldsymbol{b})$ for vectors $\boldsymbol{a}$, $\boldsymbol{b}$ and a matrix C, it holds that

$$\hat{f}_u(u, v) \cdot \hat{\nu}(u, v) = (\hat{J}(u, v) f_u(u, v)) \cdot ((\hat{J}(u, v)^T)^{-1} \nu(u, v))$$

$$= f_u(u, v) \cdot (\hat{J}(u, v)^T (\hat{J}(u, v)^T)^{-1} \nu(u, v))$$

$$= f_u(u, v) \cdot \nu(u, v) = 0,$$

and similarly $\hat{f}_v(u, v) \cdot \hat{\nu}(u, v) = 0$ holds, that is, $\hat{\nu}$ is the unit normal vector field of $\hat{f}$, and hence $\hat{f}$ is a frontal.

Next, we show that $\hat{f}$ is a wave front if f is. Since f is a wave front, the matrix $\mathbf{M}(p)$ is of rank 2 for all $p \in V$. It is sufficient to show that the matrix

$$\hat{\mathbf{M}}(p) := \begin{pmatrix} \hat{f}_u(p) & \hat{f}_v(p) \\ \hat{\nu}_u(p) & \hat{\nu}_v(p) \end{pmatrix}$$

is of rank 2. If $f_u(p)$ and $f_v(p)$ are linearly independent, that is, the matrix $(f_u(p), f_v(p))$ is of rank 2, then the result is obvious. We assume that $(f_u(p), f_v(p))$ is of rank less than or equal to 1. By Lemma 2.3.5, we may assume $f_v(p) = \mathbf{0}$. Abbreviating $\hat{J}(u, v)$ by $\hat{J}$,

$$\hat{J}_v = (f_1)_v J_x \circ f + (f_2)_v J_y \circ f + (f_3)_v J_z \circ f$$

is the zero matrix, because $f_v = \mathbf{0}$. Then[5]

$$A_v = \left((\hat{J}^T)^{-1} \right)_v = -(\hat{J}^T)^{-1} (\hat{J}_v)^T (\hat{J}^T)^{-1} = O$$

at the point p. Hence

$$\hat{\nu}_v(p) = \left(A(u, v) \nu(u, v) \right)_v \Big|_{(u,v)=p} = A(p) \nu_v(p) \tag{2.39}$$

holds. Then we have that

$$\begin{pmatrix} \hat{f}_u & \hat{f}_v \\ \hat{\nu}_u & \hat{\nu}_v \end{pmatrix} \Big|_{(u,v)=p} = \begin{pmatrix} \hat{J}(p) f_u(p) & \mathbf{0} \\ * & A(p) \nu_v(p) \end{pmatrix}.$$

---

[5]Let $A(t)$ be a regular square matrix with smooth parameter t. We set $B(t) := A(t)^{-1}$, and differentiate $A(t)B(t) = E$, where E is the identity matrix. Then the formula $B(t) = -A(t)^{-1} A'(t) A(t)^{-1}$ is obtained.

Since f is a wave front, $\nu_v(p) \neq \mathbf{0}$ holds. Moreover, since $A(p)$ is a regular matrix, $A(p)\nu_v(p) \neq \mathbf{0}$. In particular, if $f_u(p) \neq \mathbf{0}$, $\hat{f}$ is a wave front at p. So we may assume that $f_u(p) = f_v(p) = \mathbf{0}$. Then

$$\det(\hat{\nu}_u(p), \hat{\nu}_v(p), \hat{\nu}(p)) = \det A(p) \det(\nu_u(p), \nu_v(p), \nu(p)).$$

Since f is a wave front, $f_u(p) = f_v(p) = \mathbf{0}$ implies $(\nu_u(p), \nu_v(p), \nu(p))$ is a regular matrix. So we can conclude that $(\hat{\nu}_u(p), \hat{\nu}_v(p), \hat{\nu}(p))$ is also a regular matrix. Then one can easily check that $\mathbf{N}_u(p)$, $\mathbf{N}_v(p)$ are linearly independent, proving the conclusion.   □

Definition of wave fronts in terms of differentiable structure. This is a remark for readers with a knowledge of manifolds. The inner product of $\mathbf{R}^3$ is used in the proof of Theorem 2.4.1, which looks technically complicated because orthogonality is not preserved by diffeomorphisms. If one can define a notion of wave fronts only using differentiable structure of the domain and target (without use of inner product), an alternative for the proof of Theorem 2.4.1 is obtained without the complication of an inner product, in principle. Here, we give such a reformulation of wave fronts.

Consider $V := \mathbf{R}^3$ as a vector space, and let V^* be its dual space. Then we can define the bilinear map

$$V \times V^* \ni (v, \omega) \mapsto \langle v, \omega \rangle := \omega(v) \in \mathbf{R},$$

which is called a *coupling*.

Definition 2.4.2. We let V be a real vector space V of dimension n. We say that two non-zero vectors x and $y \in V$ satisfy $x \sim y$ if $\{x, y\}$ are linearly dependent. Then $\sim$ is an equivalence relation on $V \setminus \{0\}$. The set $P(V)$ of equivalent classes of $V \setminus \{0\}$ under $\sim$ is called the *projective space associated with V*.

Let $P(V^*)$ be the projective space associated with V^*. The space $P(V^*)$ can be endowed with the structure of an $(n-1)$-dimensional manifold. We denote the natural projection by

$$\pi : V^* \ni \omega \mapsto [\omega] \in P(V^*).$$

The following proposition holds, which is a special case of Corollary 10.3.11 in Chapter 10.

Proposition 2.4.3. *A C^∞-map of $f : U \to \mathbf{R}^3 (= V)$ is a wave front if and only if there exists a C^∞-map $g : U \to V^*$ satisfying the following properties:*

(1) g *does not vanish on U,*

(2) $(f, \pi \circ g) : U \to V \times P(V^*)$ *is an immersion and*

(3) *for each $p \in U$, it holds that $\langle f_u(p), g(p) \rangle = \langle f_v(p), g(p) \rangle = 0$.*

In particular, the concept of wave fronts does not depend on the inner product of $\mathbf{R}^3$.

Proof. Denote by "·" the canonical inner product of $V = \mathbf{R}^3$, and define $\Phi(v) \in V^*$ by

$$\Phi(v)(w) := v \cdot w \quad (w \in V)$$

for each $v \in V$. Then $\Phi : V \to V^*$ is a linear isomorphism. Let ν be the unit normal vector field of a wave front f, and define a map from U to V^* by $g := \Phi \circ \nu$. We show g satisfies the properties (1), (2) and (3). Since Φ is a linear isomorphism and ν never vanishes, neither does g, proving (1).

Let $S^2(\subset \mathbf{R}^3)$ be the unit sphere centered at the origin. We define a map $\hat{\pi} : S^2 \ni v \mapsto \pi \circ \Phi(v) \in P(V^*)$, which is a 2 to 1 immersion. Then

$$\mathbf{R}^3 \times S^2 \ni (x, v) \mapsto (x, \hat{\pi}(v)) \in \mathbf{R}^3 \times P(V^*)$$

is as well. Since $\hat{\pi} \circ \nu = \pi \circ g$, it holds that $(f, \pi \circ g) : U \to \mathbf{R}^3 \times P(V^*)$ is an immersion if $(f, \nu) : U \to \mathbf{R}^3 \times S^2$ is an immersion, proving (2).

By the definition of g,

$$\langle f_u(p), g(p) \rangle = g(p)(f_u(p)) = \nu(p) \cdot f_u(p) = 0$$

and $\langle f_v(p), g(p) \rangle = 0$ hold. Hence we have (3).

Conversely, let $g : U \to V^*$ be a C^∞-map satisfying the conditions (1), (2) and (3). By the condition (1), a map $\nu : U \to S^2$ can be defined by $\nu := \Phi^{-1} \circ g / |\Phi^{-1} \circ g|$. Then ν is a unit normal vector field of f by the condition (3), and $L := (f, \nu) : U \to \mathbf{R}^3 \times S^2$ is an immersion because of (2). $\quad\square$

Proposition 2.4.3 gives a condition for being a wave front using only the differentiable structures of $\mathbf{R}^3$ and $P(V^*)$, without using the inner product of $\mathbf{R}^3$. This is a reason why Theorem 2.4.1 holds. See Proposition 10.3.6 in Chapter 10 for the precise proof of the theorem (the case of $\mathcal{N}^3 = \mathbf{R}^3$ in Proposition 10.3.6 corresponds to Theorem 2.4.1).

2.5. Singularities Appearing on Surfaces

We define an equivalence relation between maps by saying two maps are equivalent when they coincide on a sufficiently small neighborhood of a point p. An equivalence class of this relation is called a *map-germ* at p. When one discusses a given map restricted to a sufficiently small neighborhood of the given point p, the map is considered as a map-germ (see Chapter 8 for details). Let $f_j(u, v)$ be a surface in $\mathbf{R}^3$ defined on a neighborhood U_j of a point $(u_j, v_j) \in \mathbf{R}^2$, for $j = 1, 2$. Then the map-germ at (u_2, v_2) of f_2 is *right-left equivalent* (or *$\mathcal{A}$-equivalent*) to that of f_1 at (u_1, v_1) if, taking U_1 and U_2 to be sufficiently small, there exist a diffeomorphism $\varphi : U_2 \to U_1$ with $\varphi(u_2, v_2) = (u_1, v_1)$ and a diffeomorphism $\Phi : \Omega_2 \to \Omega_1$, for Ω_j ($j = 1, 2$) sufficiently small neighborhoods of $f_j(u_j, v_j)$ in $\mathbf{R}^3$ such that

$$\Phi \circ f_2 = f_1 \circ \varphi.$$

Namely, two map-germs f_1 and f_2 are right-left equivalent (or $\mathcal{A}$-equivalent) if f_2 is obtained from f_1 via a suitable coordinate change of the domain (φ in this situation) and a suitable coordinate change of the target (Φ^{-1} in this situation).

Definition 2.5.1 (Nomenclature of singularities). For a given map f_0 which has a singular point named "X", a map-germ which is right–left equivalent to f_0 is said to be a *map with singularity "X"*.

We introduce here typical singularities of surfaces.

Example 2.5.2. The map $g_C(u, v) := (-3u^2, 2u^3, v)$ has singularities on the v-axis. This is a direct product of the cusp and a line (Fig. 2.4(a)). A map-germ right–left equivalent to the map-germ g_C at the origin is called a *cuspidal edge*. The intersection of the image of g_C and a plane perpendicular to the z-axis is a cusp. The unit normal vector of this cusp is the planar curve giving the unit normal vector of g_C. It is easy to check that g_C is a wave front (cf. Exercise **5** in this section).

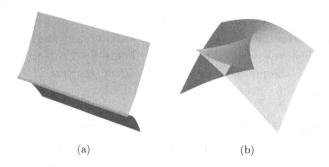

(a)   (b)

Fig. 2.4.   The cuspidal edge (a) and the swallowtail (b).

Let

$$D_{2,3} := \{(x, y, z) \in \mathbf{R}^3 \,;\, t^3 + xt + y = 0 \text{ has a multiple real root}\}.$$

Then, by the same way as in Example 1.3.1 in Chapter 1, we have

$$D_{2,3} = \{(-3t^2, 2t^3, z) \in \mathbf{R}^3 \,;\, t, z \in \mathbf{R}\} = g_C(\mathbf{R}^2),$$

where $g_C(\mathbf{R}^2)$ denotes the image of g_C. For the meaning of the notation $D_{2,3}$, see (8.23). For the sake of simplicity, we use $f_C = (u^3, u^2, v)$ for the standard cuspidal edge.

Example 2.5.3. The map $g_S(u, v) = (u, -4v^3 - 2uv, 3v^4 + uv^2)$ has a singular point at $(0, 0)$ (Fig. 2.4(b)). A map-germ right–left equivalent to the map-germ g_S at the origin is called a *swallowtail*. Since $f = g_S$ satisfies

$$f_u \times f_v = -2(6v^2 + u)(v^2, v, 1),$$

the singular point set of f is a parabola $u = -6v^2$ on the uv-plane, and

$$\nu(u, v) = \frac{1}{\sqrt{1 + v^2 + v^4}} \, (v^2, v, 1)$$

gives the unit normal vector field. Since

$$f_v(0, 0) = \mathbf{0}, \quad \nu_v(0, 0) \neq \mathbf{0} \quad \text{and} \quad f_u(0, 0) \neq \mathbf{0},$$

$f = g_S$ is a wave front on a neighborhood of $(0, 0)$. Then Theorem 2.4.1 yields that a swallowtail is a wave front on a neighborhood of the point. Note that the singular points of f, except for $(0, 0)$, are cuspidal edges.

Consider the standard form of quartic equations $t^4 + xt^2 + yt + z = 0$ (cf. the first paragraph of Example 1.3.1 in Chapter 1). Then the set

$$D_{3,3} := \{(x, y, z) \in \mathbf{R}^3 \,;\, t^4 + xt^2 + yt + z = 0 \text{ has a multiple real root}\}$$

coincides with the image of f (cf. Exercise **1** at the end of this section). For the meaning of the notation $D_{3,3}$, see (8.23). Figure 2.5 shows the cross-section of the neighborhood of the image of the swallowtail by the plane $x = c$. If $c < 0$, then the cross-section have two cusps and has a $4/3$-cusp at $c = 0$ (the third figure of Fig 2.5). If $c > 0$, then the cross-section is a regular curve. These cross-sections give a deformation of plane curves as wave fronts (cf. (1.62)). For the sake of simplicity, we use $f_S = (3u^4 + u^2 v, -4u^3 - 2uv, v)$ for the standard swallowtail.

Fig. 2.5.   Sections of the neighborhood of the swallowtail (Example 2.5.3).

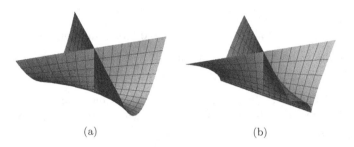

(a)   (b)

Fig. 2.6.   The image of cross cap f_W and cuspidal cross cap f_{CR}.

Fig. 2.7.   The sections of the neighborhood of the cross cap (Example 2.5.4).

Example 2.5.4. Let $f_W(u, v) := (u, uv, v^2)$. A map-germ right–left equivalent to the map-germ f_W at the origin is called a *cross cap* or *Whitney's umbrella* (Fig. 2.6(a)). This is the most frequent singularity appearing in the class of C^∞-maps from a domain of $\mathbf{R}^2$ into $\mathbf{R}^3$. Since the unit normal vector of f_W cannot be extended to the origin (cf. Exercise **2** at the end of this section), it is not a frontal. On the other hand, cuspidal edges and swallowtails are the singularities that appear most frequently on wave fronts. The self-intersection set of f_W is a half-line in $\mathbf{R}^3$ emanating from the origin $\mathbf{0} := (0, 0, 0)$. A cross-section of the image of f_W by the plane $z = x + c$ for $c < 0$, $c = 0$ and $c > 0$ are shown in Fig. 2.7. When $c = 0$, the plane passes through the singular point and the cross section becomes a cusp (as a planar curve). This figure can be considered as a deformation of curves which is not regular homotopic.

Example 2.5.5. A map-germ which is right–left equivalent to the map-germ $f_{CR}(u, v) := (u, uv^3, v^2)$ at the origin is called a *cuspidal cross cap*, whose image looks like a combination of the shapes of a cuspidal edge and a cross cap. Since there exists a smooth unit normal vector ν on a neighborhood of this singular point, it is a frontal. On the other hand, the matrix $\mathbf{M}$ in (2.31) is not of rank 2 at the origin. Hence the map is not a wave front. Like the case of a cross cap, the self-intersection set of f_{CR} is a half-line in $\mathbf{R}^3$ emanating from the origin $\mathbf{0} := (0, 0, 0)$. A cross-section

Fig. 2.8. The sections of the neighborhood of the cuspidal cross cap (Example 2.5.5)

of the image of f_{CR} by the plane $z = x + c$ for $c < 0$, $c = 0$ and $c > 0$ are shown in Fig. 2.8. When $c = 0$, the plane passes through the singular point and the cross section becomes a 5/2-cusp (as a planar curve).

For cross caps, the following is known:

Theorem 2.5.6 ([89]). *Let U be a domain on the uv-plane. Assume a C^∞-map $f : U \to \mathbb{R}^3$ satisfies $f_v(p) = \mathbf{0}$. Then the point $p \in U$ is a cross cap if and only if*

$$\det(f_u, f_{uv}, f_{vv}) \neq 0 \qquad (2.40)$$

holds at p.

Remark 2.5.7. Let p be a singular point of a given C^∞-map $f : U \to \mathbb{R}^3$. Then by Lemma 2.3.5, we can choose the coordinates so that $f_v(p) = \mathbf{0}$, using only a linear transformation. So the assumption $f_v(p) = \mathbf{0}$ of this theorem is not essential.

We shall prove Theorem 2.5.6 in Chapter 3.

Example 2.5.8. The standard cross cap $f(u, v) = (u, uv, v^2)$ actually satisfies the condition of Theorem 2.5.6. In fact,

$$f_v(0,0) = \mathbf{0}, \ f_u(0,0) = (1,0,0), \ f_{uv}(0,0) = (0,1,0), \ f_{vv}(0,0) = (0,0,2)$$

hold, and so $\det(f_u, f_{uv}, f_{vv}) \neq 0$ holds at $(u, v) = (0,0)$.

Example 2.5.9. For each real number s, consider a curve

$$\gamma_s(t) := (t^2, t(t^2 - s)) \quad (t \in \mathbb{R}).$$

If $s = 0$, the curve has a cusp. When $s < 0$, the curve has no self-intersections, and when $s > 0$, the curve has a self-intersection. If we consider the surface

$$f(t, s) = (\gamma_s(t), s) = (t^2, t(t^2 - s), s),$$

then it can be easily checked that $(0, 0)$ is a cross cap singular point.

Remark 2.5.10. Whitney [90] defined *cross cap* singularities as isolated singular points appearing in differentiable maps from a domain of R^n into R^{2n-1} and gave a criterion for such singularities, for each $n \geq 2$. (The above theorem is the case of $n = 2$.) Using them, he proved that any n-manifolds admit smooth maps into R^{2n-1} whose singular points consist only of cross caps. He then found a pairwise cancellation of cross cap singularities, and succeeded to prove that any n-manifolds can be immersed in R^{2n-1}. For example,

$$f_c(u, v) = (u, v^2, v(u^2 - v^2 + c)) \quad (c \in R)$$

gives a family of smooth maps so that $f_c : R^2 \to R^3$ ($c < 0$) has a pair of cross caps, and f_c ($c > 0$) are regular surfaces (see Fig. 2.9).

The Roman surface given in Exercise **4** at the end of this section is a realization of the projective plane P^2, and it has six cross caps (12 on S^2). These six singularities can be eliminated by applying the above cancellation of cross caps separately to three disjoint pairs of these six singularities, and P^2 can be realized as a regular surface with self-intersections in R^3 (see Fig. 2.10).

Fig. 2.9.   A pairwise cancellation ($f_{-1/2}$, f_0 and $f_{1/2}$) of cross caps.

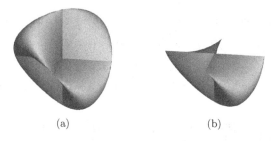

(a)   (b)

Fig. 2.10.   The entire figure of the Roman surface (a) and a piece of it (b).

Exercises 2.5

1 For a given quartic polynomial

$$f(t) = t^4 + xt^2 + yt + z,$$

show that the set of points (x, y, z) where $f(t) = 0$ has multiple real roots is congruent to the swallowtail of Example 2.5.3 (cf. Example 1.3.1 in Chapter 1).

2 Let f_W be the cross cap as in Example 2.5.4. Compute the unit normal vector $\nu(u, v)$ for $(u, v) \neq (0, 0)$, and show that the limit $\lim_{r \to 0} \nu(r \cos \theta, r \sin \theta)$ depends on the angle θ.

3 The maps

$$f_1(u, v) := (u^2, u^3, v), \quad f_2(u, v) := (3u^4 + u^2 v, -4u^3 - 2uv, v)$$

from $\mathbf{R}^2$ and $\mathbf{R}^3$ are a cuspidal edge and a swallowtail, respectively. Show that the Gaussian curvature of these maps on the regular point sets is identically zero for both surfaces. In addition, compute the mean curvature function.

4 (1) Define $F(x, y, z) = (yz, zx, xy) : \mathbf{R}^3 \to \mathbf{R}^3$, and let $f \colon S^2 \to \mathbf{R}^3$ be a map defined by restricting the domain of F to the unit sphere S^2. Show that the map f induces a smooth map $\tilde{f} \colon P^2 \to \mathbf{R}^3$ of the real projective plane $P^2 := P(\mathbf{R}^2)$ into $\mathbf{R}^3$. (The map $\tilde{f}$ is called the *Roman surface*, see Fig. 2.10.)

(2) Moreover, show that the singularities of f are all cross caps. (*Hint*: Since the Jacobian of the map $\mathbf{R}^3 \ni (x, y, z) \mapsto (yz, zx, xy)$ is xyz, the singular points appear only when $x = 0$, $y = 0$ or $z = 0$. By setting $\varphi(x, y) = (x, y, \sqrt{1 - x^2 - y^2})$, it can be easily checked that the singular points of $f \circ \varphi(x, y)$ on the set $y = 0$ appear only at $(\pm 1/\sqrt{2}, 0)$. This implies that the number of singular points of f on P^2 is six[6]. To show the assertion, it is sufficient to check that the singular point $(1/\sqrt{2}, 0)$ of $f \circ \varphi$ is a cross cap because of its symmetries.)

5 Show that the map $g_C(u, v) = (-3u^2, 2u^3, v)$ in Example 2.5.2 is a wave front.

6 Consider a surface

$$f(s, t) = (s, (1 + s \cos t) \cos t, (1 + s \cos t) \sin t) \quad (s, t \in \mathbf{R})$$

which is associated with the deformation of curves as in Example 1.6.16, and show that f has cross caps at $(s, t) = (1, \pi)$, $(-1, 0)$.

---

[6]They are $(\pm 1/\sqrt{2}, \pm 1/\sqrt{2}, 0)$, $(\pm 1/\sqrt{2}, 0, \pm 1/\sqrt{2})$ and $(\pm 1/\sqrt{2}, \pm 1/\sqrt{2}, 0)$.

2.6. Criteria for Cuspidal Edges and Swallowtail Singularities I

We introduce the criteria for cuspidal edges and swallowtails defined in the previous section. Since they are wave fronts, we start with properties of wave front singularities.

Let $f : U \to \boldsymbol{R}^3$ be a frontal defined on a neighborhood U of the origin of the uv-plane, and $\tilde{\nu}$ a non-vanishing normal vector field, which is not necessarily a unit vector. We set

$$\Lambda := \det(f_u, f_v, \tilde{\nu}),$$

where the right-hand side is the determinant of the 3×3-matrix consisting of the three column vectors. Since the zero set of the function Λ coincides with the singular points of f, we call Λ an *identifier of singularities*. (This function depends on local coordinates and choice of $\tilde{\nu}$.) As a function obtained by multiplying Λ by a non-vanishing function can play the same role as Λ, it is also called an *identifier of singularities*. When $\tilde{\nu}$ is the unit normal vector, we write λ instead of Λ, and call it the *signed area density*. This plays an important role in the Gauss–Bonnet-type formulas later (cf. Chapters 4 and 6).

A singular point $p \in U$ of a frontal $f : U \to \boldsymbol{R}^3$ is said to be *non-degenerate* if the differential $d\Lambda = \Lambda_u \, du + \Lambda_v \, dv$ of the identifier of singularities Λ does not vanish at p, that is, at least one of $\Lambda_u(p)$ and $\Lambda_v(p)$ is not 0. Since the set of singular points of f is characterized by $\Lambda = 0$, the implicit function theorem yields that the singular set is a regular curve in U on a neighborhood of p, if p is a non-degenerate singular point. We call this curve the *singular curve*, although it is regular, and its tangent vector is called the *singular vector*, while the one-dimensional vector space spanned by the singular vector is called the *singular direction*. On the other hand, the direction of a non-zero planar vector $\boldsymbol{v}$ with $df(\boldsymbol{v}) = \boldsymbol{0}$ is called a *null direction*, and the vector $\boldsymbol{v}$ is called a *null vector*. The following holds:

Lemma 2.6.1. *The vector space generated by null vectors at a non-degenerate singular point is one-dimensional.*

Proof. The identifier of singularities Λ satisfies

$$\Lambda_u = \det(f_{uu}, f_v, \tilde{\nu}) + \det(f_u, f_{uv}, \tilde{\nu}) + \det(f_u, f_v, \tilde{\nu}_u).$$

So if a non-degenerate singular point p of f satisfies $f_u(p) = f_v(p) = \boldsymbol{0}$, then $\Lambda_u(p) = 0$. Similarly, $\Lambda_v(p) = 0$ is also obtained, a contradiction to non-degeneracy. $\qquad \square$

The null direction is uniquely determined as a one-dimensional vector space at a non-degenerate singular point, by the above lemma. Then for each point $\gamma(t)$ of the singular curve, one can choose null vector $\eta(t)$ depending smoothly on t. In other words, there exists a smooth vector field $\eta(t)$ along the curve $\gamma(t)$ consisting of null directions. We call such an η the *null vector field*.

We prove the following:

Proposition 2.6.2. *Non-degeneracy of singular points of frontals does not depend on the choice of coordinates of the domain and the target space.*

Proof. When changing the normal vector field, the identifier of singularities changes by multiplication by a non-vanishing function. Since the identifier of singularities takes its value 0 at singular points, non-degeneracy is independent of choice of normal vector fields. Hence it is sufficient to prove the theorem for the signed-area element λ, that is, the identifier of singularities with respect to the unit normal vector field.

Let $f : U \to \mathbf{R}^3$ be a frontal, and consider a coordinate change of the domain

$$u = u(a, b), \quad v = v(a, b).$$

Then, by the chain rule, it holds that

$$\det(f_a, f_b, \nu) = \det(f_u u_a + f_v v_a, f_u u_b + f_v v_b, \nu)$$

$$= u_a v_b \det(f_u, f_v, \nu) + v_a u_b \det(f_v, f_u, \nu)$$

$$= (u_a v_b - u_b v_a) \det(f_u, f_v, \nu).$$

Since $u_a v_b - u_b v_a \neq 0$, the non-degeneracy does not depend on the choice of coordinate system of the domain.

Next, let $\Phi : \mathbf{R}^3 \to \mathbf{R}^3$ be a diffeomorphism of $\mathbf{R}^3$, and $\hat{f} := \Phi \circ f$. Then, as seen in (2.38), we have

$$\hat{f}_u = J f_u, \quad \hat{f}_v = J f_v,$$

where J is the Jacobi matrix of Φ. Let p be a singular point of f, and assume $f_u(p) = \mathbf{0}$, that is, we choose a local coordinate system (u, v) of the domain such that the u-direction is the null direction. By non-degeneracy, $(\lambda_u(p), \lambda_v(p)) \neq \mathbf{0}$ holds. Then there exists $(\alpha, \beta) \neq (0, 0)$ such that

$$\alpha \lambda_u(p) + \beta \lambda_v(p) \neq 0.$$

Noting that $\lambda := \det(f_u, f_v, \nu)$ and $f_u(p) = \mathbf{0}$, we have

$$0 \neq \alpha \lambda_u(p) + \beta \lambda_v(p) = \det(\alpha f_{uu}(p) + \beta f_{uv}(p), f_v(p), \nu(p)),$$

and in particular, $\alpha f_{uu}(p) + \beta f_{uv}(p)$ and $f_v(p)$ are linearly independent. Hence

$$J(\alpha f_{uu}(p) + \beta f_{uv}(p)), \quad \hat{f}_v(p)$$

are linearly independent. Moreover, since $f_u(p) = \mathbf{0}$, $\hat{f}_{uu}(p) = Jf_{uu}(p)$ and $\hat{f}_{uv}(p) = Jf_{uv}(p)$ hold, and we have that

$$\alpha \hat{f}_{uu}(p) + \beta \hat{f}_{uv}(p), \quad \hat{f}_v(p)$$

are linearly independent. Since $\hat{f}$ is also a frontal, there exists a unit normal vector field $\hat{\nu}$. Differentiating $\hat{f}_u \cdot \hat{\nu} = 0$ with respect to u and v, and noticing that $\hat{f}_u(p) = \mathbf{0}$, we find that

$$\hat{f}_{uu}(p) \cdot \hat{\nu}(p) = \hat{f}_{uv}(p) \cdot \hat{\nu}(p) = 0.$$

Hence, $\hat{f}_v(p) \cdot \hat{\nu}(p) = 0$ implies that

$$\alpha \hat{f}_{uu}(p) + \beta \hat{f}_{uv}(p), \quad \hat{f}_v(p), \quad \hat{\nu}(p)$$

are linearly independent. Let $\hat{\lambda} = \det(\hat{f}_u, \hat{f}_v, \hat{\nu})$. Then, since $\hat{f}_u(p) = \mathbf{0}$,

$$\alpha \hat{\lambda}_u(p) + \beta \hat{\lambda}_v(p) = \det(\alpha \hat{f}_{uu}(p) + \beta \hat{f}_{uv}(p), \hat{f}_v(p), \hat{\nu}(p)) \neq 0.$$

Summing up, $d\hat{\lambda}(p)$ does not vanish, that is, p is a non-degenerate singular point of the frontal $\hat{f}$.   □

The following criteria hold for cuspidal edges and swallowtails.

Theorem 2.6.3 ([44]). *Let $f: U \to \mathbf{R}^3$ be a wave front and $\gamma(t)$ a U singular curve such that each point is a non-degenerate singular point of f, and fix $t = c$.*

(1) *The map-germ of f at $\gamma(c)$ is right–left equivalent to a cuspidal edge if and only if the null direction and the singular direction are linearly independent at $t = c$.*

(2) *Let $\eta(t)(\neq \mathbf{0})$ be the null vector field, and set*

$$\Delta(t) := \det(\eta(t), \gamma'(t)), \qquad (2.41)$$

that is, $\Delta(t)$ is the determinant of the 2×2-matrix consisting of two column vectors $\eta(t)$ and $\gamma'(t)$, where $\gamma' = d\gamma/dt$. Then f is right–left equivalent to the swallowtail if and only if $\eta(c)$ and $\gamma'(c)$ are linearly dependent, and $\Delta'(t) = d\Delta(t)/dt$ does not vanish at $t = c$.

This theorem says that cuspidal edges and swallowtails can be recognized using only the information living on the domain. The first part (1) of the theorem will be proved in Section 3.4 in Chapter 3.

Remark 2.6.4. A different criterion will be introduced in Section 4.2 of Chapter 4 (Theorem 4.2.3).

We now verify the criteria in Theorem 2.6.3 for a concrete example:

Cuspidal edges and Swallowtails. We verify the criterion for the standard cuspidal edge $f(u,v) := (u^3, u^2, v)$. The function $\Lambda := u$ is an identifier of singularities, because of Example 4.2.5. Since $\Lambda_u = 1$, $d\Lambda \neq 0$ holds on the singular curve (the v-axis). And since $f_u(0,v) = \mathbf{0}$, the null vector field is $\eta = (1,0)$, which is linearly independent of the singular direction $(0,1)$. Then f satisfies (1) of Theorem 2.6.3.

To verify that the standard swallowtail satisfies the criterion (2) of Theorem 2.6.3 is an exercise left for the reader.

Criterion for cuspidal cross caps. We now introduce the known criterion for cuspidal cross caps (see [18]). Let $f \colon U \to \mathbf{R}^3$ be a frontal defined on a neighborhood U of the origin in $\mathbf{R}^2$, and let $(0,0)$ be a non-degenerate singular point. Parametrizing the singular curve as $\gamma(t) = (u(t), v(t))$ such that $\gamma(0) = (0,0)$, and letting $\eta(t) = (a(t), b(t))$ be a null vector field along γ, we set

$$\varphi(t) := \det\big(\hat{\gamma}'(t), \tilde{\nu}(u(t), v(t)), \tilde{\nu}_\eta(t)\big),$$

$$\tilde{\nu}_\eta(t) := a(t)\tilde{\nu}_u(u(t), v(t)) + b(t)\tilde{\nu}_v(u(t), v(t)),$$

where $\hat{\gamma} := f \circ \gamma$, the prime denotes differentiation with respect to t, and $\tilde{\nu}$ is a normal vector field (not necessarily unit) of f.

Fact 2.6.5 ([18]). A C^∞-map $f \colon U \to \mathbf{R}^3$ as above is right–left equivalent to the cuspidal cross cap at $(u,v) = (0,0)$ if and only if f is a frontal for a sufficiently small neighborhood of $(0,0)$, and satisfies the following conditions:

(1) the origin is a non-degenerate singular point,
(2) the null vector $\eta(0)$ and the singular vector $\gamma'(0)$ are linearly independent,
(3) $\varphi(0) = 0$ and $\varphi'(0) \neq 0$.

We verify the criterion for the standard cuspidal cross cap $f(u,v) := (u, uv^3, v^2)$. Since $f_u \times f_v = v(2v^3, -2, 3uv)$, $\{v = 0\}$ (the u-axis) is the singular set, and $\tilde{\nu} = (2v^3, -2, 3uv)$ is a non-vanishing normal vector field. Since

$$\det(f_u, f_v, \tilde{\nu}) = v(4 + 9u^2v^2 + 4v^6),$$

$\Lambda = v$ is an identifier of singularities, and since $\Lambda_v = 1(\neq 0)$, $(\Lambda_u, \Lambda_v) \neq \mathbf{0}$ holds on the u-axis.

On the other hand, the singular curve is given by $\hat{\gamma}(u) := f(u, 0)$. Since $f_v(u, 0) = \mathbf{0}$, the null vector is $\eta = (0, 1)$, which is linearly independent to $(1, 0)$, that is, the tangential direction of the singular set.

Moreover, $\hat{\gamma}'(u) = f_u(u, 0) = (1, 0, 0)$, $\tilde{\nu}(u, 0) = (0, -2, 0)$ and

$$\tilde{\nu}_\eta(u) = \tilde{\nu}_v(u, 0) = (0, 0, 3u)$$

yield

$$\varphi(u) = \det\big(\hat{\gamma}'(u), \tilde{\nu}(u, 0), \tilde{\nu}_v(u, 0)\big) = -6u.$$

Thus $\varphi(0) = 0$ and $\varphi'(0) \neq 0$, so f satisfies the criterion for a cuspidal cross cap at the origin.

Exercises 2.6

1 Verify that the swallowtail in Example 2.5.3 satisfies (2) of Theorem 2.6.3.

2.7. Special Directions of Singular Points on Frontals

We consider a frontal $f : U \to \mathbf{R}^3$ defined on a domain U in $\mathbf{R}^2$ with unit normal vector field ν. Then f is said to be *co-rank one* at p if the Jacobi matrix of f at p is of rank $1(= 2 - 1)$, where 2 is the dimension of U. Since cuspidal edges, swallowtails and cuspidal cross caps are all non-degenerate singular points on a surface, they are all co-rank one frontals.

Here, we introduce several fundamental concepts of such co-rank one singular points. We fix such a point $p \in U$. Then we can take a local coordinate system (u, v) centered at p such that $f_v(p) = \mathbf{0}$. Since p is of co-rank one, the line

$$L_1 := \{f(p) + t f_u(p) \, ; \, t \in \mathbf{R}\}$$

is defined, which is called the *tangent line* at p. The plane Π_0 passing through $f(p)$ which is perpendicular to the normal direction of f at $f(p)$ is called the *limiting tangent plane* of f at p. The line L_2 passing through $f(p)$ lying in Π_0 which is perpendicular to L_1 is called the *co-normal line* of f at p.

On the other hand, the plane Π_1 which is perpendicular to the tangent line L_1 is called the *normal plane*, and the plane Π_2 passing through $f(p)$ spanned by $\nu(p)$ and $f_u(p)$ is called the *co-normal plane* at p. By definition, the intersection of the two planes Π_0 and Π_1 is the co-normal line L_2 (see Fig. 2.11).

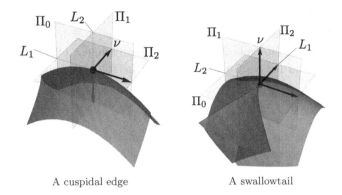

| A cuspidal edge | A swallowtail |

Fig. 2.11. The limiting tangent plane, the normal plane and the co-normal plane.

Example 2.7.1. For the standard cuspidal edge $(u, v) \mapsto (v^2, v^3, u)$ at the origin,

- the tangent line is $\{(0, 0, t); t \in \mathbf{R}\}$,
- the limiting tangent plane is $\{(s, 0, t); s, t \in \mathbf{R}\}$,
- the co-normal line is $\{(s, 0, 0); s \in \mathbf{R}\}$,
- the normal plane is $\{(s, t, 0); s, t \in \mathbf{R}\}$,
- the co-normal plane is $\{(0, s, t); s, t \in \mathbf{R}\}$.

For the standard swallowtail $(u, v) \mapsto (3u^4 + u^2 v, -4u^3 - 2uv, v)$ at the origin,

- the tangent line is $\{(0, 0, t); t \in \mathbf{R}\}$,
- the limiting tangent plane is $\{(0, s, t); s, t \in \mathbf{R}\}$,
- the co-normal line is $\{(0, s, 0); s \in \mathbf{R}\}$,
- the normal plane is $\{(s, t, 0); s, t \in \mathbf{R}\}$,
- the co-normal plane is $\{(s, 0, t); s, t \in \mathbf{R}\}$.

For cuspidal edges, the following assertion holds.

Proposition 2.7.2. *Let $f : U \to \mathbf{R}^3$ be a front, and suppose f has a cuspidal edge at $p \in U$. Let Π_1 be the normal plane of f at p. Then the slice locus of f by Π_1 is a curve in Π_1, which is a cusp (called a sectional cusp).*

Proof. By definition, there exist diffeomorphisms $\varphi : U \to \varphi(U)(\subset \mathbf{R}^2)$ and $\Phi : \Omega \to \Phi(\Omega)(\subset \mathbf{R}^3)$ such that

$$\Phi \circ f \circ \varphi = f_0,$$

where f_0 is the standard cuspidal edge given by $f_0(u,v) := (u^2, u^3, v)$ and U (resp. Ω) is a domain in $\mathbf{R}^2$ (resp. $\mathbf{R}^3$). Since Φ is a diffeomorphism, $S := \Phi(\Pi_1 \cap \Omega)$ is an embedded surface whose tangent plane at the origin does not contain the vector $e_3 := (0,0,1)$ in $\mathbf{R}^3$. It is sufficient to show that

$$f_0(\mathbf{R}^2) \cap S$$

gives a cusp on the surface at the origin. Since e_3 is not a tangent vector of S at $\mathbf{0}$, there exists a smooth function $g : D \to \mathbf{R}$ defined on a domain D containing $o := (0,0)$ such that S can be expressed as the graph of the function g, that is,

$$G(x,y) = (x, y, g(x,y))$$

gives a parametrization of S. Since

$$x = u^2, \quad y = u^3, \quad z = v.$$

parametrizes f_0, the curve $\hat{\gamma}(v) := (u^2, u^3, g(u^2, u^3))$ gives the parametrization of the image of $S \cap f_0(\mathbf{R}^2)$. Since $G^{-1} \circ \hat{\gamma}(v) = (u^2, u^3)$,

$$\Phi^{-1} \circ \hat{\gamma}(v) = \Phi^{-1} \circ G(u^2, u^3)$$

parametrizes the section $\Pi_1 \cap f(U)$. Since $\Phi^{-1} \circ G$ gives a local diffeomorphism between the xy-plane and Π_1, we obtain the assertion. $\square$

For swallowtails, we prove the following.

Proposition 2.7.3. *Let $f : U \to \mathbf{R}^3$ be a front, and suppose that f has a swallowtail $p \in U$. Let Π_0 be the limiting tangent plane of f at p, and let $\pi : \mathbf{R}^3 \to \Pi_0$ be the orthogonal projection. Then the restriction of the composition $g := \pi \circ f$ to its singular curve is a cusp in Π_0.*

Remark 2.7.4. Relating this proposition, we can show that the section of the image of f by the normal plane of f at a swallowtail point lies on a side of limiting tangent plane (see Appendix H).

To prove the assertion, we first consider the case that f coincides with the normal form $f_0(u,v) := (3u^4 + u^2 v, -4u^3 - 2uv, v)$. The singular set of f_0 is a parabola $v = -6u^2$ and

$$\hat{\gamma}(t) := (-3t^4, 8t^3, -6t^2) \tag{2.42}$$

gives a parametrization of the singular set image of f_0. In this case, the limiting tangent plane Π_0 coincides with the yz-plane, where $\hat{\gamma}(t)$ is the space curve defined by (2.42). We denote by $\pi : \mathbf{R}^3 \to \Pi_0$ the orthogonal projection. Since $\pi \circ \hat{\gamma}(t) := (8t^3, -6t^2)$, the assertion of Proposition 2.7.3 holds. We set

$$e_1 := (1, 0, 0),$$

and let Ω be a domain of $\mathbf{R}^3$ and $F : \Omega \to \mathbf{R}^2$ a smooth map satisfying $F(0, 0, 0) = (0, 0)$. We set $F(x, y, z) = (u(x, y, z), v(x, y, z))$. Then

$$J_F(x, y, z) := \begin{pmatrix} u_x(x, y, z), u_y(x, y, z), u_z(x, y, z) \\ v_x(x, y, z), v_y(x, y, z), v_z(x, y, z) \end{pmatrix}$$

is the Jacobi matrix of F at $(x, y, z) \in \Omega$. To prove the case for general f, we prepare the following lemma:

Lemma 2.7.5. *Let $\hat{\gamma}(t)$ be the curve as in (2.42). Suppose that*

$$\det \begin{pmatrix} u_y(0, 0, 0), u_z(0, 0, 0) \\ v_y(0, 0, 0), v_z(0, 0, 0) \end{pmatrix} \neq 0. \tag{2.43}$$

If there exists a non-zero vector $\xi_0 \in \mathbf{R}^3$ such that $J_F(0, 0, 0)\xi_0 = \mathbf{0}$ and $\xi_0 \cdot e_1 \neq 0$, then $F \circ \hat{\gamma}(t)$ has a cusp at $t = 0$ as a curve on the yz-plane.

Proof. We set $\sigma := F \circ \hat{\gamma}$ and $\delta := \det(\sigma''(0), \sigma'''(0))$. It is sufficient to show that $\delta \neq 0$. If we set

$$G(\boldsymbol{x}) := (F(\boldsymbol{x}), 0),$$

then we can rewrite

$$\delta = \det((G \circ \hat{\gamma})''(0), (G \circ \hat{\gamma})'''(0), \boldsymbol{e}_3) \quad (\boldsymbol{e}_3 := (0, 0, 1)). \tag{2.44}$$

We define a new map $\Psi : \mathbf{R}^3 \to \mathbf{R}^3$ by

$$\Psi(\boldsymbol{x}) := G(\boldsymbol{x}) + (\boldsymbol{x} \cdot \boldsymbol{e}_1)\boldsymbol{e}_3 \quad (\boldsymbol{x} \in \mathbf{R}^3). \tag{2.45}$$

Since the first component of $\hat{\gamma}(t)$ is $-3t^4$, we have

$$(\Psi \circ \hat{\gamma})''(0) = (G \circ \hat{\gamma})''(0), \quad (\Psi \circ \hat{\gamma})'''(0) = (G \circ \hat{\gamma})'''(0). \tag{2.46}$$

We denote by $J_\Psi(x, y, z)$ the Jacobi matrix of the map Ψ at $(x, y, z) \in \mathbf{R}^3$, that is, if $\Psi(x, y, z) = (A(x, y, z), B(x, y, z), C(x, y, z))$ then

$$J_\Psi(x, y, z) = \begin{pmatrix} A_x(x, y, z) \ A_y(x, y, z) \ A_z(x, y, z) \\ B_x(x, y, z) \ B_y(x, y, z) \ B_z(x, y, z) \\ C_x(x, y, z) \ C_y(x, y, z) \ C_z(x, y, z) \end{pmatrix}$$

holds. By (2.45), we have

$$J_\Psi(0,0,0) = \begin{pmatrix} J_F(0,0,0) \\ 1,\ 0,\ 0 \end{pmatrix}.$$

By (2.43), $J_\Psi(0,0,0)$ is a regular matrix, and

$$J_\Psi(0,0,0)\xi_0 = k e_3 \qquad (k := (\xi_0 \cdot e_1) \neq 0)$$

holds. Then we have (cf. (2.44) and (2.46))

$$\delta = \frac{1}{k} \det \left((\Psi \circ \hat{\gamma})''(0)\,,\, (\Psi \circ \hat{\gamma})'''(0),\ J_\Psi(0,0,0)\xi_0 \right).$$

We now set $\hat{\gamma}(t) = (x(t), y(t), z(t))$ and

$$\mathbf{M}(t) := J_\Psi(x(t), y(t), z(t)).$$

By the chain rule, we have

$$(\Psi \circ \hat{\gamma}(t))' = \mathbf{M}(t)\hat{\gamma}'(t).$$

So, we have that

$$(\Psi \circ \hat{\gamma})'' = (\mathbf{M}\hat{\gamma}')' = \mathbf{M}'\hat{\gamma}' + \mathbf{M}\hat{\gamma}'',$$

$$(\Psi \circ \hat{\gamma})''' = \mathbf{M}''\hat{\gamma}' + 2\mathbf{M}'\hat{\gamma}'' + \mathbf{M}\hat{\gamma}'''$$

hold. Since

$$\mathbf{M}' = (J_\Psi)_x x' + (J_\Psi)_y y' + (J_\Psi)_z z'$$

and $\mathbf{0} = \hat{\gamma}'(0) = (x'(0), y'(0), z'(0))$, $\mathbf{M}'(0)$ vanishes. So

$$(\Psi \circ \hat{\gamma})''(0) = \mathbf{M}(0)\hat{\gamma}''(0), \quad (\Psi \circ \hat{\gamma})'''(0) = \mathbf{M}(0)\hat{\gamma}'''(0).$$

Since $\mathbf{M}(0) = J_\Psi(0,0,0)$ is a regular matrix, we obtain

$$\delta = \frac{1}{k} \det(\mathbf{M}(0)\hat{\gamma}''(0), \mathbf{M}(0)\hat{\gamma}'''(0), \mathbf{M}(0)\xi_0)$$

$$= \frac{1}{k} \det \mathbf{M}(0) \det(\hat{\gamma}''(0), \hat{\gamma}'''(0), \xi_0)$$

$$= \frac{1}{k} \det \mathbf{M}(0) \left(\hat{\gamma}''(0) \times \hat{\gamma}'''(0) \right) \cdot \xi_0.$$

Since $\hat{\gamma}''(0) \times \hat{\gamma}'''(0)$ is proportional to e_1, we can conclude that $\delta \neq 0$, proving the assertion. $\square$

Proof of Proposition 2.7.3. By the definition of swallowtail, we can write

$$f \circ \varphi = \Phi \circ f_0, \tag{2.47}$$

where Φ and φ are local diffeomorphisms on $\boldsymbol{R}^3$ and $\boldsymbol{R}^2$ satisfying $\Phi(0,0,0) = \mathrm{P}(:= f(p))$ and $\varphi(0) = p$, respectively. We let $\pi : \boldsymbol{R}^3 \to \Pi_0$ be the orthogonal projection of $\boldsymbol{R}^3$ to the limiting tangent plane Π_0 of f at p. By setting $F := \pi \circ \Phi$, we have

$$\pi \circ f \circ \varphi = (\pi \circ \Phi) \circ f_0 = F \circ f_0.$$

Let $\nu_0 \in \boldsymbol{R}^3$ be the unit normal vector of f at p. By the chain rule, we have

$$J_F(0,0,0) = J_\pi(\mathrm{P})J_\Phi(0,0,0),$$

where $J_\pi(\mathrm{P})$ (respectively, $J_\Phi(0,0,0)$) is the Jacobi matrix of the map π (respectively, Φ) at P (respectively, $(0,0,0)$). Since π is a linear map, the 2×3 matrix $J_\pi(\mathrm{P})$ can be considered as a linear map from $\boldsymbol{R}^3$ to $\boldsymbol{R}^2$ and can be canonically identified with the map π itself. Then $\xi_0 := J_\Phi^{-1}\nu_0$ satisfies

$$J_F(\boldsymbol{0})\xi_0 = (J_\pi J_\Phi)(J_\Phi^{-1}\nu_0) = J_\pi\nu_0 = \pi(\nu_0) = \boldsymbol{0}.$$

Since ν_0 is a normal vector of f at P, it is linearly independent of the vectors at P lying in the limiting tangent plane Π_0. This property is preserved under diffeomorphism Φ^{-1}, and so ξ_0 is also linearly independent of the vectors in the yz-plane. So $\xi_0 \cdot e_1 \neq 0$ holds. By (2.47), $\sigma(u) := \varphi(u, -6u^2)$ is a singular curve of f passing through p, and we have

$$\pi \circ f \circ \sigma(u) = \pi \circ f \circ \varphi(u, -6u^2) = F \circ f_0(u, -6u^2) = F \circ \hat{\gamma}(u).$$

So, by applying Lemma 2.7.5, we obtain the assertion. □

Example 2.7.6. For real constants a, b and $c \neq 0$,

$$f(u,v) := (u, au^2 + v^2, bv^2 + cv^3)$$

is a wave front which has cuspidal edges along the u-axis. At the origin $(0,0)$, the normal plane Π_1 is the yz-plane. The slice locus of f by Π_1 corresponds to the v-axis and its image is

$$\{(0, v^2, bv^2 + cv^3) \, ; \, v \in \boldsymbol{R}\} \subset \Pi_1,$$

which has a cusp at the origin. See Fig. 2.12(a).

Example 2.7.7. For $a \in \boldsymbol{R}$,

$$f(u,v) := (3u^4 + u^2v + av^2, 4u^3 + 2uv, v)$$

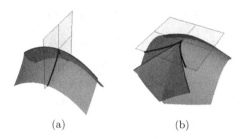

Fig. 2.12.   Examples 2.7.6 and 2.7.7

is a wave front whose singular set $\{(u, v);\ 6u^2 + v = 0\}$, and the origin $(0, 0)$ is a swallowtail. The limiting tangent plane Π_0 at $(0, 0)$ is the yz-plane. Since the image of the singular set of f is

$$\{((36a - 3)u^4, -8u^3, -6u^2)\,;\ u \in \boldsymbol{R}\},$$

its projection to the plane Π_0 gives a plane curve $u \mapsto (-8u^3, -6u^2)$ having a cusp at the origin. See Fig. 2.12(b).

Similar discussions like as in this section for cross caps are given in Appendix F.

Chapter 3

Proofs of Criteria for Singularities

In this chapter, we first explain fundamental properties of C^∞ functions. Using these properties, we prove the criterion for a cusp which is a singular point of a curve. After that we also prove the criteria for a cross cap and a cuspidal edge which are singular points of a surface.

3.1. C^∞ Functions and Whitney's Lemma

Just as for continuous functions, C^∞ functions have the following property.

Fact 3.1.1. Let A be a closed set in $\mathbf{R}^n$ ($n \geq 1$). Then there exists a C^∞ function f on $\mathbf{R}^n$ such that A is the zero set.

For the case of continuous functions, simply setting

$$f(x) := \inf_{a \in A} |x - a| \quad (x \in \mathbf{R}^n),$$

we obtain a continuous function $f : \mathbf{R}^n \to \mathbf{R}$ such that A is the zero set of f. However, for smooth functions, we need more involved techniques. See [22] for a proof. The following proposition is needed for the proofs of the criteria that we shall be demonstrating.

Proposition 3.1.2 (Termwise differentiation theorem for multivariable functions). *Let U be a domain in $\mathbf{R}^2$. Let a series of C^∞ functions $\{f_k(u, v)\}_{k=1,2,3,\dots}$ on U uniformly converge on compact sets[1] to a function $F(u, v)$ on U. Moreover, for any non-negative integers a, b, suppose*

---

[1] If a function series on U converges uniformly on any bounded closed set, it is said to be *uniformly convergent on compact sets* of U.

that the partial derivatives $\partial^{a+b} f_k / \partial u^a \partial v^b$ converge uniformly on compact sets of U to a function. Then F is a C^∞ function satisfying

$$\frac{\partial^{a+b} F}{\partial u^a \partial v^b} = \lim_{k \to \infty} \frac{\partial^{a+b} f_k}{\partial u^a \partial v^b}.$$

We give a proof by induction. Let us set $r := a + b \geq 1$ and assume that for all partial derivatives of F up to degree $r - 1$, the assertion holds. We show the assertion holds for degree r derivatives of F as well. To show differentiability of a function at a point, we need to show differentiability on a neighborhood near that point, and we may assume that neighborhood is a rectangular domain. Thus the proof of the proposition is reduced to a proof of the following lemma.

Lemma 3.1.3 (Termwise differentiation lemma for multivariable functions). *Let I, J be open intervals, and let $I \times J$ be a rectangular domain. Let a series of continuous functions $\{f_k(u, v)\}_{k=1,2,3,\ldots}$ on $I \times J$ be differentiable with respect to v. If $\{f_k\}_{k=1,2,3,\ldots}$ pointwise converges to a function F on $I \times J$, and $\partial f_k / \partial v$ converges uniformly on compact sets to a function $G(u, v)$ on $I \times J$, then F is differentiable with respect to v, and*

$$\frac{\partial F(u, v)}{\partial v} = G(u, v).$$

Proof. We show this lemma using integration. For each $u \in I$, it holds that

$$f_k(u, v) - f_k(u, v_0) = \int_{v_0}^{v} \frac{\partial f_k(u, t)}{\partial t} dt,$$

where $v_0 \in J$ is a reference point. Taking the limit $k \to \infty$, by the termwise integration theorem, we have

$$F(u, v) - F(u, v_0) = \int_{v_0}^{v} G(u, t) dt.$$

Thus, $F(u, v)$ is differentiable with respect to v, and $\partial F / \partial v = G$.   □

Here we introduce some fundamental properties of C^∞ functions.

Theorem 3.1.4 (E. Borel). *Let $f_k(u)$ $(k = 0, 1, 2, \ldots)$ be a series of C^∞ functions defined on an interval I containing the origin of $\mathbf{R}$. Then there exists a C^∞ function $F(u, v)$ defined on a neighborhood of $I \times \mathbf{R}$ containing the origin satisfying that, for any $k = 0, 1, 2, 3, \ldots$,*

$$\frac{\partial^k F}{\partial v^k}(u, 0) = f_k(u) \quad (u \in I).$$

In particular, setting the $f_k(u)$ to be constant functions, we have the following corollary.

Corollary 3.1.5. *Let* $\{a_k\}_{k=0,1,2,3,...}$ *be a sequence of real numbers. Then there exists a* C^∞ *function of one variable* $f(t)$ *defined on* $\boldsymbol{R}$ *such that*

$$f(0) = a_0, \quad f'(0) = a_1, \ldots, \quad f^{(k)}(0) = a_k, \ldots,$$

namely, there exists a C^∞ *function whose differentials at* $t = 0$ *are a given sequence of real numbers.*

Remark 3.1.6. Indeed the Taylor expansion of a real analytic function is a convergent power series. However, in the case of C^∞ functions, for any formal power series, there exists a C^∞ function whose Taylor expansion is that power series.

Proof of Theorem 3.1.4. We may assume that the series of functions $\{f_k(u)\}_{k=0,1,2,...}$ is defined on the closed interval $I = [-1, 1]$. Let $\rho : \boldsymbol{R} \to [0, 1]$ be a C^∞ function satisfying

$$\rho(v) = \begin{cases} 1 & (\text{if } |v| \leq 1/2), \\ 0 & (\text{if } |v| \geq 2/3), \end{cases}$$

and let $\{\mu_k\}_{k=0,1,2,3,...}$ be a series of numbers satisfying

- $\mu_0 \geq 1$, $\mu_{k+1} \geq \mu_k$ $(k = 0, 1, 2, \ldots)$,
- $\lim_{k \to \infty} \mu_k = \infty$.

Here, it is only needed that ρ is zero if $v \geq 1$, and there is no particular meaning to the specific number $2/3$.

We give this sequence $\{\mu_k\}$ explicitly later, and for now we assume that a sequence satisfying this condition has been given, and we set

$$F_{m,n}(u, v) := \sum_{k=m}^{n} \frac{v^k}{k!} \rho(\mu_k v) f_k(u) \quad (0 \leq m < n). \tag{3.1}$$

We want to take the limit $\lim_{n \to \infty} F_{0,n}(u, v)$ to define a smooth function on $I \times \boldsymbol{R}$. Since $k \leq n$, by monotonicity of the sequence $\{\mu_k\}$, the function $\rho(\mu_k v)$ $(k = 1, \ldots, n)$ is taking the constant value 1 for $|v| \leq 1/(2\mu_n)$. Thus

$$\frac{\partial^k F_{0,n}}{\partial v^k}(u, 0) = f_k(u)$$

holds. Hence, if the limit function $F = \lim_{n \to \infty} F_{0,n}$ exists and is termwise differentiable, then F has the desired property. To show this, we apply Proposition 3.1.2.

Differentiating (3.1) s times ($m > s$) with respect to v, we have

$$\frac{\partial^s F_{m,n}}{\partial v^s}(u, v) = \sum_{k=m}^{n} \sum_{r=0}^{s} \binom{s}{r} \frac{v^{k-r}}{(k-r)!} (\mu_k)^{s-r} \rho^{(s-r)}(\mu_k v) f_k(u),$$

where $\binom{s}{r}$ is the binomial coefficient. Differentiating this formula l times ($m > l$) with respect to u, we have

$$\frac{\partial^{s+l} F_{m,n}}{\partial v^s \partial u^l}(u, v) = \sum_{k=m}^{n} T_k, \tag{3.2}$$

$$T_k := \sum_{r=0}^{s} \binom{s}{r} \frac{v^{k-r}(\mu_k)^{s-r}}{(k-r)!} \rho^{(s-r)}(\mu_k v) \frac{d^l f_k}{du^l}(u).$$

Since any finitely many first terms do not affect the limit operations, we may assume $m > s$ and $m > l$. Furthermore, if $|v| > 1/\mu_k$, then, since $\rho^{(r)}(\mu_k v) = 0$, for evaluating the kth degree term we may assume that $|v| \leq 1/\mu_k$. We set

$$M_k := \max_{r=0,1,2,\ldots,k} \left(\sup_{u \in I} \left| \frac{d^r f_k}{du^r}(u) \right| \right).$$

Since $1/\mu_k < 1$ and $s < k$, we have

$$|T_k| \leq \sum_{r=0}^{s} \binom{s}{r} \frac{v^{k-r}}{(k-r)!} \left| \rho^{(s-r)}(\mu_k v) \right| (\mu_k)^{s-r} M_k$$

$$\leq \sum_{r=0}^{s} \binom{s}{r} \frac{M_k}{(\mu_k)^{k-s}(k-r)!} \left| \rho^{(s-r)}(\mu_k v) \right| \leq \frac{M_k}{\mu_k} \sum_{r=0}^{s} \binom{s}{r} \left| \rho^{(s-r)}(\mu_k v) \right|.$$

Setting

$$C_s := \sum_{r=0}^{s} \binom{s}{r} c_r \quad \left(c_r := \sup_{t \in \mathbf{R}} |\rho^{(r)}(t)| \right),$$

we see that

$$|T_k| \leq \frac{C_s M_k}{\mu_k}.$$

Hence we have

$$\left| \frac{\partial^{s+l} F_{s+l,n}}{\partial v^s \partial u^l}(u, v) \right| \leq C_s \sum_{k=s+l}^{n} \frac{M_k}{\mu_k}.$$

Thus, if

$$\frac{M_k}{\mu_k} \leq \frac{1}{(k+1)^2}, \tag{3.3}$$

then by the Weierstrass M-test (cf. [78, page 499]), the series on the right-hand side of (3.2) converges uniformly on $I \times \mathbf{R}$. In particular, setting

$$\mu_k = (k+1)^2 \max\{1, M_0, M_1, \ldots, M_k\},$$

(3.3) holds, and $\mu_k(> 1)$ is a diverging monotonically increasing sequence. Hence, taking the limit as $n \to \infty$, $\partial^{l+s} F_{s+l,n}/\partial v^s \partial u^l$ converges uniformly to a function on $I \times \mathbf{R}$. Since the difference between the finite sums $F_{s+l,n}$ and $F_{0,n}$ is just a first finite number of terms, the termwise differentiability for each of $F_{s+l,n}$ and $F_{0,n}$ is an equivalent condition. Thus, the limit $F(u,v) = \lim_{n\to\infty} F_{0,n}(u,v)$ is termwise differentiable l times with respect to u, and s times with respect to v. Since s and l are arbitrary, the limit is termwise differentiable infinitely many times on $I \times \mathbf{R}$. Thus, the desired property is shown. □

As an application, we show the following theorem.

Theorem 3.1.7 (Whitney's lemma [88]). *Let $f(u,v)$ be a C^∞ function defined on a neighborhood of the origin of $\mathbf{R}^2$ satisfying $f(u,v) = f(u,-v)$. Then there exists a C^∞ function $g(u,v)$ defined on a neighborhood of the origin of $\mathbf{R}^2$ such that $f(u,v) = g(u,v^2)$ holds near the origin.*

Before proving this assertion, we prepare the following three lemmas. One is the following assertion for the remainder term of Taylor's expansions.

Lemma 3.1.8. *Let $f(u,v)$ be a C^∞ function defined on a neighborhood U of the origin o of $\mathbf{R}^2$. For each non-negative integer k,*

$$R_f^{(k)}(u,v) := \frac{1}{v^{k+1}} \left(f(u,v) - \sum_{j=0}^{k} \frac{\partial^j f(u,0)}{\partial v^j} \frac{v^j}{j!} \right)$$

is a C^∞ function defined on a neighborhood $V(\subset U)$ of the origin.

Proof. It can be easily checked that

$$g(u,v) := f(u,v) - \sum_{j=0}^{k} \frac{\partial^j f(u,0)}{\partial v^j} \frac{v^j}{j!}$$

satisfies

$$g(u,0) = \frac{\partial g(u,0)}{\partial v} = \cdots = \frac{\partial^k g(u,0)}{\partial v^k} = 0.$$

So Proposition A.1 in Appendix A implies the conclusion. □

Lemma 3.1.9. *Fix k as a positive integer, and let $f(u,v)$ be a C^∞ function defined on a neighborhood $U(\subset \mathbf{R}^2)$ of the origin o. Then there exists a neighborhood V of the origin in the ut-plane and a C^∞ function η_k $(k = 1,2,\dots)$ defined on V such that the function defined by*

$$\xi_k(u,t) := t^{(2k-1)/2}\frac{\partial^k f(u,\sqrt{t})}{\partial t^k}$$

satisfies

$$\xi_k(u,t) = \eta_k(u,\sqrt{t})$$

on a neighborhood $V \cap \{t > 0\}$. In particular, $\xi_k(u,t)$ can be continuously extended to $V \cap \{t \geq 0\}$.

Proof. If we set $\varphi(t) := f(u,\sqrt{t})$, then

$$\varphi'(t) = \frac{f_v\left(u,\sqrt{t}\right)}{2\sqrt{t}}, \quad \varphi''(t) = \frac{\sqrt{t}f_{vv}\left(u,\sqrt{t}\right) - f_v\left(u,\sqrt{t}\right)}{4t^{3/2}}$$

hold. So, we can write

$$\xi_1(u,t) = \sqrt{t}\varphi'(t) = \eta_1(u,\sqrt{t}), \quad \xi_2(u,t) = \sqrt{t}^3\varphi''(t) = \eta_2(u,\sqrt{t}),$$

where

$$\eta_1(u,v) := \frac{f_v(u,v)}{2}, \quad \eta_2(u,v) := \frac{vf_{vv}(u,v) - f_v(u,v)}{4}.$$

For $k \geq 3$, one can easily prove the assertion by induction. □

Lemma 3.1.10. *Let a function $f(t)$ of one variable be continuous on a neighborhood of $t = 0$. If there exists a positive number ε such that for any t satisfying $0 < |t| < \varepsilon$, the function $f(t)$ is differentiable, and $\lim_{t\to 0} f'(t) = a$ $(a \in \mathbf{R})$, then $f(t)$ is also differentiable at $t = 0$, and $f'(0) = a$.*

Proof. By the mean value theorem, there exists a number $c(t)$ between 0 and t such that

$$\frac{f(t) - f(0)}{t} = f'(c(t)).$$

Since $c(t) \to 0$ as $t \to 0$, we have

$$f'(0) = \lim_{t\to 0}\frac{f(t) - f(0)}{t} = \lim_{t\to 0} f'(c(t)) = a.$$

□

Proof of Theorem 3.1.7. Considering u as a constant and $f(u,v)$ as a function of one variable in v, $f(u,v)$ has the formal Taylor expansion

$$\sum_{k=0}^{\infty} \frac{a_{2k}(u)}{(2k)!} v^{2k}.$$

Thus, if there exists a desired function $g(u,v)$, then considering u as a constant, it has the Taylor expansion

$$\sum_{k=0}^{\infty} \frac{a_{2k}(u)}{(2k)!} v^{k}.$$

By Theorem 3.1.4, there exists a C^{∞} function φ such that its Taylor expansion at $v=0$ is the formal power series

$$\sum_{k=0}^{\infty} \frac{a_{2k}(u)}{(2k)!} v^{k}. \tag{3.4}$$

We set

$$g(u,v) := \begin{cases} f(u, \sqrt{v}) & (\text{if } v \geq 0), \\ \varphi(u,v) & (\text{if } v < 0). \end{cases}$$

It is then clear that $f(u,v) = g(u,v^2)$ holds. So, it is sufficient to show that $g(u,v)$ is a C^{∞} function defined on a neighborhood of $(0,0)$. Since $\varphi(u,v)$ is a C^{∞} function, the assertion reduces to showing the following Lemma 3.1.11. □

Lemma 3.1.11. *For each non-negative integer r, the function*

$$h(u,v) := g(u,v) - \varphi(u,v)$$

is C^r differentiable on a neighborhood of $(0,0)$.

Proof. We prove this by induction with respect to r. Since

$$h(u,v) = 0 \quad (v \leq 0) \tag{3.5}$$

holds, it can be easily checked that h is a C^0 function defined on a neighborhood of $(0,0)$. So we suppose that the assertion holds for $r \geq 0$, that is, h is a C^r function and will prove the C^{r+1} differentiability of h. We let a, b be a pair of non-negative integers satisfying $a + b = r$ and consider the function defined by

$$\psi := D_{a,b} h, \quad D_{a,b} := \frac{\partial^{a+b}}{\partial u^a \partial v^b}.$$

It is sufficient to show that ψ_u and ψ_v exist and are continuous on a neighborhood of $(0,0)$.

For each non-negative integer k, by Lemma 3.1.8,

$$R_f^{(2k+1)}(u,v) := \frac{1}{v^{2k+1}}\left(f(u,v) - \sum_{j=0}^{k} a_{2j}(u)\frac{v^{2j}}{(2j)!}\right),$$

$$R_\varphi^{(k)}(u,v) := \frac{1}{v^{k+1}}\left(\varphi(u,v) - \sum_{j=0}^{k} a_{2j}(u)\frac{v^j}{j!}\right)$$

are C^∞ functions on a neighborhood of $(0,0)$. Since $f(u,v)$ is an even function of v, we have the following expressions:

$$f(u,\sqrt{v}) = \sum_{j=0}^{b+1} \frac{a_{2j}(u)}{(2j)!}v^j + v^{b+2}R_f^{(2b+3)}(u,\sqrt{v})$$

and

$$h(u,v) = v^{b+2}\left(R_f^{(2b+3)}(u,\sqrt{v}) - R_\varphi^{(b+1)}(u,v)\right)$$

for $v > 0$, which induces the identity

$$\psi(u,v)(= D_{a,b}h(u,v)) = v^2 A_0(u,v) + v^{5/2}A_1(u,v), \tag{3.6}$$

where

$$A_0 := \frac{(b+2)!}{2}\frac{\partial^a}{\partial u^a}(R_f^{(2b+3)}(u,\sqrt{v}) - R_\varphi^{(b+1)}(u,v))$$

$$- \sum_{j=0}^{b-1}\binom{b}{j}\frac{(b+2)!v^{b-j}}{(b+2-j)!}\frac{\partial^{a+b-j}}{\partial u^a \partial v^{b-j}}R_\varphi^{(b+1)}(u,v),$$

$$A_1 := \sum_{j=0}^{b-1}\binom{b}{j}\frac{(b+2)!v^{b-j-1/2}}{(b+2-j)!}\frac{\partial^{a+b-j}}{\partial u^a \partial v^{b-j}}R_f^{(2b+3)}(u,\sqrt{v}).$$

Obviously, we can write

$$A_0(u,v) = B_0(u,\sqrt{v}),$$

where $B_0(u,v)$ is a C^∞ function defined on a neighborhood of $(0,0)$. On the other hand, applying Lemma 3.1.9 to the functions

$$\xi_{b-j} := t^{b-j-1/2}\frac{\partial^{b-j}}{\partial t^{b-j}}\rho(u,\sqrt{t}), \quad \rho := \frac{\partial^a R_f^{(2b+3)}(u,v)}{\partial u^a} \quad (j = 0,\ldots,b-1),$$

we can show the existence of a C^∞ function $B_1(u,v)$ defined on a neighborhood of $(0,0)$ satisfying $A_1(u,v) = B_1(u,\sqrt{v})$. So, by a suitable choice of a neighborhood V of $(0,0)$, we have the following identity

$$\psi(u,v) = v^2 B_0(u, \sqrt{v}) + v^{5/2} B_1(u, \sqrt{v}) \tag{3.7}$$

on $V \cap \{v > 0\}$. Since h is a C^r function, $\psi(u,v)$ is a continuous function on V. So, (3.7) holds on $V \cap \{v \geq 0\}$, and the following assertions follow:

- $\psi_u(u,v)$ exists on $V \cap \{v \geq 0\}$ satisfying $\psi_u(u,0) = 0$, and
- $\psi_u(u,v)$ is continuous on $V \cap \{v \geq 0\}$.

On the other hand, $\psi(u,v) = 0$ holds on $V \cap \{v < 0\}$ because of (3.5). In particular, $\psi_u(u,v)$ exists and is identically equal to zero on $V \cap \{v < 0\}$. Consequently, $\psi_u(u,v)$ is a continuous function on V.

We next show that ψ_v exists and is a continuous function on V. In fact, differentiating (3.7) by v, we can observe that $C(u,v) := \psi_v(u,v)$ can be continuously extended on $V \cap \{v \geq 0\}$ satisfying $C(u,0) = 0$. On the other hand, (3.5) implies that $\psi_v(u,v)$ exists on $V \cap \{v < 0\}$ and is identically zero. By Lemma 3.1.10, $\psi_v(u,0)$ also exists and is equal to 0 for each $(u,0) \in V$. In particular, ψ_v is a continuous function on V, proving the assertion. $\square$

As a corollary of Theorem 3.1.7, we prove the following variant of Whitney's lemma for the odd function $f(u,v)$ with respect to the variable v:

Corollary 3.1.12. *Let $f(u,v)$ be a C^∞ function defined on a neighborhood of the origin of $\mathbf{R}^2$ satisfying $f(u,-v) = -f(u,v)$. Then there exists a C^∞ function $g(u,v)$ defined on a neighborhood of the origin of $\mathbf{R}^2$ such that $f(u,v) = vg(u,v^2)$.*

Proof. Because $f(u,-v) = -f(u,v)$, it holds that $f(u,0) = 0$. By the division lemma (Proposition A.1 in Appendix A), there exists a C^∞ function defined on a neighborhood of the origin such that $f(u,v) = vh(u,v)$. Since

$$vh(u,v) = f(u,v) = -f(u,-v) = vh(u,-v),$$

the continuity of h implies that $h(u,v) = h(u,-v)$ on a neighborhood of the origin. Then by Theorem 3.1.7, there exists a C^∞ function $g(u,v)$ defined on a neighborhood of the origin such that $h(u,v) = g(u,v^2)$. This proves the assertion. $\square$

3.2. Proof of the Criterion for Cusps

Let ε_1, ε_2 be sufficiently small positive numbers. We consider two planar curves

$$\gamma_j(t) \quad (|t - c_j| < \varepsilon_j, \; j = 1, 2)$$

defined on neighborhoods of $t = c_j (\in \mathbf{R})$ $(j = 1, 2)$, respectively. We say that $\gamma_1(t)$ and $\gamma_2(t)$ are *right-left equivalent* (or *$\mathcal{A}$-equivalent*) if there exists a diffeomorphism $\varphi : (c_1 - \varepsilon_1, c_1 + \varepsilon_1) \to (c_2 - \varepsilon_2, c_2 + \varepsilon_2)$ satisfying $\varphi(c_1) = c_2$, and a diffeomorphism $\Phi : U_1 \to U_2$ from an open set U_1 of $\mathbf{R}^2$ containing $\gamma_1(c_1)$ to an open set U_2 containing $\gamma_2(c_2)$ such that

$$\gamma_2 \circ \varphi(t) = \Phi \circ \gamma_1(t) \quad (|t - c_1| < \varepsilon_1).$$

On the other hand, $\gamma_0(t) = (t^2, t^3)$ gives the standard cusp at $t = 0$. If a curve $\gamma(t)$ has a singular point at $t = c$, and is right–left equivalent to the above $\gamma_0(t)$, then $t = c$ is a cusp. Here we give a proof for the criterion (Theorem 1.3.2) given in Chapter 1.

Proof of Theorem 1.3.2. Without loss of generality, we may assume $c = 0$ and $\gamma(0) = (0, 0)$. We already gave a proof that the condition does not depend on the right–left equivalence in Proposition 1.3.5. We will now show that a singular point on a curve satisfying (1.20) (cf. Theorem 1.3.2) is right–left equivalent to the standard cusp.

Consider a 2×2 matrix $T := (\gamma''(0), \gamma'''(0))$, which is a regular matrix because γ satisfies (1.20). Then $\hat{\gamma}(t) := T^{-1}\gamma(t)$ satisfies

$$\hat{\gamma}''(0) = (1, 0), \quad \hat{\gamma}'''(0) = (0, 1).$$

If we set $\hat{\gamma}(t) = (\hat{x}(t), \hat{y}(t))$, then

$$\hat{x}(0) = \hat{x}'(0) = 0, \ \hat{x}''(0) = 1,$$

$$\hat{y}(0) = \hat{y}'(0) = \hat{y}''(0) = 0, \ \hat{y}'''(0) = 1$$

hold. By the division lemma (cf. Corollary A.3 in Appendix A), we can write

$$\hat{x}(t) = t^2 a(t), \quad \hat{y}(t) = t^3 b(t),$$

where $a(t)$ and $b(t)$ are smooth functions defined on an open interval containing $t = 0$ such that $a(0) = 1/2$ and $b(0) = 1/6$. We then set $s := t\sqrt{a(t)}$, which can be taken as a new parameter of $\hat{\gamma}$. Then we can write

$$\hat{\gamma}(s) = \left(s^2, \frac{s^3 b(s)}{a(s)^{3/2}} \right).$$

So without loss of generality we may assume that $\gamma(t)$ has a singular point at $t = 0$ and is right–left equivalent to

$$\gamma_1(s) := (s^2, s^3 \psi(s)),$$

where $\psi(s)$ is a C^∞ function of a variable s satisfying $\psi(0) \neq 0$. By applying a reflection if necessary, we may assume $\psi(0) > 0$. We set

$$u(s) := \frac{\psi(s) + \psi(-s)}{2}, \quad v(s) := \frac{\psi(s) - \psi(-s)}{2}.$$

Then, since $u(s)$ is an even function and $v(s)$ is an odd function, by Whitney's lemma (Theorem 3.1.7 and Corollary 3.1.12), we can write

$$u(s) = \hat{u}(s^2), \quad v(s) = s\hat{v}(s^2),$$

and by $\psi(0) \neq 0$, it holds that $\hat{u}(0) \neq 0$. Using these formulas, we can write

$$\gamma_1(s) = (s^2, s^3(u(s) + v(s))) = (s^2, s^3\hat{u}(s^2) + s^4\hat{v}(s^2)).$$

Considering the map $\Phi_2(x, y) := (x, y\hat{u}(x) + x^2\hat{v}(x))$, defined on a neighborhood of the origin of $\boldsymbol{R}^2$, the fact that $\hat{u}(0) \neq 0$ implies that the Jacobian of Φ_2 does not vanish at $(x, y) = (0, 0)$. Thus it is a diffeomorphism of a neighborhood of the origin of $\boldsymbol{R}^2$. Then, because $\Phi_2(s^2, s^3) = \gamma_1(s)$, we see that γ has a cusp at $t = 0$, since $s \mapsto (s^2, s^3)$ gives the standard cusp. $\square$

3.3. Proof of the Criterion for Cross Caps

As defined in Chapter 2, a cross cap is a C^∞ map which is right–left equivalent to the standard cross cap $f(u, v) = (u, uv, v^2)$. In this section, following Whitney [89], we give a proof of this criterion for cross caps (Theorem 2.5.6 in Chapter 2).

First we show that the condition in Theorem 2.5.6 does not depend on the choice of coordinate system (u, v) satisfying $f_v(p) = \boldsymbol{0}$. Let

$$u = u(\xi, \eta), \quad v = v(\xi, \eta)$$

be a coordinate transformation. Without loss of generality, we may set $u(p) = v(p) = 0$, $\xi(p) = \eta(p) = 0$ and suppose $f_\eta(p) = \boldsymbol{0}$. Since

$$\boldsymbol{0} = f_\eta(p) = f_u(p)u_\eta(p) + f_v(p)v_\eta(p) = u_\eta(p)f_u(p)$$

we have $u_\eta(p) = 0$. It holds that

$$f_\xi(p) = f_u(p)u_\xi(p) + f_v(p)v_\xi(p) = u_\xi(p)f_u(p). \tag{3.8}$$

Since the Jacobian of the coordinate transformation does not vanish,

$$0 \neq u_\xi(p)v_\eta(p) - u_\eta(p)v_\xi(p) = u_\xi(p)v_\eta(p) \tag{3.9}$$

holds. By a direct calculation, we have

$$f_{\eta\eta} = (f_u u_\eta + f_v v_\eta)_\eta$$

$$= f_u u_{\eta\eta} + f_v v_{\eta\eta} + (f_{uu}u_\eta + f_{uv}v_\eta)u_\eta + (f_{vu}u_\eta + f_{vv}v_\eta)v_\eta$$

and by $u_\eta(p) = 0$ and $f_\eta(p) = \mathbf{0}$, it holds that

$$f_{\eta\eta}(p) = u_{\eta\eta}(p)f_u(p) + v_\eta(p)^2 f_{vv}(p). \qquad (3.10)$$

Similarly, we have

$$f_{\xi\eta}(p) = u_{\xi\eta}(p)f_u(p) + u_\xi(p)v_\eta(p)f_{uv}(p) + v_\eta(p)v_\xi(p)f_{vv}(p).$$

By this with (3.8) and (3.10), we have

$$\det(f_\xi, f_{\xi\eta}, f_{\eta\eta}) = (v_\eta)^3 (u_\xi)^2 \det(f_u, f_{uv}, f_{vv})$$

at $p(= (0,0))$. By (3.9), we obtain the independence of the condition under a coordinate transformation on the source space.

On the other hand, let the image of f be contained in a domain D of $\mathbf{R}^3$. For a diffeomorphism into the image $\Phi : D \to \mathbf{R}^3$, regarding the differential $J := (\Phi_x, \Phi_y, \Phi_z)$ (namely, the Jacobi matrix) as a 3×3 regular matrix-valued C^∞ map from D into $GL(3, \mathbf{R})$, the chain rule (2.38) tells us

$$\left((\Phi \circ f)_u, (\Phi \circ f)_v\right) = (\hat{J}f_u, \hat{J}f_v) = \hat{J}(f_u, f_v), \qquad (3.11)$$

where $\hat{J} := J \circ f$. Let us set $f(u,v) = \big(x(u,v), y(u,v), z(u,v)\big)^T$. Since $f_v = (x_v, y_v, z_v)^T$, $f_v(p) = \mathbf{0}$ and

$$\hat{J}_v = (J_x \circ f)x_v + (J_y \circ f)y_v + (J_z \circ f)z_v = O$$

at p (O denotes the zero matrix), it holds that

$$\left((\Phi \circ f)_{uv}, (\Phi \circ f)_{vv}\right) = \hat{J}_v(f_u, f_v) + \hat{J}(f_{uv}, f_{vv}) = \hat{J}(f_{uv}, f_{vv})$$

at p. Thus, we have

$$\det((\Phi \circ f)_u, (\Phi \circ f)_{uv}, (\Phi \circ f)_{vv}) = (\det \hat{J}) \det(f_u, f_{uv}, f_{vv})$$

at p. This proves independence from the choice of coordinate system on the target space.

We next prepare the following lemma:

Lemma 3.3.1. *Let $\gamma(t)$ $(a < t < b)$ be a regular curve in $\mathbf{R}^3$ and fix $c \in (a,b)$ arbitrarily. Then there exist a local coordinate transform Φ on $\mathbf{R}^3$, an open interval $I(\subset \mathbf{R})$ containing 0, and a local parameter change $\varphi : I \to (a,b)$ satisfying $\varphi(0) = c$ such that*

$$\Phi \circ \gamma \circ \varphi(s) = (s, 0, 0) \quad (s \in I).$$

Proof. Without loss of generality, we may assume that $c = 0$ and $\gamma(0) = \mathbf{0}$. We then set

$$\gamma(t) = (a(t), b(t), c(t)).$$

Changing the order of a, b, c if necessary, we may assume that $a(0) = 0$ and $a'(0) \neq 0$, because $t = 0$ is a regular point of $\gamma(t)$. Moreover, replacing $\gamma(t)$ by $-\gamma(t)$, we may assume that $a'(0) > 0$. Then $s = a(t)$ can be taken as a new parameter of $\gamma(t)$ near $t = c$. So replacing $\gamma(t(s))$ by $\gamma(s)$, we can write

$$\gamma(s) = (s, b(s), c(s)) \quad (s \in I),$$

where I is a sufficiently small interval containing $s = 0$. We next consider a map Φ defined by

$$\Phi(x, y, z) = (x, y - b(x), z - c(x))$$

on a neighborhood of the origin in $\mathbf{R}^3$. Then the Jacobian of the map Φ at $(x, y, z) = \mathbf{0}$ does not vanish, and so Φ gives a diffeomorphism which maps a neighborhood Ω of the origin of $\mathbf{R}^3$ to a neighborhood $\Phi(\Omega)$ of the origin in $\mathbf{R}^3$. Taking I to be sufficiently small, we may assume $\gamma(I) \subset \Omega$. Then we have

$$\Phi \circ \gamma(s) = \Phi(s, b(s), c(s)) = (s, 0, 0) \quad (s \in I),$$

proving the assertion. $\square$

Now f is written in the form $f(u, v) = (x(u, v), y(u, v), z(u, v))$. By Lemma 3.3.1 and (3.11), we may assume that

$$f(u, 0) = (u, 0, 0) \quad \text{and} \quad f_v(0, 0) = \mathbf{0}$$

hold without loss of generality. We set

$$\tilde{u} := x(u, v), \quad \tilde{v} := v.$$

Since $x_u(0, 0) \neq 0$ and $x_v(0, 0) = 0$, $(\tilde{u}, \tilde{v})$ can be chosen as a new parametrization satisfying $f_{\tilde{v}}(0, 0) = \mathbf{0}$. Replacing $(\tilde{u}, \tilde{v})$ by (u, v), we can write

$$f(u, v) = (u, y(u, v), z(u, v)).$$

We set

$$T := \big(f_u(0, 0), \, f_{uv}(0, 0) \, f_{vv}(0, 0) \big),$$

which is a regular matrix because of the assumption of the theorem. Then we have

$$T e_1 = f_u(0, 0) = e_1, \quad T e_2 = f_{uv}(0, 0), \quad T e_3 = f_{vv}(0, 0),$$

where

$$e_1 = \begin{pmatrix} 1 \\ 0 \\ 0 \end{pmatrix}, \quad e_2 = \begin{pmatrix} 0 \\ 1 \\ 0 \end{pmatrix}, \quad e_3 = \begin{pmatrix} 0 \\ 0 \\ 1 \end{pmatrix}.$$

Since $Te_1 = e_1$, we have $Tf(u,0) = (u,0,0)^T$. So, by replacing f with $T^{-1}f$, we have

$$f(u,0) = (u,0,0), \quad f_{uv}(0,0) = (0,1,0), \quad f_{vv}(0,0) = (0,0,1).$$

In particular, it holds that $y(u,0) = z(u,0) = 0$. By the division lemma (Proposition A.1) in Appendix A, y and z can be written as

$$y(u,v) = va(u,v), \quad z(u,v) = vb(u,v).$$

Since $f_v(0,0) = \mathbf{0}$, it holds that $a(0,0) = b(0,0) = 0$. On the other hand, by the division lemma, we can write $a(u,v) - a(u,0) = v\hat{a}(u,v)$. Since $a(0,0) = 0$, again by the division lemma, we can write $a(u,0) = ua_1(u)$. In particular,

$$y(u,v) = uva_1(u) + v^2\hat{a}(u,v).$$

Similarly, we may assume that

$$z(u,v) = uvb_1(u) + v^2\hat{b}(u,v).$$

Then, by $f_{uv}(0,0) = (0,1,0)$, it holds that $a_1(0) \neq 0$. Thus, by the coordinate transformation

$$\Psi(x,y,z) := \left(x, y, z - \frac{b_1(z)}{a_1(z)} y \right)$$

on the target space, we may assume that f has the form

$$g(u,v) := \Psi \circ f(u,v) = (u, va(u,v), v^2 c(u,v)),$$

where

$$c(u,v) := \hat{b}(u,v) - \frac{b_1(u)}{a_1(u)} \hat{a}(u,v).$$

Since $a(0,0) = 0$, it holds that $g_v(0,0) = \mathbf{0}$. Thus, by the coordinate independence of the condition (2.40) which was proven above in the first half of this proof, it holds that $g_{vv}(0,0) \neq \mathbf{0}$. In particular, by a reflection with respect to the xy-plane if necessary, we may assume $c(0,0) > 0$. By the coordinate transformation $s = u$, $t = v\sqrt{c(u,v)}$ on the source space, it holds that $va(u,v) = t\,a(u,v)/\sqrt{c(u,v)}$. Since the function $a(u,v)/\sqrt{c(u,v)}$ is a function of s,t, we can write

$$g(s,t) = \big(s, tA(s,t), t^2\big).$$

Here, $A(s,t)$ is a C^∞ function of two variables. Furthermore, by Theorem 3.1.7 and Corollary 3.1.12, there exist two C^∞ functions A_0, A_1 such that

$$A(s,t) = \frac{A(s,t) + A(s,-t)}{2} + \frac{A(s,t) - A(s,-t)}{2}$$

$$= A_0(s,t^2) + tA_1(s,t^2).$$

Thus,

$$g(s,t) = \left(s, tA_0(s,t^2) + t^2 A_1(s,t^2), t^2\right).$$

Here we remark that it holds that $A_0(0,0) = A_1(0,0) = 0$ because $a(0,0) = 0$. Then by the coordinate transformation

$$\Psi_1(x,y,z) := (x, y - zA_1(x,z), z)$$

on the target space $\boldsymbol{R}^3$, we have

$$h(s,t) := \Psi_1 \circ g = (s, tA_0(s,t^2), t^2).$$

Since $h_t(0,0) = \boldsymbol{0}$, by the independence of the condition for the criterion with respect to right–left equivalence, it holds that $h_{st}(0,0) \neq \boldsymbol{0}$, and we have

$$\left. \frac{\partial A_0}{\partial s} \right|_{(s,t)=(0,0)} \neq 0. \tag{3.12}$$

We consider a function

$$F(x,y,z) := y - A_0(x,z).$$

Then (3.12) yields that $F_x(0,0,0) \neq 0$. By the implicit function theorem[2], there exists a smooth function $C(y,z)$ of two variables such that $C(0,0) = 0$ and

$$y - A_0(C(y,z), z) = 0. \tag{3.13}$$

Substituting $y = \xi$ and $z = \eta^2$, we have

$$\xi - A_0(C(\xi,\eta^2), \eta^2) = 0. \tag{3.14}$$

Differentiating this by ξ, we see that (cf. (3.12))

$$\left. \frac{\partial C(\xi,\eta^2)}{\partial \xi} \right|_{(\xi,\eta)=(0,0)} \neq 0.$$

---

[2] If a C^∞ function $F(x,y,z)$ satisfies $F(0,0,0) = 0$ and $F_x(0,0,0) \neq 0$, then there exists a C^∞ function $G(y,z)$ of two variables satisfying $F(G(y,z),y,z) = 0$. Furthermore, any points (a,b,c) near the origin satisfying $F(a,b,c) = 0$ satisfy $a = G(b,c)$.

By the inverse function theorem, we can use the new coordinate system (ξ, η) determined by

$$s := C(\xi, \eta^2), \quad t := \eta$$

near the origin $(0, 0)$. Then

$$h(s, t) = (s, tA_0(s, t^2), t^2) = (C(\xi, \eta^2), \eta A_0(C(\xi, \eta^2), \eta^2), \eta^2)$$

$$= (C(\xi, \eta^2), \xi\eta, \eta^2).$$

So, if we consider the coordinate transformation $\Phi(x, y, z) := (C(x, z), y, z)$ near the origin of $\boldsymbol{R}^3$, then differentiating the identity (cf. (3.14))

$$\xi - A_0(C(\xi, \eta^2), \eta^2) = 0$$

with respect to ξ, and noticing (3.12), we see that $\partial C(\xi, \eta)/\partial \xi$ also does not vanish at $(\xi, \eta) = (0, 0)$. Thus, Φ is a local diffeomorphism satisfying

$$h(\xi, \eta) = \Phi(\xi, \xi\eta, \eta^2).$$

3.4.   Proof of the Criterion for Cuspidal Edges

Proof of (1) **in Theorem 2.6.3.** Theorem 2.4.1 tells us that f being a front or not is independent of the choices of coordinate systems on the source and target spaces, and Proposition 2.6.2 tells us that whether a singular point is non-degenerate or not is also independent of these choices. Moreover, the linear independence of the singular vector and the null vector also does not depend on the choice of coordinate systems on the source space and the target space. We already saw in Section 2.6 that the standard cuspidal edge satisfies the condition (1) of Theorem 2.6.3. So the above facts imply that any cuspidal edges satisfy the condition (1) of Theorem 2.6.3. So it is sufficient to show that a front f which satisfies the condition (1) of Theorem 2.6.3 is right–left equivalent to the cuspidal edge $(u, v) \mapsto (u^2, u^3, v)$ by coordinate transformations on the source space and the target space. (It should be remarked that the property that f satisfies the condition (1) of Theorem 2.6.3 does not depend on choices of coordinate changes.)

First, we take a coordinate system (u, v) such that the singular set is the v-axis, and the null direction is the u-direction along the v-axis. (We show existence of this coordinate system as in Lemma 5.2.10 in Section 5.) Since the singular direction and the null direction are linearly independent, the image of the singular curve $v \mapsto f(0, v)$ is a regular curve in $\boldsymbol{R}^3$. By Lemma 3.3.1, we may assume that

$$f(0, v) = (0, 0, v) \tag{3.15}$$

holds for sufficient small v. We set

$$f(u, v) = (f_1(u, v), f_2(u, v), f_3(u, v)).$$

Since $f_3(0, v) - v = 0$, the division lemma (Lemma A.1 in Appendix A) implies there exists a C^∞ function $h(u, v)$ defined on a neighborhood of the origin such that $f_3(u, v) = v + uh(u, v)$. We apply the coordinate transformation $\xi := u$, $\eta := v + uh(u, v)$. Since $f_u(0, v) = \mathbf{0}$, we have $h(0, v) = 0$. By the inverse function theorem, (ξ, η) defines a new coordinate system on the source space. Since $\xi = u$, the η-axis, namely $\{\xi = 0\}$, is still the singular curve, and since on the singular set,

$$\mathbf{0} = f_u = f_\xi \xi_u + f_\eta \eta_u = f_\xi + f_\eta h = f_\xi$$

holds if $u = 0$, we see that $f_\xi = \mathbf{0}$ on the singular set. Replacing (ξ, η) by (u, v), we may assume that f can be written as $f(u, v) = (f_1(u, v), f_2(u, v), v)$ such that $f_u(0, v) = \mathbf{0}$. In particular,

$$f_i(0, v) = (f_i)_u(0, v) = 0 \quad (i = 1, 2).$$

Thus, by the division lemma, there exist C^∞ functions $a(u, v)$ and $b(u, v)$ such that

$$f_1(u, v) = u^2 a(u, v), \quad f_2(u, v) = u^2 b(u, v).$$

Hence f is written as $f(u, v) = (u^2 a(u, v), u^2 b(u, v), v)$. Since $f_u(0, v) = \mathbf{0}$, we have $\lambda_v(0, v) = 0$. Since the singular point is non-degenerate, $d\lambda \neq \mathbf{0}$ holds. In particular, we have $\lambda_u(0, 0) \neq 0$. Hence

$$0 \neq \lambda_u(0, 0) = \det(f_u, f_v, \nu)_u|_{(u,v)=(0,0)} = \det(f_{uu}, f_v, \nu)|_{(u,v)=(0,0)},$$

and we can conclude that $f_{uu}(0, 0) \neq \mathbf{0}$, namely, $a(0, 0)$ and $b(0, 0)$ do not vanish at the same time. Then we may assume $a(0, 0) > 0$ by interchanging the x-axis and the y-axis in $\mathbf{R}^3$ and perhaps also reversing the direction of the x-axis if necessary. By the inverse function theorem,

$$\varphi : (u, v) \mapsto (u\sqrt{a(u, v)}, v)$$

is a local diffeomorphism near the origin. Let us take a new coordinate system

$$s = u\sqrt{a(u, v)}, \quad t = v.$$

Since s is a scalar multiple of u, the singular set is $\{s = 0\}$, namely, the t-axis is the singular curve on the (s, t) coordinate system. Noticing that $t = v$, we have

$$\mathbf{0} = f_u(0, v) = s_u(0, v)f_s(0, v) + t_u(0, v)f_t(0, v) = \sqrt{a(0, v)}f_s(0, v)$$

on the singular curve. By this formula, $f_s = \mathbf{0}$ holds on the t-axis.

Taking (s,t) to be (u,v), then $f(u,v) = (u^2, c(u,v)u^2, v)$, where $c :=$ b/a. Setting

$$\alpha(u,v) := \frac{c(u,v) + c(-u,v)}{2}, \qquad \beta(u,v) := \frac{c(u,v) - c(-u,v)}{2},$$

then α is an even function with respect to u, and β is an odd function with respect to u, and $c(u,v) = \alpha(u,v) + \beta(u,v)$ holds. By Whitney's lemma (Theorem 3.1.7 and Corollary 3.1.12), there exist C^∞ functions $\hat{\alpha}(u,v), \hat{\beta}(u,v)$ defined on a neighborhood of the origin such that $\alpha(u,v) = \hat{\alpha}(u^2,v)$, $\beta(u,v) = u\hat{\beta}(u^2,v)$. Thus f can be written as

$$f(u,v) = (u^2, \hat{\alpha}(u^2,v)u^2 + \hat{\beta}(u^2,v)u^3, v).$$

Since the map $\Psi_1 : (x,y,z) \mapsto (x, y - \hat{\alpha}(x,z)x, z)$ is a local diffeomorphism on a neighborhood of the origin of $\boldsymbol{R}^3$, considering $\Psi_1 \circ f$ instead of f, we may assume that f is written as $f(u,v) = (u^2, \hat{\beta}(u^2,v)u^3, v)$. By a direct calculation,

$$\left(\tilde{\nu} := \right)\frac{f_u(u,v) \times f_v(u,v)}{u} = \left(3u\hat{\beta}(u^2,v), -2, 0\right) + u^2\boldsymbol{v}(u,v),$$

and this gives a normal vector of $f(u,v)$, where $\boldsymbol{v}(u,v)$ is a vector valued C^∞ function defined near $(0,0)$. Setting $\nu := \tilde{\nu}/|\tilde{\nu}|$, this gives a unit normal vector field of f. Since $f_u(0,0) = \boldsymbol{0}$ and f is a front near $(0,0)$, it holds that $\nu_u(0,0) \neq \boldsymbol{0}$. Then we see $\hat{\beta}(0,0) \neq 0$. Hence the map

$$\Psi_2 : (x,y,z) \mapsto (x, \hat{\beta}(x,z)y, z)$$

is a local diffeomorphism on a neighborhood of the origin of $\boldsymbol{R}^3$. Setting $f_0 := (u^2, u^3, v)$, we have $\Psi_2 \circ f_0 = f$. Thus, f coincides with f_0 under coordinate transformations of the source and target spaces. $\qquad\square$

Chapter 4

Applications of Criteria for Singularities

In this chapter, we explain vector fields and use them for finding criteria for cuspidal edges and swallowtails that are different from those in Chapter 2, and give applications of them. We also explain criteria for "folds" and "Whitney cusps", appearing in maps between domains in $\boldsymbol{R}^2$, which are very similar to those of cuspidal edges and swallowtails.

4.1. Vector Fields on Regions in $\boldsymbol{R}^2$

Here we explain vector fields, which are frequently used in this text, for readers who are not familiar with manifold theory.

The exterior derivative. We fix a domain U in the uv-plane $\boldsymbol{R}^2$. The formal sum

$$d\varphi := \varphi_u(u, v) \, du + \varphi_v(u, v) \, dv \qquad (4.1)$$

for a C^∞-function of two variables $\varphi(u, v)$ on U is called the *exterior derivative* or *total derivative* of the function φ. Let

$$U \ni (u, v) \longmapsto \big(x(u, v), y(u, v)\big) \in \tilde{U}$$

be a diffeomorphism from a domain U in the uv-plane to a domain $\tilde{U}$ in the xy-plane. Regarding this as a coordinate transformation and substituting

$$dx = x_u \, du + x_v \, dv, \quad dy = y_u \, du + y_v \, dv$$

into (4.1), the chain rule tells us that

$$d\varphi = \varphi_u \, du + \varphi_v \, dv = (\varphi_x x_u + \varphi_y y_u) \, du + (\varphi_x x_v + \varphi_y y_v) \, dv$$

$$= \varphi_x(x_u \, du + x_v \, dv) + \varphi_y(y_u \, du + y_v \, dv) = \varphi_x \, dx + \varphi_y \, dy,$$

and then we see that

$$\varphi_u \, du + \varphi_v \, dv = \varphi_x \, dx + \varphi_y \, dy,$$

which is the invariance of the exterior derivative under coordinate transformations. The exterior derivative contains the same information as the gradient vector

$$\nabla \varphi(u, v) := \big(\varphi_u(u, v), \varphi_v(u, v) \big)$$

of a function, and by adding in the symbols du, dv, it becomes a coordinate-independent notion. For any functions φ, ψ, we have

(1) $d(\varphi + \psi) = d\varphi + d\psi$,
(2) $d(\varphi\psi) = \varphi \, d\psi + \psi \, d\varphi$, and
(3) $d(\tau \circ \varphi) = \tau'(\varphi)d\varphi$ for any differentiable function of one variable $\tau(x)$.

See [84, §7] for proofs.

The directional derivative. Let v be a vector situated at the point $p = (a, b) \in U$ which is obtained by parallel translation to $p = (a, b)$ of the vector (α, β) situated at the origin $(0, 0)$. Then, v is parallel to the position vector (α, β). Since the vector v is determined by two pieces of information, the choice of point p and the choice of vector (α, β), we can write it as

$$v := \alpha \left(\frac{\partial}{\partial u} \right)_p + \beta \left(\frac{\partial}{\partial v} \right)_p . \tag{4.2}$$

The advantage of this notation is that it exhibits how this is interlocked with directional derivatives. We define the *directional derivative of a function φ by v at p* by

$$\alpha\varphi_u(a, b) + \beta\varphi_v(a, b),$$

and we write it as

$$\varphi_v \quad \text{or} \quad d\varphi(v).$$

In this text, we use both of these notations. In particular, the notation $d\varphi(v)$ means restricting $d\varphi$ to the special direction v and assigning to it the rate of change of that function with respect to v. In fact, considering v as a differential operator as written above, we have

$$v(\varphi) = \left(\alpha \left(\frac{\partial}{\partial u} \right)_p + \beta \left(\frac{\partial}{\partial v} \right)_p \right) \varphi = d\varphi(v).$$

So the directional derivative can be considered to be "applying an operator $\boldsymbol{v}$", and the three notations

$$\varphi_{\boldsymbol{v}}, \quad d\varphi(\boldsymbol{v}), \quad \boldsymbol{v}(\varphi)$$

all represent the same directional derivative. Let $T_p\boldsymbol{R}^2$ stand for the set of vectors whose starting points are p, namely,

$$T_p\boldsymbol{R}^2 := \left\{ \alpha\left(\frac{\partial}{\partial u}\right)_p + \beta\left(\frac{\partial}{\partial v}\right)_p ; \alpha, \beta \in \boldsymbol{R} \right\}.$$

$T_p\boldsymbol{R}^2$ is a real two-dimensional vector space called the tangent space of $\boldsymbol{R}^2$ at p. For a given vector $\boldsymbol{v}$ at the point p, considering a planar curve γ satisfying $\gamma(0) = p$ and $\gamma'(0) = \boldsymbol{v}$, $\boldsymbol{v}$ can be regarded as the tangent vector of the curve at p. The following proposition provides motivation for the notation (4.2).

Proposition 4.1.1. *Let U be a domain of the uv-plane containing the point p, and let $\gamma(t)$ be a curve in the uv-plane satisfying $\gamma(0) = p$ and*

$$(\boldsymbol{v} =)\,(\alpha, \beta) := \gamma'(0) \quad \left(\,' = \frac{d}{dt}\right).$$

Then for any C^∞-function $\varphi : U \to \boldsymbol{R}$, the derivative of the composition $\varphi \circ \gamma(t)$ at $t = 0$ coincides with $d\varphi(\boldsymbol{v})$. Namely,

$$\frac{d\varphi \circ \gamma(0)}{dt} = \left(\alpha\left(\frac{\partial}{\partial u}\right)_p + \beta\left(\frac{\partial}{\partial v}\right)_p \right)\varphi.$$

Proof. Setting $\gamma(t) = (u(t), v(t))$,

$$\text{the left-hand side} = \varphi_u(u(0), v(0))u'(0) + \varphi_v(u(0), v(0))v'(0)$$

$$= \varphi_u(p)\alpha + \varphi_v(p)\beta = \text{the right-hand side},$$

by the chain rule. $\qquad\square$

Let $\Psi : U \to \tilde{U}$ be a diffeomorphism from a domain in the uv-plane U to a domain $\tilde{U}$ in the xy-plane. For $p \in U$, we define a linear map $d\Psi_p : T_p\boldsymbol{R}^2 \to T_{\Psi(p)}\boldsymbol{R}^2$ by

$$d\Psi_p(\boldsymbol{v}) := dx_p(\boldsymbol{v})\left(\frac{\partial}{\partial x}\right)_{\Psi(p)} + dy_p(\boldsymbol{v})\left(\frac{\partial}{\partial y}\right)_{\Psi(p)}, \qquad (4.3)$$

where $\Psi(u, v) = \big(x(u,v), y(u,v)\big)$. This map $d\Psi_p$ is called the *differential* of the map Ψ at p. Since $\boldsymbol{v}$ has the expression (4.2), we identify $\boldsymbol{v}$

with the vector (α, β), and the vector $d\Psi_p(v)$ is identified with the vector $\big(dx_p(v), dy_p(v)\big)$. Furthermore,

$$dx_p(v) = \alpha x_u(p) + \beta x_v(p), \quad dy_p(v) = \alpha y_u(p) + \beta y_v(p). \tag{4.4}$$

Thus, substituting this into (4.3), $d\Psi_p(v)$ can be expressed as

$$d\Psi_p(v) = (\alpha x_u(p) + \beta x_v(p)) \left(\frac{\partial}{\partial x}\right)_{\Psi(p)}$$

$$+ (\alpha y_u(p) + \beta y_v(p)) \left(\frac{\partial}{\partial y}\right)_{\Psi(p)}. \tag{4.5}$$

Rewriting this, we have

$$d\Psi_p(v) = J_\Psi \begin{pmatrix} \alpha \\ \beta \end{pmatrix}, \quad J_\Psi := \begin{pmatrix} x_u(p) & x_v(p) \\ y_u(p) & y_v(p) \end{pmatrix}. \tag{4.6}$$

Namely, the operation $d\Psi_p$ to v amounts to multiplying the Jacobi matrix J_Ψ by (α, β) from the right-hand side, regarding (α, β) as a column vector. For this reason, $d\Psi$ can be identified with the Jacobi matrix J_Ψ. The next proposition explains how the two vectors v and $d\Psi_p(v)$ are identified when Ψ is a diffeomorphism.

Proposition 4.1.2. *Let $\Psi : U \to \tilde{U}$ be a diffeomorphism from a domain U in the uv-plane to a domain $\tilde{U}$ in the xy-plane. Let $\gamma(t)$ be a curve in the uv-plane satisfying $\gamma(0) = p$ and $\gamma'(0) = (\alpha, \beta)$, namely, the curve $\gamma(t)$ has the velocity vector $v = (\alpha, \beta)$ at p $(t = 0)$. Then the velocity vector*

$$\hat{\gamma}(t) := \Psi \circ \gamma(t)$$

of the curve at $t = 0$ in the xy-plane at $\Psi(p)$ coincides with $d\Psi_p(v)$.

Proof. Let v be expressed as in (4.2). Setting $\gamma(t) = \big(u(t), v(t)\big)$, we have

$$\alpha = u'(0), \quad \beta = v'(0). \tag{4.7}$$

We set $\Psi(u, v) = (x(u, v), y(u, v))$. Then

$$\frac{d}{dt}\Psi \circ \gamma(t) = \frac{d}{dt}(x(u(t), v(t)), y(u(t), v(t)))$$

$$= \Big(x_u(u(t), v(t))u'(t) + x_v(u(t), v(t))v'(t),$$

$$y_u(u(t), v(t))u'(t) + y_v(u(t), v(t))v'(t) \Big),$$

and substituting $t = 0$, (4.7) leads to

$$\frac{d}{dt}\Psi \circ \gamma(t)\bigg|_{t=0} = (\alpha x_u(p) + \beta x_v(p)) \left(\frac{\partial}{\partial x}\right)_{\Psi(p)}$$

$$+ (\alpha y_u(p) + \beta y_v(p)) \left(\frac{\partial}{\partial y}\right)_{\Psi(p)}$$

$$= d\Psi_p(\boldsymbol{v}).$$

This shows the assertion. □

Since Ψ is a diffeomorphism, $\Psi(u, v) = (x(u, v), y(u, v))$ gives a coordinate transformation. The curve $\hat{\gamma}(t)$ is the composition of $\gamma(t)$ with this coordinate transformation. Then the vector $\boldsymbol{v} = \gamma'(0)$ in the uv-plane corresponds to the vector $\hat{\gamma}'(0)$ in the xy-plane. Using this new coordinate system (x, y), we get different expressions for the same geometric objects, so we expect that, when identifying $\gamma(t)$ with $\hat{\gamma}(t)$, then $\gamma'(0)$ will be identified with $\hat{\gamma}'(0)$. In fact, for a function $\varphi(x, y)$ on the xy-plane,

$$d\varphi(d\Psi_p(\boldsymbol{v})) = d\varphi(\hat{\gamma}'(0)) = \frac{d\varphi \circ \hat{\gamma}}{dt}\bigg|_{t=0}$$

$$= \frac{d\varphi \circ \Psi \circ \gamma}{dt}\bigg|_{t=0} = d(\varphi \circ \Psi)(\gamma'(0)) = d(\varphi \circ \Psi)(\boldsymbol{v}).$$

Since, when applying the coordinate transformation, the function φ is identified with $\varphi \circ \Psi$, and the two vectors determine the same operator as directional derivatives. This provides a justification for the identification. We set

$$\left(\frac{\partial}{\partial u}\right)_p = \frac{\partial x}{\partial u}\left(\frac{\partial}{\partial x}\right)_{\Psi(p)} + \frac{\partial y}{\partial u}\left(\frac{\partial}{\partial y}\right)_{\Psi(p)},$$

$$\left(\frac{\partial}{\partial v}\right)_p = \frac{\partial x}{\partial v}\left(\frac{\partial}{\partial x}\right)_{\Psi(p)} + \frac{\partial y}{\partial v}\left(\frac{\partial}{\partial y}\right)_{\Psi(p)}.$$

Then, by substituting this into (4.2), we have the expression (4.5) for $d\Psi_p(\boldsymbol{v})$, giving us an identification of the vectors $\boldsymbol{v}$ and $d\Psi_p(\boldsymbol{v})$ via the coordinate transformation. In this sense, the expression (4.2) can be regarded as a coordinate independent expression for a vector. (In manifold theory, this identification corresponds to the coordinate independence of tangent vectors.)

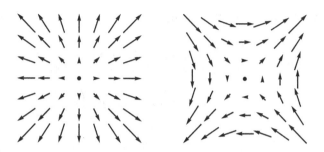

Fig. 4.1.   Vector fields X_1 and X_2.

Vector fields. A correspondence $X : U \ni p \mapsto X_p \in T_p \mathbf{R}^2$ from each point p in the domain U in the uv-plane to the vector X_p with base point p that changes smoothly with respect to smooth movement of p is called a *vector field* on U. Let X be such a vector field. Then there exist C^∞-functions α and β on U such that

$$X_p = \alpha(p) \left(\frac{\partial}{\partial u} \right)_p + \beta(p) \left(\frac{\partial}{\partial v} \right)_p \quad (p \in U).$$

For this reason, a vector field can be written as $X = \alpha \frac{\partial}{\partial u} + \beta \frac{\partial}{\partial v}$. For example, setting

$$X_1 := u\frac{\partial}{\partial u} + v\frac{\partial}{\partial v}, \quad X_2 := v\frac{\partial}{\partial u} + u\frac{\partial}{\partial v},$$

we obtain the vector fields depicted in Fig. 4.1, and are parallel translations of (u, v), (v, u) to the base point (u, v).

Letting $\Psi : U \to \tilde{U}$ be a diffeomorphism from a domain U in the uv-plane to a domain $\tilde{U}$ in the xy-plane, we write $\Psi(u, v) = \big(x(u, v), y(u, v) \big)$. Regarding the map Ψ as a coordinate transformation, a vector field

$$X := \alpha \frac{\partial}{\partial u} + \beta \frac{\partial}{\partial v}$$

on U can be identified with the vector field

$$d\Psi(X) = dx(X)\frac{\partial}{\partial x} + dy(X)\frac{\partial}{\partial y}$$

$$= (\alpha x_u + \beta x_v)\frac{\partial}{\partial x} + (\alpha y_u + \beta y_v)\frac{\partial}{\partial y}$$

on $\tilde{U}$ in the xy-plane. We can also consider a vector field that is not defined on the entire domain. Let $\gamma(t)$ $(a < t < b)$ be a regular curve in a domain U

in the uv-plane. If, for each t, a vector

$$X_t := \alpha(t) \left(\frac{\partial}{\partial u} \right)_{\gamma(t)} + \beta(t) \left(\frac{\partial}{\partial v} \right)_{\gamma(t)}$$

with base point $\gamma(t)$ is given, where $\alpha(t)$ and $\beta(t)$ are C^∞-functions on (a, b), then the correspondence $X : (a, b) \ni t \mapsto X_t$ is called a *vector field along* γ. We already used the phrase "unit normal vector field" in Chapter 1 and "null vector field" along a singular curve in Chapter 2, and these are examples of vector fields along a curve. We sometimes identify X with the vector valued function $(\alpha(t), \beta(t))$. For example, the velocity vector $\gamma'(t) = (u'(t), v'(t))$ of a curve γ can be written as

$$u'(t) \left(\frac{\partial}{\partial u} \right)_{\gamma(t)} + v'(t) \left(\frac{\partial}{\partial v} \right)_{\gamma(t)},$$

and this expression demonstrates how $\gamma'(t)$ is a vector field along γ, but we usually write

$$\gamma'(t) = (u'(t), v'(t)),$$

which we interpret as having base point $\gamma(t)$.

4.2. Criteria for Cuspidal Edges and Swallowtail Singularities II

In Chapter 2, we already introduced criteria for cuspidal edges and swallowtails (Theorem 2.6.3), and here we introduce another criteria for these two types of singularities, using the vector fields introduced in the previous section.

Let $p = (0, 0) \in U$ be a non-degenerate singular point of a frontal $f : U \to \mathbf{R}^3$, and let $\gamma(t)$ be the singular curve passing through p, and assume that $\eta(t)$ is a null vector field along γ, that is, $df_{\gamma(t)}(\eta(t))$ vanishes for each t.

Definition 4.2.1. A vector field $\tilde{\eta}$ on U is called an *extended null vector field* if, for each t satisfying $\gamma(t) \in U$, we have

$$\tilde{\eta}(\gamma(t)) = \eta(t).$$

The following lemma guarantees the existence of $\tilde{\eta}$ for frontals.

Lemma 4.2.2. *Let $f : U \to \mathbf{R}^3$ be a frontal, with $p \in U$ a non-degenerate singular point. Then for a sufficiently small neighborhood $V (\subset U)$ of p,* *there exists a local coordinate system[1] (x, y) on V centered at p such that $\partial_x := \partial/\partial x$ is an extended null vector field.*

Proof. We may assume that U lies in the uv-plane $(\mathbf{R}^2; u, v)$, and the singular curve γ satisfies $\gamma(0) = p$.

Let $\mathbf{n}(t)$ be a unit normal vector field along the singular curve $\gamma(t)$. Then, since $\gamma'(0)$ and $\mathbf{n}(0)$ are linearly independent, a coordinate system (t, s) centered at p can be defined by the map

$$(t, s) \longmapsto \gamma(t) + s\mathbf{n}(t) \in \mathbf{R}^2.$$

In this coordinate system, the t-axis is the singular curve, and we let $\eta(t)$ be a null vector field along that axis (i.e., the singular curve γ). Then

$$\tilde{\eta}(s, t) := \eta(t),$$

gives an extended null vector field on a neighborhood of the origin in the st-plane. Using Theorem B-5-3 of [84], we have a new coordinate system (x, y) centered at p such that ∂_x is a scalar multiple of $\tilde{\eta}$, which is the desired coordinate system. □

By the above lemma, the existence of an extended null vector field is guaranteed for non-degenerate singular points on frontals. Using it, we can state criteria for cuspidal edges and swallowtails that are different from those in Chapter 2.

Theorem 4.2.3. *Let $f : U \to \mathbf{R}^3$ be a front, and let $p \in U$ be a non-degenerate singular point of f, that is, $d\Lambda$ does not vanish at p, where Λ is an identifier of the singularities (see Section 2.6). Let $\tilde{\eta}$ be an extended null vector field. Then:*

(1) *p is a cuspidal edge if and only if the directional derivative of Λ with respect to $\tilde{\eta}$, namely $\Lambda_\eta := d\Lambda(\tilde{\eta}) : U \to \mathbf{R}$, does not vanish at p. (Note that $\Lambda_\eta \neq 0$ implies p is non-degenerate.)*
(2) *p is a swallowtail if and only if $\Lambda_\eta(p) = 0$ and the directional derivative $\Lambda_{\eta\eta} := d\Lambda_\eta(\tilde{\eta})$ of Λ_η with respect to $\tilde{\eta}$ does not vanish at p.*

We show these criteria by proving that they are equivalent to the criteria (cf. Theorem 2.6.3) for cuspidal edges and swallowtails introduced in Chapter 2. We first show the following:

---

[1]Here, a *local coordinate system* means a parametrization of f by two variables defined on V.

Proposition 4.2.4. *The conditions in Theorem 4.2.3 depend neither on the choice of identifier Λ of the singularity of f nor on the choice of extended null vector field $\tilde{\eta}$.*

Proof. We show the conditions remain the same when multiplying Λ by a non-zero function. Let k be a nowhere-vanishing C^∞-function defined on a neighborhood of p. We assume $\Lambda(p) = 0$. By $d(k\Lambda) = \Lambda\,dk + k\,d\Lambda$, the condition $d\Lambda_p = 0$ is equivalent to $d(k\Lambda)_p = 0$. Furthermore, since $(k\Lambda)_\eta = k_\eta\,\Lambda + k\Lambda_\eta$, the condition $\Lambda_\eta(p) = 0$ is equivalent to $(k\Lambda)_\eta(p) = 0$. Next, we assume $\Lambda(p) = \Lambda_\eta(p) = 0$. By

$$(k\Lambda)_{\eta\eta} = k_{\eta\eta}\Lambda + 2k_\eta\Lambda_\eta + k\Lambda_{\eta\eta},$$

the condition $\Lambda_{\eta\eta}(p) = 0$ is equivalent to $(k\Lambda)_{\eta\eta}(p) = 0$, under the assumption $\Lambda(p) = \Lambda_\eta(p) = 0$.

We show that condition (2) does not depend on the choice of extended null vector field. Let k be a nowhere-vanishing C^∞-function. We set $\tilde{\xi} := k\tilde{\eta}$. Then $\tilde{\xi}$ is also an extended null vector field, and it satisfies

$$\Lambda_{\xi\xi}(p) = k(p)(k_\eta(p)\Lambda_\eta(p) + k(p)\Lambda_{\eta\eta}(p)) = k(p)^2\Lambda_{\eta\eta}(p).$$

Thus the condition (2) does not change when multiplying the extended null-vector field by a nowhere-vanishing function.

Let $\tilde{\zeta}$ be another extended null vector field. Multiplying $\tilde{\eta}$ by a nowhere-vanishing function of t, we may assume $\tilde{\eta}(\gamma(t)) = \tilde{\zeta}(\gamma(t))$ for each t. Setting $h := \Lambda_{\tilde{\zeta}} - \Lambda_{\tilde{\eta}}$, we have $h(\gamma(t)) = 0$. On the other hand, since $\Lambda_{\tilde{\eta}}(p) = 0$ and $\tilde{\eta}(p) = \gamma'(0)$,

$$\Lambda_{\zeta\zeta}(p) - \Lambda_{\eta\eta}(p) = d\Lambda_\zeta(\tilde{\zeta}(p)) - d\Lambda_\eta(\tilde{\eta}(p)) = d\Lambda_\zeta(\tilde{\eta}(p)) - d\Lambda_\eta(\tilde{\eta}(p))$$

$$= dh(\tilde{\eta}(p)) = dh(\gamma'(0))$$

holds. Since $h(\gamma(t)) = 0$, we have

$$0 = (h \circ \gamma)'(0) = dh(\gamma'(0)) = \Lambda_{\zeta\zeta}(p) - \Lambda_{\eta\eta}(p).$$

This proves $\Lambda_{\zeta\zeta}(p) = \Lambda_{\eta\eta}(p)$. In particular, the condition (2) for the extended null vector field $\tilde{\eta}$ implies the same for $\tilde{\zeta}$, proving the assertion. $\square$

In the next examples, we confirm that the standard cuspidal edge and the standard swallowtail satisfy the respective conditions in Theorem 4.2.3.

Example 4.2.5. We check here that the cuspidal edge $f(u, v) := (u^2, u^3, v)$ satisfies the condition in the given criterion. Since $f_u \times f_v = u\,(3u, -2, 0)$, the set $\{u = 0\}$ (namely, the v-axis) is the set of singularities of f. We can take $\tilde{\nu} = (3u, -2, 0)$ as a normal vector field. We established that f is a front in Example 2.5.2. Since

$$\det(f_u, f_v, \tilde{\nu}) = u(4 + 9u^2),$$

we can take $\Lambda := u$ as an identifier of the singularities. As noted, the set of singular points of f is $\{u = 0\}$, and we can see that $f_u = \mathbf{0}$ holds on $\{u = 0\}$, so we can take $\partial_u := \partial/\partial u$ as an extended null vector field. Since $\Lambda_u = 1$, the condition in the criterion for cuspidal edges in Theorem 4.2.3 is satisfied.

Example 4.2.6. Here, we check that the standard swallowtail given in Example 2.5.3 satisfies the condition in the given criterion. We can take $\tilde{\nu} := (1, u, u^2)$ as a normal vector, and we established that f is a front in Example 2.5.3. An identifier of the singularities can be taken as $\Lambda = 6u^2 + v$. Since $f_u = \mathbf{0}$ on the set of singular points $\{6u^2 + v = 0\}$, we may take ∂_u as the extended null vector field. Since

$$(\Lambda_u(0, 0), \Lambda_v(0, 0)) = (0, 1)$$

is not zero, $(0, 0)$ is non-degenerate. Moreover, since $\Lambda_u = 12u$ and $\Lambda_{uu} = 12$, the condition in the criterion for swallowtails in Theorem 4.2.3 is satisfied.

Equivalency of Theorems 4.2.3 and 2.6.3. We now show that the criteria in Theorems 4.2.3 and 2.6.3 in Section 2.6 are equivalent.

We first consider the case of cuspidal edges. By Lemma 4.2.2, there exists a coordinate system (u, v) centered at p such that ∂_v is an extended null vector field.

We write the singular curve as $\gamma(t) = (u(t), v(t))$. Since $\Lambda(u(t), v(t)) = 0$, taking its derivative along γ, we have

$$u'\Lambda_u + v'\Lambda_v = 0, \tag{4.8}$$

and the vector (Λ_u, Λ_v) is perpendicular to the singular direction $\gamma'(= d\gamma/dt) = (u', v')$. Linear independence of the singular direction and the null direction is equivalent to ∂_v (which is equal to $(0, 1)$ by the identification of (4.2)) not being perpendicular to (Λ_u, Λ_v). This is equivalent to $\Lambda_v \neq 0$. Hence the equivalency of the two criteria is shown.

Next, we consider the case of swallowtails. Since swallowtails do not satisfy the criteria for cuspidal edges, the singular direction and the null direction coincide. Thus, $\Lambda_v(p) = 0$, and by non-degeneracy, $\Lambda_u(p) \neq 0$

follows. Then, by the implicit function theorem, the singular curve can be parametrized as $(u(t), t)$. Since this is the special case $v(t) = t$ in (4.8), it holds that

$$u' \Lambda_u(u(t), t) + \Lambda_v(u(t), t) = 0. \tag{4.9}$$

Substituting $t = 0$ into this equation, $u'(0) = 0$ holds, since $\Lambda_v(p) = 0$. Differentiating (4.9), we have

$$\Lambda_{uu}(u(t), t)u'(t)^2 + 2\Lambda_{uv}(u(t), t)u'(t)$$
$$+ \Lambda_{vv}(u(t), t) + \Lambda_u(u(t), t)u''(t) = 0,$$

and by $u'(0) = 0$ we have $u''(0) = -\Lambda_{vv}(0, 0)/\Lambda_u(0, 0)$. Since $\eta = \partial_v$ along γ, the function Δ in Theorem 2.6.3 is

$$\Delta(t) = \det(\eta(t), \gamma'(t)) = \det \begin{pmatrix} 0 & u' \\ 1 & v' \end{pmatrix} = -u'(t).$$

Thus, $(u''(0) =)\Delta'(0) \neq 0$ is equivalent to $\Lambda_{vv}(p) \neq 0$. Hence the equivalency of the criterion for swallowtails in Theorem 2.6.3 and that of Theorem 4.2.3 is shown.

4.3. Singularities of Parallel Surfaces

As an application of the criteria for cuspidal edges and swallowtails, we consider what the conditions are for a singular point of a parallel surface to be a cuspidal edge or swallowtail.

Let $f : U \to \mathbf{R}^3$ be a regular surface, with ν a unit normal vector of f. For a real number $t(\neq 0)$, we consider the parallel surface $f^t := f + t\nu$. In Section 2.2, we have seen that ν is a unit normal vector field of f^t for each t. Let $p \in U$ be a non-umbilic point of f. The parallel surface f^t has a singular point at p if and only if $t = 1/\lambda_1(p)$ or $1/\lambda_2(p)$ (see (2.28) in Chapter 2).

Suppose that $\lambda_1(p) \neq 0$ and set $t_0 := 1/\lambda_1(p)$. Let us examine the singularity of f^{t_0} at p by using criteria for singularities. Let (u, v) be a curvature line coordinate system (cf. Fact 2.1.19). Then, by (2.27),

$$\det(f_u^t, f_v^t, \nu) = (1 - t\lambda_1)(1 - t\lambda_2) \det(f_u, f_v, \nu)$$

holds. Since p is not an umbilic point, if $t = t_0$, then we have that $1 - t\lambda_2$ and $\det(f_u, f_v, \nu)$ do not vanish on a sufficiently small neighborhood $V(\subset U)$ of p, and we may choose $\Lambda := 1 - t_0\lambda_1$ as an identifier of the singularities for f^{t_0} on V. Then the null vector of f^{t_0} at p can be taken to be the principal vector ∂_u. Applying Theorem 4.2.3 with $\Lambda = 1 - t_0\lambda_1$, we obtain the following assertion:

Theorem 4.3.1. *Let $f : U \to \mathbf{R}^3$ be a regular surface, and p a non-umbilic point. Let (u, v) be a curvature line coordinate system centered at p, and λ_1, λ_2 the principal curvatures of f with respect to the u-curves and the v-curves, respectively. Suppose that $\lambda_1(p) \neq 0$. Then the parallel surface f^{t_0} $(t_0 := 1/\lambda_1(p))$ of f has a singular point at p, and by taking a sufficiently small neighborhood $V(\subset U)$ of p, the following hold:*

(1) a singular point of f^{t_0} on V is a cuspidal edge if and only if $(\lambda_1)_u(p) \neq 0$,

(2) a singular point of f^{t_0} on V is a swallowtail if and only if $(\lambda_1)_u(p) = 0$, $(\lambda_1)_v(p) \neq 0$ and $(\lambda_1)_{uu}(p) \neq 0$ (the condition $(\lambda_1)_v(p) \neq 0$ corresponds to the condition that p is non-degenerate).

In this theorem, a point p satisfying $(\lambda_1)_u(p) = 0$ is called a *ridge point* of f with respect to the principal curvature λ_1 (cf. [64]).

4.4.   Maps Between Planes and Singularities of the Gauss Map

We first introduce two new singularities appearing on maps between planes and their criteria. Let U be a domain in the uv-plane, and let $f : U \to \mathbf{R}^2$ be a C^∞-map into the xy-plane. A point $(u, v) \in U$ where the rank of the Jacobi matrix

$$J_f = \begin{pmatrix} x_u(u, v) & x_v(u, v) \\ y_u(u, v) & y_v(u, v) \end{pmatrix}$$

of the map $f = (x(u, v), y(u, v))$ is less than or equal to 1 is called a *singular point* of f.

An example of a map with singularities is the *standard fold map*

$$f_0(u, v) := (u, v^2), \tag{4.10}$$

whose Jacobi matrix is computed as

$$J_{f_0} = \begin{pmatrix} 1 & 0 \\ 0 & 2v \end{pmatrix}. \tag{4.11}$$

Thus, $v = 0$, namely the u-axis, is the set of singular points of f_0.

Another example of a map with singularities is the *standard Whitney cusp*

$$f_1(u, v) := (u^3 - 3uv, v). \tag{4.12}$$

In this case, the Jacobi matrix is

$$J_{f_1} = \begin{pmatrix} 3u^2 - 3v & -3u \\ 0 & 1 \end{pmatrix}. \tag{4.13}$$

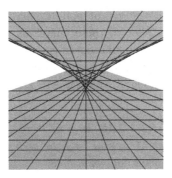

Fig. 4.2. Standard Whitney cusp.

Since the determinant of this matrix is $3(u^2 - v)$, the parabola $v = u^2$ on the uv-plane is the set of singular points of f_1. The origin $(0, 0)$ is the standard Whitney cusp singularity (see Fig. 4.2).

We call a map which is right–left equivalent to the above standard fold singular point, respectively the standard Whitney cusp, a *fold*, respectively, a *Whitney cusp*. The following assertion holds:

Proposition 4.4.1. *The singular set image of a fold (respectively, Whitney cusp) is a regular curve (respectively, a cusp) in the plane.*

Proof. To prove this assertion, we may check it for the standard fold and the standard Whitney cusp. Then $u \mapsto f_0(u, 0) = (u, 0)$ and $u \mapsto f_1(u, u^2) = (-2u^3, u^2)$ give parametrization of the singular set image of f_0 and f_1, respectively. $\square$

We state the criteria for these singularities which are similar to those for cuspidal edges and swallowtails.

Suppose $f : U \to \mathbf{R}^2$ has a singular point at $p \in U$. We can take the Jacobian $\det(J_f)$ as an identifier of the singularities. Multiplying $\det(J_f)$ by a nowhere-vanishing function, it will still identify singular points, so we call a nowhere-vanishing functional multiple of $\det(J_f)$ an *identifier of singularities*, and we denote it by Λ.

A singular point p is *non-degenerate* if

$$\big(\Lambda_u(p), \Lambda_v(p)\big) \neq (0, 0)$$

for a coordinate system (u, v) of U. This definition does not depend on the choice of coordinate system, nor the choice of identifier of singularities Λ. Like as in the case of non-degenerate singular points of fronts, each connected component of the set of non-degenerate singular points is a regular curve on U. We refer to these curves as *singular curves*. Taking a parametrization $\gamma(t)$ of a singular curve passing through p, there exists a nowhere-vanishing vector field $\eta(t)$ along γ such that $df(\eta(t)) = 0$. This is called a *null vector field* along γ. Analogous to Theorem 4.2.3, the following theorem holds.

Theorem 4.4.2. *Let $f : U \to \mathbf{R}^2$ be a C^∞-map, with $p = (0,0)$ a non-degenerate singular point of f, and let $\tilde{\eta}$ be an extended null vector field (see Lemma 4.2.2). Then the following hold:*

(1) *The directional derivative $\Lambda_\eta := d\Lambda(\tilde{\eta}) : U \to \mathbf{R}$ with respect to $\tilde{\eta}$ of an identifier Λ of singularities does not vanish at p if and only if p is a fold singular point.*

(2) *When $\Lambda_\eta(p) = 0$, the directional derivative $\Lambda_{\eta\eta} := d\Lambda_\eta(\eta)$ of Λ_η does not vanish at p if and only if p is a Whitney cusp (see Exercise 3 in this section).*

Imitating the proof of Proposition 4.2.4, one can easily show that (1) and (2) depend neither on the choice of identifier Λ nor on the choice of extended null vector field $\tilde{\eta}$. Proving the criterion for a fold singular point is left as an exercise (Problem 3). One can also obtain criteria of folds and cusps like as in Theorem 2.6.3 by modifying the arguments in "Equivalency of Theorems 4.2.3 and 4.2.6" on page 116. See Appendix D for a proof of the criterion for a Whitney cusp.

Example 4.4.3. Since the Jacobian of the standard fold map is $\Lambda = 2v$, the origin is a non-degenerate singular point. Furthermore, by (4.11), we can take a null direction as $\tilde{\eta} = (0, 1)$. Since $\Lambda_\eta(0,0) = 2$, the standard fold map satisfies the condition in the criterion for a fold singular point.

Example 4.4.4. Next, we apply the criterion for the standard Whitney cusp. Since the Jacobian is $\Lambda = 3(u^2 - v)$, the origin is a non-degenerate singular point. Furthermore, by (4.13), we can take a null direction as $\tilde{\eta} = (1, 0)$. Since $\Lambda_\eta(0,0) = 0$ and $\Lambda_{\eta\eta}(0,0) = 6$, the standard Whitney cusp satisfies the condition in the criterion for a Whitney cusp.

We proved that Theorems 4.2.3 and 2.6.3 are equivalent, in Section 4.2. With the same arguments, we can show a criterion using information in the source space as follows:

Corollary 4.4.5. *Let $f : U \to \mathbf{R}^2$ have a non-degenerate singular point at $p \in U$. Let $\gamma(t)$ be a parametrization of the singular curve passing through*

p with $\gamma(c) = p$. Let $\eta(t)$ be a null vector field along γ, and set

$$\Delta(t) := \det(\eta(t), \gamma'(t)).$$

Then

- p is a fold singular point if and only if $\Delta(c) \neq 0$,
- p is a Whitney cusp if and only if $\Delta(c) = 0$ and $\Delta'(c) \neq 0$.

Verifying that the standard fold singular point and standard Whitney cusp satisfy these conditions is left as exercises (see Exercise 1 in this section).

Singularities of Gauss maps. As an application, let us consider singular points of Gauss maps of regular surfaces. Let $f : U \to \mathbf{R}^3$ be a regular surface, and $\nu : U \to S^2$ a unit normal vector field. Let ν have a singular point at $p \in U$, and then, by the Weingarten formula (Proposition 2.1.5), we have $K(p) = 0$. Let us assume that ν has a rank one singular point at p. Then we can see that p is not an umbilic point. By Theorem 2.1.19, there exists a curvature line coordinate system, and we assume $(U; u, v)$ is such a curvature line coordinate system of f. Let λ_1 and λ_2 be the principal curvatures with respect to the u-curves and v-curves, respectively. We may assume $\lambda_1(p) = 0$ and $\lambda_2(p) \neq 0$. Then, again by the Weingarten formula (Proposition 2.1.5),

$$\nu_u = -\lambda_1 f_u, \quad \nu_v = -\lambda_2 f_v,$$

and the Jacobian of ν vanishes if and only if $\lambda_1 \lambda_2$ does as well. Since $\lambda_2(p) \neq 0$, the function $\Lambda = \lambda_1$ can be taken as an identifier of singularities. Furthermore, since the set of singular points of ν is $\Sigma_\nu = \{\lambda_1 = 0\}$, the vector field ∂_u, namely, the principal direction corresponding to λ_1, can be taken as a null vector field. By Corollary 4.4.5, we have the following theorem.

Theorem 4.4.6. *Let* $f : U \to \mathbf{R}^3$ *be a regular surface with unit normal vector field* ν. *Let* λ_1, λ_2 *be the principal curvatures of* f, *and* $\boldsymbol{v}_i$ $(i = 1, 2)$ *vector fields on* U *giving the principal directions with respect to the principal curvatures* λ_i. *Suppose that* $\lambda_1(p) = 0$ *and* $\lambda_2(p) \neq 0$. *Then,*

- p *is a fold singular point of* ν *if and only if* $\boldsymbol{v}_1 \lambda_1(p) \neq 0$, *namely, the directional derivative of* λ_1 *at* p *in the direction* $\boldsymbol{v}_1$ *does not vanish.*
- p *is a Whitney cusp of* ν *if and only if*

$$\boldsymbol{v}_1 \lambda_1(p) = 0, \quad \boldsymbol{v}_1 \boldsymbol{v}_1 \lambda_1(p) \neq 0, \quad \boldsymbol{v}_2 \lambda_1(p) \neq 0,$$

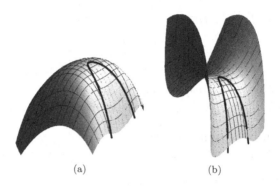

(a)   (b)

Fig. 4.3. Positive and negative godron points.

where $v_1 v_1 \lambda_1(p)$ is the directional derivative of $v_1 \lambda_1$ at p in the direction v_1. (The condition $v_2 \lambda_1(p) \neq 0$ corresponds to the condition that p is non-degenerate.)

Remark 4.4.7. If we take curvature line coordinates (u, v) so that $\partial_u (= \partial/\partial u)$ (respectively, ∂_v) points in the principal direction with respect to the principal curvature λ_1 (respectively, λ_2), then $v_1 \lambda_1 = (\lambda_1)_u$, $v_2 \lambda_1 = (\lambda_1)_v$ and $v_1 v_1 \lambda_1 = (\lambda_1)_{uu}$.

A point at which the Gauss map ν is a Whitney cusp is called a *godron point*, or A_3-*parabolic point* [46]. Godron points have a plus-minus property determined by the positivity or negativity of the Gaussian curvature over the exterior domain of the cusp. Figure 4.3 shows positive and negative godron points. The dark curves in the figures stand for the points where the Gauss curvature is zero, namely, the singular points of the Gauss map.

Exercises 4.4

1 Verify that the standard fold singular point and standard Whitney cusp satisfy the conditions in Corollary 4.4.5.

2 The surfaces in Fig. 4.3 are defined by the formula

$$(u, v) \mapsto (u, v, -\varepsilon u^4 + u^2 v - v^2) \quad (\varepsilon = \pm 1)$$

((a): $\varepsilon = +1$, and (b): $\varepsilon = -1$). Calculate the Gauss map of these surfaces, and verify that $(u, v) = (0, 0)$ is a godron point of each surface.

3 Give a proof of the criterion for a fold singular point in Theorem 4.4.2 given, referring to the proof in Section 3.4 in Chapter 3. (Hint: By Lemma 3.3.1, we may set $f(0, v) = (0, v)$. Like as in the proof of Theorem 2.6.3 in Chapter 3, we can write $f(u, v) = (u^2 a(u, v), v)$, where $a(0, 0) > 0$. By setting $s := u\sqrt{a(u, v)}$ and $t := v$, we obtain the expression $f(s, t) = (s^2, t)$.)

4.5. Koenderink's Theorem

As an application of Corollary 4.4.5, we introduce a theorem concerning singular points of a projection of a surface.

Let Π be a plane in $\mathbf{R}^3$, and let $\pi : \mathbf{R}^3 \to \Pi$ be the orthogonal projection. For a regular surface $f : U \to \mathbf{R}^3$, we now define the "apparent contour" (as a plane curve lying in Π) made by projecting this surface into Π. Let I be an interval. A planar curve $\sigma : I \to \Pi$ as a C^∞-map is an *apparent contour* of $\pi \circ f$ if the image of $\pi \circ f$ lies to one side of σ within Π. The apparent contour $\sigma(t)$ is the image of the singular set of $\pi \circ f$.

For instance, the projection into the xz-plane of the cylinder

$$f(u, v) := (\cos u, \sin u, v) \quad ((u, v) \in [0, 2\pi) \times \mathbf{R}) \qquad (4.14)$$

of radius one is expressed by the map $h(u, v) := (\cos u, v)$ from the uv-plane. The Jacobian $-\sin u$ of h is an identifier of the singularities. Thus, the apparent contour passing through $(\pm 1, 0)$ (which is a projection of $(\pm 1, 0, 0)$), is given by $\sigma(t) := (\pm 1, t)$ $(t \in \mathbf{R})$. These two lines give the set of boundaries of the image of the projection of the cylinder in the xz-plane.

Theorem 4.5.1 (Koenderink's theorem [43]). *For a regular surface* $f : U \to \mathbf{R}^3$, *fix* $p \in U$ *and set* $\mathrm{P} := f(p)$. *Let* $\nu(p)$ *be the unit normal vector of* f *at* P. *Let* $\mathbf{v}$ *be a tangent vector of* f *at* P, *and assume that the normal curvature* $\kappa_n(p)$ *of* f *with respect to* $\mathbf{v}$ *(cf. (2.19)) does not vanish. Let* $\Pi(\subset \mathbf{R}^3)$ *be the plane passing through* P *and perpendicular to* $\mathbf{v}$, *and let* $\pi : \mathbf{R}^3 \to \Pi$ *be the orthogonal projection of* $\mathbf{R}^3$ *into* Π. *In this setting, the image of the apparent contour* σ *passing through* P *coincides with the set of fold singular points of* $\pi \circ f$ *in the plane* Π. *Moreover, we give an orientation to this plane and the apparent contour* σ *at* P *respectively so that* $\nu(p)$ *is the leftward unit normal vector of* σ. *We let* $\mu(p)$ *be the curvature of the planar curve* σ *at* P *with respect to these orientations. Then the Gaussian curvature* K_p *of* f *at* p *is* $K_p = \mu(p)\kappa_n(p)$.

The above theorem implies the following: Under the orthogonal projection of a surface to a plane containing the normal vector, the product of the curvature of the apparent contour and the normal curvature with respect to the collapsed direction coincides with the Gaussian curvature.

Remark 4.5.2. Let the apparent contour of a projection into a plane Π of the image of a surface f be parametrized by $\sigma(t)$ as a planar curve, and let $\sigma(0)$ correspond to the projected point of P into Π. Then

$$\mu(p) = \frac{\sigma''(0) \cdot \nu(0)}{\sigma'(0) \cdot \sigma'(0)} \qquad (4.15)$$

is the curvature of σ with respect to the orientation for which the direction of ν is the leftward normal vector. The vector $\nu(0)$ is a normal vector of the surface at P, and is also a unit normal vector of σ at $\sigma(0)$ in the plane Π. Choosing ν so that κ_n is positive, then μ coincides with the "singular curvature", which will be defined in Chapter 5, of a singular point of the first kind (see Proposition 5.4.6).

Example 4.5.3. Let f be the cylinder given in (4.14). We have seen that the apparent contour $\sigma(t)$ of the orthogonal projection of f is a pair of disjoint straight lines. Since the curvature of these line is identically zero, the fact that the Gaussian curvature of f vanishes identically follows from Koenderink's theorem.

Example 4.5.4. Projecting the unit sphere (Fig. 4.4(a))

$$S^2 := \{(x, y, z) \in \mathbf{R}^3 \, ; \, x^2 + y^2 + z^2 = 1\}$$

to the xz-plane, to compute the apparent contour of its orthogonal projection, we set

$$f = (u, v, \sqrt{1 - u^2 - v^2}),$$

which parametrizes the upper hemisphere of S^2, and $h(u, v) := (u, \sqrt{1 - u^2 - v^2})$ gives its orthogonal projection to the xz-plane. The Jacobi matrix of the map h is

$$J_h = \frac{1}{\sqrt{1 - u^2 - v^2}} \begin{pmatrix} \sqrt{1 - u^2 - v^2} & 0 \\ -u & -v \end{pmatrix},$$

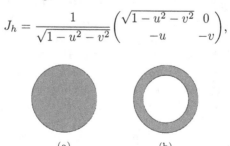

(a)   (b)

Fig. 4.4.   The image of the orthogonal projection sphere (a) and rotationally symmetric torus (b).

and its determinant is $\det(J_h) := -v/\sqrt{1 - u^2 - v^2}$. Its zero set is the set of singular points, which is the u-axis. The image of the set of singular points is $h(u, 0) = (u, \sqrt{1 - u^2})$, and is a half circle. This corroborates the fact that the apparent contour of an orthogonal projection of a sphere into a plane is a circle. This circle is the set of singular points of the map h. In this case, the normal curvature of the sphere is everywhere 1, and since the apparent contour of the projection is a circle of radius 1, its curvature is also 1. Then the product of these two curvatures is 1. Namely we have verified Koenderink's theorem directly in this case (cf. Fig. 4.4(a)).

Proof of Theorem 4.5.1. Taking a point $P = f(p)$ on the image of the surface f to be $(0, 0, 0)$ and set the tangent plane as the xy-plane, then the image of f can be locally expressed as the graph $z = \varphi(x, y)$ in a neighborhood of P. Since the xy-plane is the tangent plane of the surface,

$$\varphi_x(0, 0) = \varphi_y(0, 0) = 0 \tag{4.16}$$

holds, and also the surface can be expressed as

$$f(x, y) = (x, y, \varphi(x, y)) \tag{4.17}$$

near P (here we are changing the parametrization f of the surface). Since, by the formulas for the fundamental quantities of a surface that is expressed as a graph (Example 2.1.3), we have

$$E(0, 0) = G(0, 0) = 1, \qquad F(0, 0) = 0,$$

$$L(0, 0) = \varphi_{xx}(0, 0), \qquad M(0, 0) = \varphi_{xy}(0, 0), \qquad N(0, 0) = \varphi_{yy}(0, 0),$$

and the Gaussian curvature at the point $P = f(p) = (0, 0, 0)$ is given by

$$K_p := \varphi_{xx}(0, 0)\varphi_{yy}(0, 0) - \varphi_{xy}(0, 0)^2. \tag{4.18}$$

On the other hand, $\nu_0 := (0, 0, 1)$ gives a unit normal vector at P. Rotating the xy-plane if necessary, we may assume that the positive direction of the y-axis is the same as the direction of v. In particular,

$$\pi(x, y, z) := (x, z). \tag{4.19}$$

Then, substituting $(\alpha, \beta) = (0, 1)$ into Definition 2.1.9, the normal curvature of f at P with respect to the direction v is

$$\kappa_n(p) = \frac{N(0, 0)}{G(0, 0)} = N(0, 0) = \varphi_{yy}(0, 0). \tag{4.20}$$

Since $\kappa_n(p) \neq 0$, by the assumption of the theorem, it holds that

$$\varphi_{yy}(0,0) \neq 0. \tag{4.21}$$

Projecting f into the plane which is perpendicular to $\boldsymbol{v}$, namely, the xz-plane, we have the map (cf. (4.17))

$$h(x,y) := (x, \varphi(x,y)).$$

The Jacobi matrix of h is

$$J = \begin{pmatrix} 1 & 0 \\ \varphi_x & \varphi_y \end{pmatrix}, \tag{4.22}$$

and the Jacobian of h is

$$\Lambda := \det J = \varphi_y(x,y),$$

and so (4.21) implies p is a non-degenerate singular point. Since $(0,1)^T$ is the eigenvector of J at $p = (0,0)$ associated with the eigenvalue 0, $\partial_y(= (0,1))$ gives the null direction at p. The set of singular points of h can be expressed by the implicit function $(\Lambda =)\varphi_y(x,y) = 0$. By (4.21), $(\varphi_{yy}(0,0) =)\Lambda_y(0,0) \neq 0$ holds. Thus, the set $\Sigma(\subset U)$ of singular points of h near $(0,0)$ can be expressed as the graph $y = y(x)$. Since the singular curve

$$\gamma(x) := (x, y(x))$$

gives a parametrization of Σ, it is called the *contour generator*. The planar curve

$$\sigma(x) := h(x, y(x))$$

gives a parametrization of the apparent contour. Since

$$\varphi_y(x, y(x)) = 0 \tag{4.23}$$

holds, (4.22) implies that $(0,1)$ is the null direction along the singular curve σ of h.

Differentiating both sides of (4.23) by x, we have

$$\varphi_{yx}(x, y(x)) + \varphi_{yy}(x, y(x))y'(x) = 0,$$

where $y' = dy/dx$. Thus

$$y' = -\frac{\varphi_{xy}(x, y(x))}{\varphi_{yy}(x, y(x))}.$$

In particular, the vector

$$(1, y'(0)) = \left(1, -\frac{\varphi_{xy}(0,0)}{\varphi_{yy}(0,0)}\right)$$

at $x = 0$ points in the singular direction of h at $(0,0)$ and is linearly independent of the null vector $(0,1)$. Thus, by Corollary 4.4.5, the map h has fold singular points on the image of the contour generator σ. Hence $\sigma := \pi \circ f \circ \gamma$ and (cf. (4.19))

$$\sigma(x) := (x, \varphi(x, y(x)))$$

holds. In particular, σ is a regular curve, and we now calculate its curvature at $x = 0$. Since

$$\sigma'(x) = (1, \varphi_x + \varphi_y y'), \tag{4.24}$$

differentiating once again, we have

$$\sigma''(x) = (0, \varphi_{xx} + 2\varphi_{xy} y' + \varphi_y y'' + \varphi_{yy}(y')^2). \tag{4.25}$$

Substituting $x = 0$ into (4.24) gives

$$\sigma'(0) = (1, 0), \tag{4.26}$$

and substituting $x = 0$ into (4.25), we have

$$\sigma''(0) = (0, \varphi_{xx} + 2\varphi_{xy} y' + \varphi_{yy}(y')^2)$$

$$= \left(0, \varphi_{xx} + 2\varphi_{xy}\left(-\frac{\varphi_{xy}}{\varphi_{yy}}\right) + \varphi_{yy}\left(-\frac{\varphi_{xy}}{\varphi_{yy}}\right)^2\right).$$

Thus,

$$\sigma''(0) = \frac{1}{\varphi_{yy}(0,0)}\left(0, \varphi_{xx}(0,0)\varphi_{yy}(0,0) - \varphi_{xy}(0,0)^2\right). \tag{4.27}$$

Projecting $\nu_0 = (0,0,1)$ into the xz-plane, we obtain the unit normal vector $\boldsymbol{n} = (0,1)$ of the apparent contour σ at the origin. By (4.15), the curvature $\mu(p)$ of the apparent contour at Q, noting that $\sigma'(0) \cdot \sigma'(0) = 1$ by (4.26), is

$$\mu(p) = \frac{\sigma''(0) \cdot \boldsymbol{n}}{\sigma'(0) \cdot \sigma'(0)} = \sigma''(0) \cdot \boldsymbol{n}.$$

Since $\boldsymbol{n} = (0,1)$, the right-hand side is the second component of $\sigma''(0)$. So, by (4.27), it holds that

$$\mu(p) = \frac{\varphi_{xx}(0,0)\varphi_{yy}(0,0) - \varphi_{xy}(0,0)^2}{\varphi_{yy}(0,0)} = \frac{K_p}{\kappa_n(p)}.$$

In the last equality above, we used (4.18) and (4.20). This shows all assertions of the theorem. $\qquad\qquad\Box$

Exercise 4.5

1   By projecting the torus (2.22) in the direction of the z-axis (Fig. 4.4(b)), find the Gaussian curvature corresponding to the two apparent contours, one is $(a - b)^{-1}b^{-1}$ and the other is $(a + b)^{-1}b^{-1}$.

4.6.   A Gauss–Bonnet Type Formula for Closed Wave Fronts

We will deal with the Gauss–Bonnet type formulas in Chapter 6, using the notion of manifolds; here, we initially introduce an overview without relying on the notion of manifolds. The discussion here is less mathematically strict, but rather more intuitive.

Geodesic curvature and the Gauss–Bonnet theorem. Let U be a domain in the uv-plane $\boldsymbol{R}^2$, and $f : U \to \boldsymbol{R}^3$ a regular surface. We let $\gamma : [a, b] \to U$ be a regular curve on the source space of f, which can be regarded as a curve on the surface itself (its image lies in the image of f). More precisely, when we regard it as a curve on a surface, we consider the composition

$$\hat{\gamma} := f \circ \gamma : [a, b] \longrightarrow \boldsymbol{R}^3. \tag{4.28}$$

In this text, regarding curves on a surface, sometimes we do not identify them as lying specifically on the source or on the target, but in the following discussion here we will be clear about making that distinction. For this purpose, we use the notation as in (4.28), using the overhead "hat" for the curve that lies on the surface in $\boldsymbol{R}^3$.

We let $\boldsymbol{n}_L(t)$ be the *left-ward co-normal vector field* of $\gamma(t)$, that is, $\boldsymbol{n}_L(t)$ points to the left hand side of γ in the uv-plane and $\hat{\boldsymbol{n}}_L(t) := df(\boldsymbol{n}_L(t))$ is a unit tangent vector which is perpendicular to $\hat{\gamma}'(t)$. Then the geodesic curvature κ_g of the curve γ with respect to the left-ward co-normal vector field is defined by

$$\kappa_g(t) = \frac{\hat{\gamma}''(t) \cdot \hat{\boldsymbol{n}}_L(t)}{\hat{\gamma}'(t) \cdot \hat{\gamma}'(t)}. \tag{4.29}$$

If we take the unit normal vector field ν so that the signed area density function

$$\lambda(u, v) = \det(f_u, f_v, \nu)$$

is positive, then $\hat{\boldsymbol{n}}_L = \nu \times \hat{\gamma}$ holds along γ, and $\kappa_g(t)$ can be expressed as

$$\kappa_g(t) = \frac{\det(\hat{\gamma}'(t), \hat{\gamma}''(t), \hat{\nu}(t))}{|\hat{\gamma}'(t)|^3},$$

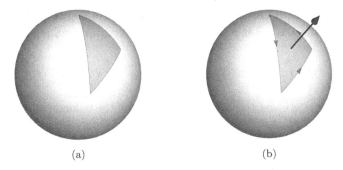

(a)   (b)

Fig. 4.5. A triangle on the sphere (a), with orientation (b).

using the restriction $\hat{\nu}(t) := \nu(\gamma(t))$ of the unit normal vector field ν of f on U to the curve (see (2.15) in Chapter 2). The integration of $\kappa_g(t)$ is denoted by

$$\Theta_\gamma := \int_a^b \kappa_g(t)\,ds \quad (ds := |\hat{\gamma}'(t)|dt),$$

and is called the *total geodesic curvature* with respect to the left-ward co-normal vector field of the curve $\hat{\gamma}$.

Considering a triangle $\triangle$ABC on U whose edges are three regular curves AB, BC, CA which connect the three points A, B, C on U, we assume the domain bounded by these three curves is simply connected. We call this domain the *interior of* $\triangle$ABC, and $\triangle$ABC is defined to be the closed domain containing the interior of $\triangle$ABC with the above curves as boundary. For the sake of brevity, in the later discussions we also denote the image $f(\triangle$ABC$)$ of the domain triangle $\triangle$ABC in the uv-plane simply by $\triangle$ABC.

We fix a unit normal vector ν_0 at a point in the interior of $\triangle$ABC$(\subset \mathbf{R}^3)$, and give an orientation along the edges of the triangle so that motion in that direction would result in a screw moving in the direction of ν_0 (see the (b) Fig. 4.5). We call this *counterclockwise orientation*. Let $\mathbf{n}_L$ be the left-ward co-normal vector as in (4.29) pointing toward the interior of the triangle. We denote the total geodesic curvature with respect to $\mathbf{n}_L$ for each edge by

$$\Theta_{\text{AB}}, \quad \Theta_{\text{BC}}, \quad \Theta_{\text{CA}},$$

respectively, and we call

$$dA := |\det(f_u, f_v, \nu)|\,du \wedge dv$$

the *area element* of the surface, which does not depend on the choice of orientation preserving parametrization of the surface. We have the following theorem.

Fact 4.6.1 (Local Gauss–Bonnet formula for a regular surface [84, Proposition 14.1]). Let K be the Gaussian curvature of a regular surface. Then

$$\int_{\triangle ABC} K \, dA - \Theta_{AB} - \Theta_{BC} - \Theta_{CA} = -\pi + \angle A + \angle B + \angle C, \qquad (4.30)$$

where $\angle A$, $\angle B$, $\angle C$ are the angles (in radians) of the triangle at each vertex, as a surface in $\mathbf{R}^3$.

We say that a triangle $\triangle ABC$ is a *geodesic triangle* if all edges AB, BC and CA are images of geodesics. In this case, since

$$\Theta_{AB} = \Theta_{BC} = \Theta_{CA} = 0,$$

it holds that

$$\angle A + \angle B + \angle C = \pi + \int_{\triangle ABC} K \, dA.$$

In particular, if the Gaussian curvature K has non-changing sign, the sum of the interior angles of the geodesic triangle $\triangle ABC$ is greater than π when $K > 0$, and is less than π when $K < 0$. Using the formula (4.30), we can show the global Gauss–Bonnet theorem, as in what follows.

So far we considered a regular surface in $\mathbf{R}^3$ by taking a parametrization $f : U \to \mathbf{R}^3$ from a domain U in the uv-plane. However, in general, to treat closed surfaces, we will need smoothly glued parametrized surfaces, and we sometimes need to consider a closed surface with self-intersections. To meet these needs, we consider a triangulation of a closed surface as follows. We consider N distinct triangles on a regular surface S in $\mathbf{R}^3$:

$$T_1, T_2, \ldots, T_N. \qquad (4.31)$$

For each triangle T_i $(i = 1, \ldots, N)$, we assume that there exists a regular subdomain of the surface

$$f_i : U_i \to \mathbf{R}^3 \quad (i = 1, \ldots, N)$$

such that T_i is a triangle on $f_i(U_i)$ in the sense described above. We assume that the triangles are glued in pairs along edges, and that together they constitute the surface S, possibly with boundary. Namely, S satisfies the following conditions (see Fig. 4.6):

(o) the boundary ∂S of S consists of a union of finitely many spacial regular C^∞-curves,

(i) S is a union of triangles of (4.31), and there exists a connected path along edges between any two vertices,

Fig. 4.6. Triangulations of immersed surfaces.

(ii) each individual triangle T_i does not have self-intersections,

(iii) if an edge of a triangle T_i is also an edge of another triangle T_j, then the two triangles share the two vertices of the shared edge, and $f_i(T_i) \cup f_j(T_j)$ is glued smoothly as a surface along that edge,

(iv) calling an edge of a triangle which is not an edge of a second triangle a *boundary edge*, then the union of all boundary edges coincides with the boundary ∂S of S,

(v) if the surface has self-intersections, they are not considered as edges, and they are interpreted simply as triangles overlapping.

Although we are constructing S here by gluing triangles, in practice we start with a smooth surface S (as an image of an immersion from a two-dimensional manifold), and then determine a triangulation of S to arrive at the above decomposition. Later, when we rigorously prove the Gauss–Bonnet theorem, we use a triangulation of the domain that is a two-dimensional manifold, and not of the image of the surface, and, as a result, consideration of self-intersections is unnecessary. See Chapter 6 for details.

Let P_i ($i = 1, 2, \ldots, n$) be vertices lying on the boundary ∂S of S, and let $\angle P_i$ be the sum of the interior angles at P_i of all the triangles having vertex P_i. We will also consider the case that there is no boundary of S, namely ∂S is empty, in which case S is a closed surface without boundary. The Euler number of S is given by

$$\chi(S) := (\text{number of vertices}) - (\text{number of edges}) + (\text{number of faces}),$$

and is determined by the triangulation, although it is a topological invariant of S, that is, it is independent of the choice of triangulation.

Furthermore, we assume that S is orientable. More precisely,

(vi) giving orientations to all triangles: The orientations of two triangles which share an edge are said to be *compatible* (see Fig. 4.7) if the

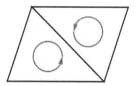

Fig. 4.7.   Triangles with compatible orientations.

orientation along the common edge is opposite. If one can give orientations for all triangles that are everywhere compatible, then S is said to be *orientable*, and the orientation for the triangles is called the *orientation* of S. On the other hand, when such an orientation is not possible, S is said to be *non-orientable*.

When we reverse the orientations of all the triangles, the new orientations are again all compatible. Since a single triangle has only two orientations, and since each choice of orientation fixes the orientations of its neighboring triangles in order for compatibility to hold, employing condition (i), there are only two orientations for the entirety of S. We say that we *fix an orientation of S* when we assign one of those two orientations to S.

With these terminologies in place, we now fix an orientation of S. Now the boundary ∂S can be considered as a smooth path with an orientation. We have the following proposition.

Proposition 4.6.2 (Global Gauss–Bonnet formula for closed regular surfaces). *For an orientable surface S with boundary ∂S and with a triangulation as in* (4.31), *denoting by $\Theta_{\partial S}$ the total geodesic curvature of the boundary ∂S, we have*

$$\int_S K\,dA + \Theta_{\partial S} = 2\pi\chi(S) - \sum_{i=1}^{n}(\pi - \angle\mathrm{P}_i). \tag{4.32}$$

In particular, when ∂S is empty

$$\int_S K\,dA = 2\pi\chi(S), \tag{4.33}$$

where $\int_S K\,dA := \sum_{j=1}^{N} \int_{T_j} K\,dA$.

Proof. Let k be the number of vertices in the triangulation, and we denote those vertices by $\mathrm{V}_1, \ldots, \mathrm{V}_k$. Let m_i be the number of triangles adjacent to each vertex V_i $(i = 1, \ldots, k)$, and denote the triangles adjacent to V_i by $T_{i,1}, \ldots, T_{i,m_i}$.

Taking the sum of all the integral formulas (4.30) for each triangle, we see cancellation of the two total geodesic curvatures along any edge shared by two adjacent triangles. (More explicitly, since the acceleration vector of a curve does not depend on the orientation of the curve, and since the conormals of two triangles which share an edge are opposite along that edge, it follows that the two geodesic curvatures will be equal but will have opposite sign.) Also, the sum of the interior angles (at V_i) of the triangles connecting to a vertex V_i not lying on the boundary is 2π. We denote by $\{P_1, \ldots, P_n\}$ ($n < k$) the vertices on the boundary ∂S. ($\{P_1, \ldots, P_n\}$ is a proper subset of $\{V_1, \ldots, V_k\}$). With $\angle P_i$ ($i = 1, \ldots, n$) as defined above, we have

$$2\pi k - \sum_{l=1}^{n}(2\pi - \angle P_l) = \sum_{i=1}^{k}\sum_{j=1}^{m_i}(\text{the interior angle of } T_{i,j} \text{ at } V_i)$$

$$= (\text{the sum of all interior angles of triangles})$$

$$= \int_S K \, dA + \pi N + \Theta_{\partial S}.$$

On the other hand, an edge that is not on the boundary ∂S is the shared boundary of two geodesic triangles, and each triangle is surrounded by three edges. Since the number of edges on the boundary is n, we have

$$3N = 3 \cdot (\text{the number of triangles}) = 2 \cdot (\text{the number of edges}) - n.$$

Since the number of edges is $(3N + n)/2$, and the number of the vertices is k, we have

$$\int_S K \, dA = -\Theta_{\partial S} + 2\pi k - \pi \cdot (\text{the number of triangles}) - \sum_{l=1}^{n}(2\pi - \angle P_l)$$

$$= -\Theta_{\partial S} + 2\pi \cdot (\text{the number of vertices})$$

$$- 2\pi \cdot (\text{the number of edges})$$

$$+ 2\pi \cdot (\text{the number of triangles}) + \pi n - \sum_{l=1}^{n}(2\pi - \angle P_l)$$

$$= -\Theta_{\partial S} + 2\pi \chi(S) - \sum_{l=1}^{n}(\pi - \angle P_l).$$

This proves the assertion. □

A Gauss–Bonnet type formula for closed wave fronts. Next, we show two Gauss–Bonnet type formulas for closed wave fronts (see Fig. 4.8(a) and (c)). Here we also consider finitely many distinct triangles $T_1, \ldots, T_N$,

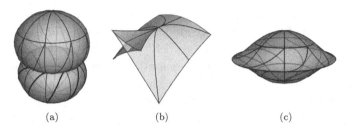

Fig. 4.8.   (a) Singular points at an edge or at a vertex. (b) Swallowtail at a vertex.
(c) is an example of a closed front.

but the meaning of the word "triangles" is now a bit different. We assume
that for each triangle T_i, there exists a front $f_i : U_i \to \mathbf{R}^3$ $(i = 1, \ldots, N)$
such that $T_i \subset f_i(U_i)$. A subset S is called a *triangulated closed front*
(*without boundary*) if it satisfies the conditions (i)–(vi) above with the following additional conditions.

(vii) The interiors of the triangles consist only of regular points, and when
an edge has at least one singular point, then the entire edge consists
of singular points. Moreover, any singular point other than a cuspidal
edge must be a swallowtail, which will only appear at a vertex of the
triangulation.

(viii) The boundary ∂S of S is empty.

(ix) Similarly to the case of regular surfaces, each individual triangle has
no self-intersections. If two triangles intersect, we interpret this as
occurring in the target space, and intersections are not detected on
the source manifold. For example, in the case of a swallowtail, the
swallowtail singular point is a vertex of the triangulation, and the
cuspidal edges connecting to it will be edges of the triangulation, and
then we consider the self-intersections near the swallowtail as just an
intersection of different triangles, and we do not consider them as an
edge of the triangulation (see Fig. 4.8(b)).

For the existence of triangulations satisfying the above conditions, see
Chapter 6.

Moreover, we assume S is orientable, namely (vi) holds, like as in the case
of regular surfaces. Each triangle is a subset of a front, and one can choose a
smooth unit normal vector field. We further assume the following condition:

(x) Properly choosing the directions of the unit normal vectors for all triangles, those normal vectors continuously extend to all of S, giving a
well-defined normal direction on S.

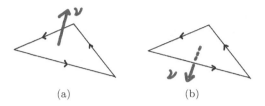

(a) (b)

Fig. 4.9. A positive triangle (a) and a negative triangle (b).

We denote the global unit normal vector field as above by ν. The existence of such a ν is referred to as the *co-orientability* of S. From now on, we assume that S is co-orientable.[2]

When each triangle has an orientation, we can choose the normal vector field such that the orientation is counterclockwise oriented. We call this the *unit normal vector field induced by oriented triangles*. If the surface has no singular points, then all unit normal vector fields induced by oriented triangles can be connected smoothly, and orientability and co-orientability become equivalent notions. However, if we treat a surface with singular points, we need to distinguish these two notions. By co-orientability, a unit normal vector field ν is defined on the entirety of S. On the other hand, fixing an orientation of S, each triangle has an orientation, and the unit normal vector field induced by that orientation is determined. Where that unit normal vector field coincides with ν, the triangle is called a *positive triangle*, and if that unit normal vector field is the opposite of ν, then it is called a *negative triangle* (see Fig. 4.9).

The union of all positive triangles is denoted by S_+, and the union of all negative triangles is denoted by S_-, which we call the *positive part* and *negative part*, respectively. In particular, $S = S_+ \cup S_-$. Setting

$$\Sigma := S_+ \cap S_-,$$

then Σ coincides the set of singular points of S.

Definition of singular curvature. Next, we explain the notion of "singular curvature". Considering a front $f \colon U \to \mathbf{R}^3$ ($U \subset \mathbf{R}^2$ is a domain) so that $f(U) \subset S$ and a C^∞-curve $\gamma \colon [a, b] \to U$ on U, the composition $\hat{\gamma} = f \circ \gamma \colon [a, b] \to \mathbf{R}^3$ is a space curve lying in the image of the surface. We assume $\gamma(t)$ is a parametrization of a cuspidal edge of the surface. Since the singular direction is different from the null direction, the image of the cuspidal edge is regular, namely, $\hat{\gamma}'(t) \neq \mathbf{0}$.

---

[2]An orientable surface may not be co-orientable in general, see (2.32) in Chapter 2.

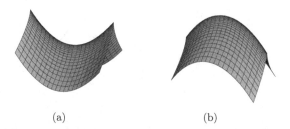

(a)   (b)

Fig. 4.10. Cuspidal edges with negative singular curvature (a) and positive singular curvature (b).

Recall that the positivity and negativity of the Gaussian curvature on a surface reflect geometric behavior: if the shape of the surface is saddle-shaped, then the Gaussian curvature is negative, and if the shape of the surface is bowl-shaped, then the Gaussian curvature is positive (see Fig. 2.1 in Chapter 2). With this in mind, we now consider how to define a new curvature for cuspidal edges having an analogous geometric property to the one above. We introduce an invariant called "singular curvature": the cuspidal edge on Fig. 4.10(a) will have negative singular curvature, and the cuspidal edge on the Fig. 4.10(b) will have positive singular curvature. Let $\hat{\nu}(t)$ be a smooth unit normal vector field of S along $\hat{\gamma}(t)$. Intuitively, we can define the *singular curvature* (function) by

$$\kappa_s(t) := \begin{cases} \Delta(t) & \text{(for a bowl-shaped image of the cuspidal edge),} \\ -\Delta(t) & \text{(for a saddle-shaped image of a cuspidal edge),} \end{cases} \tag{4.34}$$

where (see (5.17) in Chapter 5 for a rigorous definition)

$$\Delta(t) := \left| \frac{\det(\hat{\gamma}'(t), \hat{\gamma}''(t), \hat{\nu}(t))}{|\hat{\gamma}'(t)|^3} \right|.$$

Comparing with the definition of the geodesic curvature (4.29), one can see that this definition is, up to choice of sign, the absolute value of "a function corresponding to the geodesic curvature" along the curve of singular points. In fact, this is an invariant of a singular point of a surface which does not depend on the choice of the parametrization (Proposition 5.4.1 in Chapter 5).

We give here a geometric formula for singular curvature κ_s. We let

$$\gamma : (a, b) \to \mathbf{R}^3$$

be a regular space curve parametrizing a cuspidal edge on a surface S. Here, we consider the case that $\gamma(t)$ ($a < t < b$) is a space curve whose curvature function $\kappa(t)$ does not have zeros. Then the unit tangent vector $e(t)$, the unit principal normal vector $n(t)$ and the unit bi-normal vector $b(t)$ are defined. We consider the normal plane $\Pi(t)$ at $\gamma(t)$ spanned by $n(t)$ and $b(t)$. The slice of S by the normal plane $\Pi(t)$ is a cusp C_t on the plane $\Pi(t)$, as shown in Proposition 2.7.3 in Chapter 2. Then we can take the unique unit vector $v(t)$ at $\gamma(t)$ which points in the direction[3] of the cusp of C_t. Since the plane $\Pi(t)$ is spanned by $n(t)$ and $b(t)$, we can write

$$v(t) = \cos\theta(t)n(t) + \sin\theta(t)b(t), \tag{4.35}$$

where $\theta(t)$ ($a < t < b$) is called the *cuspidal angle* and $\kappa_s(t)$ can be written as

$$\kappa_s(t) = \kappa(t)\cos\theta(t), \tag{4.36}$$

which can be considered as a variant of Meusnier's theorem (see Corollary 5.5.2 in Section 5).

The integral of this curvature is denoted by

$$\Theta_\gamma := \int_a^b \kappa_s(t)\,ds \qquad (ds := |\hat\gamma'(t)|\,dt), \tag{4.37}$$

and is called the *total singular curvature* of γ, with ds called the *line element* of the curve $\hat\gamma(t)$. Upon choosing an orientation of the curve, we can obtain this value as a limit of the total geodesic curvature, as explained below, using the same notation as that for the notion of total geodesic curvature.

Let U be a domain in the uv-plane $\mathbf{R}^2$, $\varepsilon(> 0)$ a real number, and let $\gamma_w : [a,b] \to U$ be a family of regular curves defined for each w ($|w| < \varepsilon$) satisfying $\gamma_0(t) = \gamma(t)$. Letting $f(t,w) := f(\gamma_w(t))$ be a parametrization of a surface f, then γ_w does not intersect the set of singular points of the surface when $w \neq 0$. We have the following proposition.

Proposition 4.6.3. *Suppose that γ_w lies in S_+ if $w > 0$ and lies in S_- if $w < 0$. Then the geodesic curvature*

$$\kappa_g^w(t) := \frac{\hat\gamma_w''(t) \cdot \hat{n}_L(t)}{\hat\gamma_w'(t) \cdot \hat\gamma_w'(t)}$$

with respect to the left-ward co-normal vector field $n_L(t)$ satisfies

$$\lim_{w\to 0^+} \kappa_g^w(t) = \kappa_s(t), \qquad \lim_{w\to 0^-} \kappa_g^w(t) = -\kappa_s(t). \tag{4.38}$$

---

[3]For example, for the standard cusp $\sigma(t) = (t^2, t^3)$, the unit vector v should be $(1,0)$.

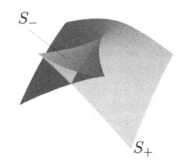

Fig. 4.11. Positive swallowtail (the interior domain of the cusped curve is the domain where $\lambda < 0$).

The fact that the limit $\lim_{w \to 0^-} \kappa_g^w(t)$ is not $\kappa_s(t)$, but rather $-\kappa_s(t)$, arises from the property that the surface is folded to one side of the image of the singular curve along the cuspidal edges. This proposition is essentially the same as Proposition 5.4.3 in Chapter 5 (see also Remark 5.4.4 in Chapter 5), and we give a proof there.

A Gauss-Bonnet type formula. Suppose that a closed front S has only cuspidal edges and swallowtails as singular points. Then, since the number of swallowtails is finite, we can denote them by $P_1, \ldots, P_n$. The set of singular points Σ is a piecewise smooth curve on S, and each interior angle at P_i $(i = 1, \ldots, n)$ from the side of S_+ is either 2π or 0, the first case being called a *positive swallowtail* and the latter a *negative swallowtail* (Fig. 4.11).

Theorem 4.6.4. *Let S be a closed front, with cuspidal edges and swallowtails for singular points. Then $K\, dA$ can be extended continuously to the boundary Σ of each of S_+ and S_-, and the following two formulas hold:*

$$\int_{S_\pm} K\, dA + \Theta_\Sigma = 2\pi\chi(S_\pm) \pm \pi \sum_{i=1}^{n} \operatorname{sgn}(P_i), \qquad (4.39)$$

with $\pm$ all correspondingly the same for each of the two formulas. Here Σ is oriented by regarding both S_+ and S_- as lying to the left-hand side from the standpoint of the direction in which ν points, and then Θ_Σ is the total singular curvature with respect to this orientation. The sign $\operatorname{sgn}(P_i)$ stands for the sign of the swallowtail P_i.

Proof. Since this theorem is a special case of Proposition 6.7.3 and Corollary 6.7.4 in Chapter 6, one can see Section 6.7 for a rigorous proof, and we give only an intuitive proof here. This formula essentially follows by applying the cases of $S = S_+$ and $S = S_-$ to (4.32). In these two formulas, the reason why the two signs of the term $\pi \sum_{i=1}^{n} \mathrm{sgn}(\mathrm{P}_i)$ are opposite is that when setting θ_i to be the interior angle to the side of S_+ at P_i, the interior angle to the side of S_- is $2\pi - \theta_i$, and the reason why the sign of the term Θ_Σ in the left-hand side of (4.39) is the same is the discontinuity of the singular curvature κ_s in the formula (4.38). □

We consider the following total sums of the Gaussian curvature:

$$\int_S K \, dA := \int_{S_+} K \, dA + \int_{S_-} K \, dA,$$

$$\int_S K \, d\hat{A} := \int_{S_+} K \, dA - \int_{S_-} K \, dA.$$

Taking the sum and difference of the two formulas in (4.39), we have the following corollary.

Corollary 4.6.5. *The following two formulas hold:*

$$\frac{1}{2\pi} \left(\int_S K \, dA + 2\Theta_\Sigma \right) = \chi(S), \tag{4.40}$$

$$\frac{1}{2\pi} \int_S K \, d\hat{A} = \chi(S_+) - \chi(S_-) + \sum_{i=1}^{n} \mathrm{sgn}(\mathrm{P}_i). \tag{4.41}$$

This corollary is a special case of Theorem 6.7.5 in Chapter 6. The formula (4.40) is a Gauss–Bonnet type formula for fronts shown by Kossowski [47] in 2002, and one can see that the singular curvature is deeply related to the topology of the surface.[4] In the proof of the global Gauss–Bonnet formula (Proposition 4.6.2), pairs of contributions of geodesic curvatures cancel for the two triangles sharing an edge. However, in the case of cuspidal edges, the surface is folded and S_+ and S_- lie on the same side, and so the geodesic curvatures which originally canceled are now changed to the singular curvature, and not canceling, but rather being counted twice. This contribution to the topology of the surface is seen from the discontinuity (4.38) of the geodesic curvature.

---

[4]In the paper [47], $\kappa_s \, ds$ is treated as a 1-form. The singular curvature is not defined in [47].

The formula (4.41) was pointed out by Langevin *et al.* [49] in 1995. In fact, these two formulas hold even for frontals, not just fronts, as seen in Chapter 6. Regarding the unit normal vector field ν of f as the Gauss map of the closed front, ν can be considered as a frontal. Then we have two Gauss–Bonnet type formulas for ν. Thus, we have four Gauss–Bonnet type formulas in total, and we can obtain various applications. See Chapter 6 for details.

In the case where the surface has a cross cap singular point, then, although it is not even a frontal at that point, the cross cap does not affect the integration of the Gaussian curvature [28].

Chapter 5

Singular Curvature

In this chapter, we give precise definitions of the singular curvature and the limiting normal curvature for cuspidal edges, and introduce fundamental properties of those curvatures. In the latter half of this chapter, we investigate behaviors of these curvatures in relation to the Gaussian curvature near singular points.

5.1. Area Elements

We enter this chapter now assuming that the reader already has a foundation in manifold theory. First, we give a most general form for the definition of singular points:

Definition 5.1.1. A point $p \in \mathcal{M}^m$, in an m-dimensional manifold $\mathcal{M}^m$, is a *singular point* of a map $f : \mathcal{M}^m \to \mathcal{N}^n$ into an n-dimensional manifold $\mathcal{N}^n$ if the rank of the linear map $df_p : T_p\mathcal{M}^m \to T_{f(p)}\mathcal{N}^n$ as the differential of f is strictly less than $\min\{m, n\}$, namely, the smaller value amongst m and n. (This definition is equivalent to the rank of the Jacobi matrix being strictly less than $\min\{m, n\}$, when regarding f as a map from a domain of $\mathbf{R}^m$ into $\mathbf{R}^n$, by using local coordinate systems of $\mathcal{M}^m$ and $\mathcal{N}^n$.) A point $p \in \mathcal{M}^m$ which is not a singular point is called a *regular point*.

Singular points which have been treated in this text so far are all singular points in the sense of the above definition.

Since we treat surfaces in this chapter, we assume that $\mathcal{M}^m$ is an oriented two-dimensional manifold $\mathcal{M}^2$. For a given C^∞-map $f : \mathcal{M}^2 \longrightarrow \mathbf{R}^3$, any of its singular points is, following the above definition, a point p where the rank of the linear map $df_p : T_p\mathcal{M}^2 \to T_{f(p)}\mathbf{R}^3$ is strictly less than 2.

A C^∞-map without any singular points is called an *immersion*. In this chapter, a *"regular surface"* means that we have an immersion from a two-dimensional manifold into $\mathbf{R}^3$.

Here, we give a definition of fronts whose source spaces are manifolds, via first defining frontals. A C^∞-map $f : \mathcal{M}^2 \to \mathbf{R}^3$ is a *frontal* if for any $p \in \mathcal{M}^2$, there exist a neighborhood U of p and a C^∞-map

$$\nu : U \to S^2 \tag{5.1}$$

into the unit sphere $S^2 := \{(x, y, z) \in \mathbf{R}^3 \, ; \, x^2 + y^2 + z^2 = 1\}$ such that for any $q \in U$,

$$df(\mathbf{v}) \cdot \nu(q) = 0$$

holds for any $\mathbf{v} \in T_q U$, where $T_q U$ is the tangent space of U at $q \in U$. The map ν is called the *Gauss map* of f on U. The vector $\nu(q)$, determined for each $q \in U$, is called a *unit normal vector field* if $\nu(q)$ belongs to $T_{f(q)}\mathbf{R}^3$ for each $q \in U$. (In this case, ν can be considered as a vector field of $\mathbf{R}^3$ along the map f.) If the unit normal vector field can be extended smoothly so that it is defined on the entirety of $\mathcal{M}^2$, then f is said to be *co-orientable*.

A frontal $f : \mathcal{M}^2 \to \mathbf{R}^3$ is a *wave front* or *front*, if for any $p \in \mathcal{M}^2$, there exists a neighborhood U of p such that

$$L := (f, \nu) : U \longrightarrow \mathbf{R}^3 \times S^2$$

is an immersion, where ν is a unit normal vector field. In the above definition of frontals, the unit normal vector field does not need to be defined on the entirety of $\mathcal{M}^2$, but for each $p \in \mathcal{M}^2$, there exists a neighborhood U of p such that the restriction map $f|_U$ gives a co-orientable front (i.e., there exists a smooth unit normal vector field of f on U). So, at least, to investigate local properties of wave fronts, we may assume that f is co-orientable. For the sake of simplicity, we treat co-orientable fronts in this text, unless otherwise stated.[1] Moreover, we fix a unit normal vector field ν of f defined on $\mathcal{M}^2$.

For a given frontal $f : \mathcal{M}^2 \to \mathbf{R}^3$, the real-valued function defined by

$$\lambda := \det(f_u, f_v, \nu) \tag{5.2}$$

for a coordinate system $(U; u, v)$ of $\mathcal{M}^2$ is called a *signed area density* on U.

---

[1]Like as in the case of orientability, taking a suitable double covering of the source manifold, we may assume that a given wave front is co-orientable.

As we stated in Section 2.6, in Chapter 2, λ is an identifier of singularities of the map f. We let (u, v) be positively oriented, that is, the coordinates is compatible with respect to the orientation of $\mathcal{M}^2$. The two 2-forms

$$d\hat{A} := \lambda \, du \wedge dv, \quad dA := |\lambda| \, du \wedge dv = |f_u \times f_v| \, du \wedge dv$$

on U are called the *signed area element* and (*unsigned*) *area element* (on U) of f, respectively. These differential forms have the following independence of choice of coordinate systems and choice of unit normal vector field ν:

Proposition 5.1.2. *The definition of $d\hat{A}$ (respectively dA) does not depend on the choice of local coordinate system of $\mathcal{M}^2$ (respectively local coordinate system of $\mathcal{M}^2$ with compatible orientation). In particular, $d\hat{A}$ defines a smooth 2-form on $\mathcal{M}^2$, and dA is continuous on $\mathcal{M}^2$. Moreover, $d\hat{A}$ (respectively dA) changes sign (respectively is unchanged) if we replace ν by $-\nu$.*

Proof. This is equivalent to showing the coordinate independence of the differential form $(d\hat{A} =)\lambda \, du \wedge dv$, and we leave it to the reader as an exercise. Taking the absolute value of a smooth function λ, the differentiability may be lost at points where $\lambda = 0$, and $|\lambda|$ is then merely a continuous function. The last assertion is obvious from (5.2). $\qquad\square$

Proposition 5.1.3. *Let K be the Gaussian curvature of a frontal f which is defined on the set $\mathcal{R}$ of regular points of f. Suppose that $\mathcal{R}$ is open dense in $\mathcal{M}^2$. Then the differential form $K d\hat{A}$ can be smoothly extended to a 2-form defined on all of $\mathcal{M}^2$, including the set of singular points.*

Proof. Regarding ν as a position vector in $\mathbf{R}^3$, it can be identified with the Gauss map $\nu : \mathcal{M}^2 \to S^2$ of f. Taking a coordinate system $(U; u, v)$ that agrees with the orientation of $\mathcal{M}^2$, the Weingarten formula (Proposition 2.1.5 in Chapter 2) implies we can write

$$(\nu_u, \nu_v) = -(f_u, f_v)W, \tag{5.3}$$

where W is the Weingarten matrix. In particular, regarding ν as a map into the unit sphere, ν itself can be regarded as a unit normal vector field of ν, so ν is a frontal. Thus the signed area element for ν is defined by

$$d\hat{A}_\nu := \det(\nu_u, \nu_v, \nu) \, du \wedge dv. \tag{5.4}$$

By (5.3), we have

$$\nu_u \times \nu_v = \det W \det(f_u, f_v, \nu)\nu. \tag{5.5}$$

Thus,

$$d\hat{A}_\nu = \det(\nu_u, \nu_v, \nu)\, du \wedge dv = (\nu_u \times \nu_v) \cdot \nu\, du \wedge dv$$
$$= \det W \det(f_u, f_v, \nu)\, du \wedge dv = \det W\, d\hat{A} \qquad (5.6)$$

holds. Since the definition of the Gaussian curvature is $K = \det W$, we have

$$d\hat{A}_\nu = K d\hat{A} \qquad (5.7)$$

on the set of regular points of f. Since the differential form $d\hat{A}_\nu$ itself is, by Proposition 5.1.2, a smooth form extendable to all of $\mathcal{M}^2$, the form $K\, d\hat{A}$ itself also extends to all of $\mathcal{M}^2$. □

By the above proposition, $d\hat{A}$ is of class C^∞ on $\mathcal{M}^2$, and dA is of class C^0 and might not be differentiable at a singular point. We give here a typical such example:

Example 5.1.4. Consider a cuspidal edge $f(u,v) = (u^2, u^3, v)$. Then $\nu(u,v) := (4 + 9u^2)^{-1/2}(3u, -2, 0)$ is a unit normal vector field, and the two area elements are given by

$$d\hat{A} = u\sqrt{4 + 9u^2}\, du \wedge dv, \qquad dA = |u|\sqrt{4 + 9u^2}\, du \wedge dv.$$

Definition 5.1.5. We denote by $\hat{\Omega}$ the smooth 2-form defined on $\mathcal{M}^2$ as an extension of $K\, d\hat{A}$, and call it the (signed) *Euler form* of f.

The name "Euler form" comes from the fact that the differential form $\hat{\Omega}$ corresponds to the Euler class of the tangent bundle. We set

$$\mathcal{M}^+ := \{\mathrm{P} \in \mathcal{M}^2 \,;\, dA_\mathrm{P} = d\hat{A}_\mathrm{P}\}, \quad \mathcal{M}^- := \{\mathrm{P} \in \mathcal{M}^2 \,;\, dA_\mathrm{P} = -d\hat{A}_\mathrm{P}\}.$$

Then the intersection

$$\Sigma(f) := \mathcal{M}^+ \cap \mathcal{M}^- = \partial\mathcal{M}^+ = \partial\mathcal{M}^-$$

is the singular set of f, where $\partial\mathcal{M}^+$ and $\partial\mathcal{M}^-$ are the boundaries of $\mathcal{M}^+$ and $\mathcal{M}^-$, respectively. From now on, we assume that the singular points of f are all non-degenerate (cf. Section 2.6 in Chapter 2). Then $\Sigma(f)$ is a union of embedded curves on $\mathcal{M}^2$. Each of these curves is called a *singular curve*.

Definition 5.1.6. We consider the 2-form $\Omega := K\, dA$ defined on $\mathcal{M}^2 \backslash \Sigma(f)$, which is called the *unsigned Euler form* of f.

The following example shows that Ω is discontinuous on $\Sigma(f)$, in general.

Example 5.1.7. A map defined by $f(u,v) := (u^2, u^3 + v^2/4, v)$ is a front defined on $\mathcal{M}^2 := \mathbf{R}^2$ whose singular set $\{(0,v)\}$ consists of cuspidal edges. In fact, the unit normal vector field can be chosen as

$$\nu(u,v) := (3u, -2, v)/\Delta(u,v) \qquad (\Delta(u,v) := \sqrt{4 + 9u^2 + v^2}).$$

Since the signed area density function is $\lambda(u,v) = u\Delta(u,v)$, we have $\mathcal{M}^+ = \{(u,v)\,;\, u \geq 0\}$ and $\mathcal{M}^- = \{(u,v)\,;\, u \leq 0\}$. The Gaussian curvature K is computed as $K = 6\Delta(u,v)^{-4}/u$, the Euler form and the unsigned Euler form are computed as

$$\hat{\Omega} = 6\Delta(u,v)^{-3}\, du \wedge dv, \quad \Omega = \begin{cases} 6\Delta(u,v)^{-3}\, du \wedge dv & (\text{on } \mathcal{M}^+ \setminus \Sigma(f)), \\ -6\Delta(u,v)^{-3}\, du \wedge dv & (\text{on } \mathcal{M}^- \setminus \Sigma(f)), \end{cases}$$

where $\Delta(u,v)^{-1}$ is a smooth function on $\mathbf{R}^2$. Since $\Delta(u,0) \neq 0$, the unsigned Euler form Ω is discontinuous at each point of the v-axis.

Remark 5.1.8. As seen in the above example, Ω cannot be extended across $\Sigma(f)$ in general. In fact, the Euler form $\hat{\Omega}$ does not vanish at a generic cuspidal edge, see Theorem 5.2.7 and Proposition 5.2.9.

The following is obtained by definition.

Proposition 5.1.9. *The unsigned Euler form Ω satisfies*

$$\Omega = \begin{cases} K\, d\hat{A}(= \hat{\Omega}) & (on\ \mathcal{M}^+ \setminus \Sigma(f)), \\ -K\, d\hat{A}(= -\hat{\Omega}) & (on\ \mathcal{M}^- \setminus \Sigma(f)). \end{cases} \tag{5.8}$$

As a consequence, the following assertion holds:

Corollary 5.1.10. *The restriction of Ω to $\mathcal{M}^+$ (respectively $\mathcal{M}^-$) can be extended to a smooth 2-form defined on a neighborhood of $\mathcal{M}^+$ (respectively $\mathcal{M}^-$).*

Proof. In fact, the restriction of $\hat{\Omega}$ (respectively $-\hat{\Omega}$) to $\mathcal{M}^+$ (respectively $\mathcal{M}^-$) is Ω. $\qquad\square$

Remark 5.1.11. The point $p \in \Sigma(f)$ where Ω is discontinuous if and only if the Euler form $\hat{\Omega}$ does not vanish at p. For example, a generic cuspidal edge (where $\kappa_\nu \neq 0$ holds) is a such point, (cf. Theorem 5.2.7 and Proposition 5.2.9).

We prove the following assertion as a consequence of Proposition 5.1.3.

Proposition 5.1.12. *The mapping degree[2] $\deg(\nu)$ of the Gauss map ν of the frontal $f : \mathcal{M}^2 \to \mathbf{R}^3$ from a compact orientable two-dimensional manifold $\mathcal{M}^2$ satisfies*

$$2\deg(\nu) = \frac{1}{2\pi} \int_{\mathcal{M}^2} \hat{\Omega} \in 2\mathbf{Z}. \tag{5.9}$$

Proof. Since the surface area of the unit sphere is 4π,

$$\omega := \frac{1}{4\pi} d\hat{A}_\nu,$$

for $d\hat{A}_\nu$ as given by (5.4), is the pull-back of the fundamental 2-form of S^2 (i.e., the 2-form on S^2 that can be integrated to give 1). Thus, by (5.7), we have

$$\deg(\nu) = \int_{\mathcal{M}^2} \omega = \frac{1}{4\pi} \int_{\mathcal{M}^2} \hat{\Omega}. \qquad \square$$

On the other hand, by (5.6), we have the following corollary.

Proposition 5.1.13. *Let $f : \mathcal{M}^2 \to \mathbf{R}^3$ be a frontal, and $\nu : \mathcal{M}^2 \to S^2$ its Gauss map, defined on a compact orientable two-dimensional manifold $\mathcal{M}^2$. Then the signed area element $d\hat{A}_\nu$ and the area element dA_ν of ν as a frontal are expressed as*

$$d\hat{A}_\nu = Kd\hat{A} = \hat{\Omega}, \qquad dA_\nu = |K|\,dA,$$

using the signed area element $d\hat{A}$ and the area element dA of f. In particular, $|K|\,dA$ is a continuous differential form on $\mathcal{M}^2$, where K is the Gauss curvature of f.

Exercises 5.1

1 Let $\mathcal{M}^n$ and $\mathcal{N}^n$ be compact orientable manifolds of the same dimension n. A *regular value* of a smooth map $f : \mathcal{M}^n \to \mathcal{N}^n$ is a point q in $\mathcal{N}^n$ such that all points belonging to the inverse image $f^{-1}(q)$ are regular points (namely, non-singular points) of f. Confirm that the value given by the formula (5.9) for the mapping degree is the same as the sum of the signs of the Jacobian of f at each point in $f^{-1}(q)$.

---

[2]The mapping degree is an integer which expresses how many times the manifold covers the target manifold. See [26] or [55] for details.

5.2. Limiting Normal Curvature

Let U be a domain in the uv-plane $\mathbf{R}^2$, $f : U \to \mathbf{R}^3$ a frontal, and ν its unit normal vector field. Using the signed area density λ given by (5.2), a point p is a singular point of f if and only if $\lambda(p) = 0$. Moreover, p is non-degenerate (cf. Section 2.6 in Chapter 2) if and only if

$$\big(\lambda_u(p), \lambda_v(p)\big) \neq (0,0).$$

By the implicit function theorem, there exists a regular curve $\gamma : (-\varepsilon, \varepsilon) \to U$ ($\varepsilon > 0$) such that $\gamma(0) = p$ and $\lambda(\gamma(t)) = 0 (|t| < \varepsilon)$ hold. As in Section 2.6 in Chapter 2, we call γ a *singular curve* and its tangent vector $\gamma'(t)$ a *singular vector*. Also, the one-dimensional space spanned by the singular vector is called the *singular direction*. Also, as defined in Chapter 2, a non-zero tangent vector $\boldsymbol{v} \in T_p U$ satisfying $df(\boldsymbol{v}) = \boldsymbol{0}$ is called a *null vector*, and the direction defined by $\boldsymbol{v}$ called the *null direction*. Since, at a non-degenerate singular point, the null vector is uniquely determined up to a scalar multiplication, one can take a nowhere-vanishing vector field $\eta(t)$ along $\gamma(t)$ pointing in the null direction. We called this a *null vector field*. Writing

$$\eta(t) = (a(t), b(t)) \left(= a(t)\frac{\partial}{\partial u} + b(t)\frac{\partial}{\partial v} \right),$$

by definition we have

$$a(t)f_u(\gamma(t)) + b(t)f_v(\gamma(t)) = \boldsymbol{0} \qquad (|t| < \varepsilon). \tag{5.10}$$

From now on, we assume that all singular points of f are non-degenerate. A non-degenerate singular point $p \in U$ is said to be of the *first kind* if the null direction and the singular direction are different, and of the *second kind* at p if the null direction and the singular direction coincide. A singular point of the second kind is said to be *admissible* if it is not an accumulation point of the set of singular points of the second kind on the source space. Namely, there are no singular points other than singular points of the first kind near admissible singular points of the second kind. We remark that the definitions of non-degeneracy and of the singular point of the first and second kind do not depend on the choice of coordinate systems on the source and target spaces. So, we can check whether a non-degenerate singular point is of the first or second kind by using the standard forms. We list such examples here, leaving their confirmations to the reader.

Example 5.2.1. A cuspidal edge (the standard form is given in Example 2.5.2 in Chapter 2) is a singular point of the first kind on a front.

A swallowtail (cf. Example 2.5.3 in Chapter 2 for its standard form) is an admissible singular point of the second kind on a front.

Example 5.2.2. A cuspidal cross cap (Example 2.5.5 in Chapter 2 for its standard form) is a singular point of the first kind on a frontal, but is not a front.

Example 5.2.3. The standard form of the cone

$$f(u, v) = (v \cos u, v \sin u, v) \qquad (v \in \mathbf{R}, \ u \in [0, 2\pi))$$

can be considered as a wave front (see Example 2.3.7 in Chapter 2), and has non-degenerate singular points on the set $\{v = 0\}$. Since a null vector field is $\eta = \partial_u$, all singular points are of the second kind, and they are not admissible. The image of the set of singular points $\{v = 0\}$ is the origin $\{(0, 0, 0)\}$. So an arbitrarily given cone is also of the second kind, and is not admissible.

Let p be a singular point of the first kind. Then the null direction and the singular direction are linearly independent. In particular, the space curve

$$\hat{\gamma}(t) := f \circ \gamma(t) \qquad (\gamma(0) = p)$$

as the image of the singular curve γ is regular. We now define the limiting normal curvature along this singular curve. Let $f : U \to \mathbf{R}^3$ be a frontal with a unit normal vector field ν. A real number $\kappa_\nu(t)$ defined by

$$\kappa_\nu(t) := \frac{\hat{\gamma}''(t) \cdot \hat{\nu}(t)}{\hat{\gamma}'(t) \cdot \hat{\gamma}'(t)} \tag{5.11}$$

is called the *limiting normal curvature*, where $\hat{\nu}(t) := \nu \circ \gamma(t)$. (This definition is analogous to the definition of the normal curvature (2.16) in Chapter 2.)

Proposition 5.2.4. *The limiting normal curvature does not depend on the choice of the parameter t of the singular curve.*

Proof. The independence of $\kappa_\nu(t)$ from the choice of parameter can be easily shown by the chain rule. $\qquad \square$

The limiting normal curvature can be defined for singular points of the second kind as well (cf. [54]). Preparing for this, we fix a coordinate system (u, v) centered at a non-degenerate singular point of the first kind p satisfying $f_v(p) = \mathbf{0}$. Since $p = (0, 0)$ is non-degenerate, by Lemma 2.6.1 in Chapter 2 we have $f_u(p) \neq \mathbf{0}$. Then, since

$$f_v \cdot \nu_u = (f_v \cdot \nu)_u - f_{vu} \cdot \nu = -f_{vu} \cdot \nu = -f_{uv} \cdot \nu = f_u \cdot \nu_v, \tag{5.12}$$

the fact that $f_v(p) = \mathbf{0}$ yields

$$f_u(p) \cdot \nu_v(p) = 0. \tag{5.13}$$

Setting $\gamma(t) = (u(t), v(t))$ and $\hat{\gamma}(t) = f(u(t), v(t))$, we have

$$\hat{\gamma}'(0) = u'(0)f_u(p) + v'(0)f_v(p) = u'(0)f_u(p).$$

Since $\hat{\nu}(t) = \nu(u(t), v(t))$, we have

$$\hat{\nu}'(0) = u'(0)\nu_u(p) + v'(0)\nu_v(p).$$

Moreover, we have

$$\hat{\gamma}'' \cdot \hat{\nu} = (\hat{\gamma}' \cdot \hat{\nu})' - \hat{\gamma}' \cdot \hat{\nu}' = -\hat{\gamma}' \cdot \hat{\nu}',$$

and also (5.13), we have $\hat{\gamma}'(0) \cdot \hat{\nu}'(0) = u'(0)^2 f_u(p) \cdot \nu_u(p)$. Hence, we have the expression

$$\frac{\hat{\gamma}''(0) \cdot \hat{\nu}(0)}{\hat{\gamma}'(0) \cdot \hat{\gamma}'(0)} = -\frac{\hat{\gamma}'(0) \cdot \hat{\nu}'(0)}{\hat{\gamma}'(0) \cdot \hat{\gamma}'(0)} = -\frac{f_u(p) \cdot \nu_u(p)}{f_u(p) \cdot f_u(p)}.$$

Continuing from the results above, we can define the limiting normal curvature as

$$\kappa_\nu(p) := -\frac{f_u(p) \cdot \nu_u(p)}{f_u(p) \cdot f_u(p)} \tag{5.14}$$

for a non-degenerate singular point with a coordinate system satisfying $f_u(p) \neq \mathbf{0}$. This coincides with the original definition for a singular point of the first kind. The following proposition guarantees the validity of this definition.

Proposition 5.2.5. *Let p be a non-degenerate singular point. The right-hand side of* (5.14) *does not depend on the choice of coordinate system satisfying* $f_v(p) = \mathbf{0}$.

Thus, the limiting normal curvature can also be defined for a singular point of the second kind.

Proof. We take a coordinate transformation $u = u(\xi, \eta)$, $v = v(\xi, \eta)$, and assume $f_\xi(p) \neq \mathbf{0}$ and $f_\eta(p) = \mathbf{0}$. Because of

$$\mathbf{0} = f_\eta(p) = f_u(p)u_\eta(p) + f_v(p)v_\eta(p) = f_u(p)u_\eta(p),$$

it holds that $u_\eta(p) = 0$. Computing the Jacobi matrix of this coordinate transformation, we have

$$0 \neq u_\xi(p)v_\eta(p) - u_\eta(p)v_\xi(p) = u_\xi(p)v_\eta(p).$$

Noticing that $f_v(p) = \mathbf{0}$, we have

$$f_\xi(p) = f_u(p)u_\xi(p) + f_v(p)v_\xi(p) = f_u(p)u_\xi(p),$$
$$\nu_\xi(p) = \nu_u(p)u_\xi(p) + \nu_v(p)v_\xi(p).$$

Because $f_u(p) \cdot \nu_v(p) = 0$ (cf. (5.13)), we have

$$\frac{f_\xi(p) \cdot \nu_\xi(p)}{f_\xi(p) \cdot f_\xi(p)} = \frac{f_u(p) \cdot \nu_u(p)}{f_u(p) \cdot f_u(p)}.$$

This shows the assertion. $\square$

Extending the definition of the limiting normal curvature in this way, it has the following important geometric meaning.

Proposition 5.2.6. *Let $p \in U$ be a non-degenerate singular point of a front $f : U \to \mathbf{R}^3$. Then the Gauss map $\nu : U \to \mathbf{R}^3$ of f is regular at p if and only if the limiting normal curvature does not vanish at p.*

Proof. We take a local coordinate system (u, v) at p such that $f_v(p) = \mathbf{0}$. Then $\kappa_\nu(p) \neq 0$ is equivalent to

$$f_u(p) \cdot \nu_u(p) \neq 0. \tag{5.15}$$

Since f is a front, the fact $f_v(p) = \mathbf{0}$ implies $\nu_v(p) \neq \mathbf{0}$. Then by (5.12), we have $\nu_v(p) \cdot f_u(p) = 0$. By this with (5.15), we can conclude that $\{\nu_u(p), \nu_v(p)\}$ are linearly independent. $\square$

Furthermore, we show the following theorem.

Theorem 5.2.7. *If p is a non-degenerate singular point of a front, then the following three conditions are equivalent:*

(1) *$\kappa_\nu(p) = 0$,*
(2) *p is a singular point of the Gauss map ν,*
(3) *the Euler form $\hat{\Omega}$ vanishes at p.*

Proof. The equivalency of (1) and (2) was already shown in Proposition 5.2.6. Here we show the equivalency of (2) and (3). By the proof of Proposition 5.1.3,

$$\det(\nu_u, \nu_v, \nu) = K \ \det(f_u, f_v, \nu) \tag{5.16}$$

holds, where K is the Gaussian curvature of f. Thus, the condition that ν is an immersion is equivalent to the left-hand side not vanishing, providing us with the assertion. $\square$

Remark 5.2.8. When p is a regular point, it is well known that the following three assertions are equivalent:

(1) The Gaussian curvature does not vanish at p,
(2) p is a singular point of the Gauss map ν,
(3) the Euler form $\hat{\Omega} = K dA$ vanishes at p.

So, Theorem 5.2.7 can be considered as a generalization of this fact.

As an application, we study the behavior of the Gaussian curvature near a cuspidal edge. For a regular space curve $\hat{\gamma}$, the plane spanned by the tangent vector and the principal normal vector at $\hat{\gamma}(t)$ is called the *osculating plane* of the space curve.

In the case of a singular point of the first kind, the composition $\hat{\gamma}(t) := f \circ \gamma(t)$ of f with the singular curve $\gamma(t)$ on the source space is a regular space curve, and at each point on the curve, a plane which is perpendicular to the normal direction of the surface is determined. We call this plane the *limiting tangent plane* (cf. Section 2.7 in Chapter 2). We consider the case that $\gamma(t)$ consists only of cuspidal edge points. A cuspidal edge is said to be *generic* if its tangential plane and the osculating plane of $\hat{\gamma}(t)$ as a space curve are different, in which case the following holds.

Proposition 5.2.9. *A cuspidal edge is generic if and only if the limiting normal curvature does not vanish.*

Proof. We set $\hat{\nu}(t) := \nu \circ \gamma(t)$. By definition, the limiting normal curvature is zero if and only if $\hat{\gamma}''(t)$ and $\hat{\nu}(t)$ are perpendicular. Since $\hat{\gamma}'(t)$ and $\hat{\nu}(t)$ are perpendicular, and the plane spanned by $\hat{\gamma}'(t)$ and $\hat{\gamma}''(t)$ is the osculating plane, the assertion is clear. $\qquad\square$

Near a non-degenerate singular point, there exists a coordinate system that is convenient for calculating the invariants.

Lemma 5.2.10. *Near a non-degenerate singular point p of a frontal, there exists a coordinate system (u, v) centered at p which satisfies the following:*

(1) *the u-axis is the singular curve, and*
(2) *if p is of the first kind, then $\partial_v := \partial/\partial v$ is the null direction along the u-axis.*

Proof. Let $f(u, v)$ be a frontal defined on a neighborhood of p. Let $\gamma(t)$ be the singular curve on U satisfying $p = \gamma(0)$, and let $\xi(t)$ be a vector field in the uv-plane along $\gamma(t)$ that is linearly independent of $\gamma'(t)$. Then by

$$(u(x,y), v(x,y)) := \gamma(x) + y\xi(x),$$

a coordinate transformation from the xy-plane to the uv-plane can be defined. Then the x-axis corresponds to the singular curve. We would like to show that (x, y) is the desired coordinates. In fact, by the linear independence of $\xi(t)$ and $\gamma'(t)$, we can use the inverse function theorem to see that the map $\varphi(x, y) := (u(x, y), v(x, y))$ is a local diffeomorphism on a neighborhood of the origin. If p is a singular point of the first kind, then $\xi(t)$ can be taken as a null vector field (along $\gamma(t)$), and

$$(f \circ \varphi)_y = df(d\varphi(\partial/\partial y)) = df(\xi) = \mathbf{0}$$

if $y = 0$. Resetting $u := x$ and $v := y$, we obtain the assertion.   □

Theorem 5.2.11 ([69]). *On a generic cuspidal edge, the Gaussian curvature diverges to $+\infty$ on one side of the singular curve, while it diverges to $-\infty$ on the other side of the singular curve.*

Proof. Taking a local coordinate system (u, v) as in Lemma 5.2.10, since $(0, 0)$ gives a generic cuspidal edge of $f(u, v)$, the limiting normal curvature does not vanish. By Theorem 5.2.7, the two vectors $\{\nu_u, \nu_v\}$ are linearly independent on the u-axis. Then the left-hand side of (5.16) does not vanish, and the signed area density $\det(f_u, f_v, \nu)$ changes sign across the u-axis. Thus, we have the assertion.   □

Since the Gaussian curvature of the standard cuspidal edge and the standard swallowtail are identically zero (see Exercise **3** in Section 2.5 of Chapter 2), the two cuspidal edges appearing on these two maps are non-generic. Figure 5.1 shows a generic cuspidal edge. Taking a close look and

Fig. 5.1.   A cuspidal edge whose Gaussian curvature is unbounded.

noting where the surface is concave and convex, the reader can recognize the sign difference of the Gaussian curvature on the opposing sides of the cuspidal edge.

Exercises 5.2

1 The map
$$f(u, v) = (u^2 + v^2, av^2 + v^3, u) \quad (a \in \mathbf{R})$$
from $\mathbf{R}^2$ into $\mathbf{R}^3$ has cuspidal edge points on the u-axis. Show that if $a \neq 0$, then the Gaussian curvature $K(u, v)$ is unbounded near $(0, 0)$. Furthermore, compute $\lim_{v \to 0^+} K$ and $\lim_{v \to 0^-} K$, to verify that the Gaussian curvature diverges with different signs across the singular curve.

5.3. More on Relationships Between Parallel Surfaces and Curvature Lines

At a non-umbilic point on a regular surface, there exist two directions called the principal directions, and they are perpendicular to each other. The principal curvatures are the normal curvatures with respect to these directions (Fact 2.1.12 in Chapter 2). The product of the two principal curvatures is the Gaussian curvature, and their average is the mean curvature. We start by showing the following non-existence of umbilic points on parallel surfaces of a neighborhood of a non-degenerate singular point in a front.

Proposition 5.3.1. *Let U be a domain in the uv-plane. Let $f : U \to \mathbf{R}^3$ be a front, and let $p \in U$ be a non-degenerate singular point. Then p is not an umbilic point of the parallel surface f^t, for all $t \in \mathbf{R}$.*

In fact, by Proposition 2.3.4 in Chapter 2, p is a regular point of f^t for sufficiently small t. Thus, interchanging the role of f and f^t, the proof of the proposition reduces to the proof of the following lemma.

Lemma 5.3.2. *Let $f : U \to \mathbf{R}^3$ be a regular surface, and let $p \in U$ be an umbilic point. If the parallel surface $f^t : U \to \mathbf{R}^3$ of f has a singular point at p, then p is not a non-degenerate singular point.*

Proof. Since p is an umbilic point, the principal curvature $\lambda_0 \in \mathbf{R}$ is uniquely determined. By (2.27) in Chapter 2, f^t has a singular point at p if and only if $\lambda_0 \neq 0$ and $t = 1/\lambda_0$. Then we have $f_u^c(p) = f_v^c(p) = \mathbf{0}$ ($c := 1/\lambda_0$). If p is non-degenerate, then it is impossible for both the u-direction and v-direction to simultaneously be the null direction. This contradiction proves the result. □

Corollary 5.3.3. *A non-degenerate singular point of a front cannot become a non-singular umbilic point on parallel surfaces. In particular, umbilic points cannot accumulate at a non-degenerate singular point.*

Proof. Let $f : U \to \mathbf{R}^3$ be a front and $p \in U$ a non-degenerate singular point. We assume that a sequence $\{p_n\}_{n=1}^{\infty}$ of umbilic points accumulates to the point p. The non-degenerate singular point p becomes a regular point of the parallel surfaces f^t for some values of $t \in \mathbf{R}$ (Proposition 2.3.4 in Chapter 2). By Theorem 2.2.1 in Chapter 2, $\{p_n\}_{n=1}^{\infty}$ is also a set of umbilics of f^t. Then the limit p is also an umbilic point of the regular surface f^t. By Lemma 5.3.2, p cannot be a non-degenerate singular point of $f = (f^t)^{-t}$, giving a contradiction. □

By this corollary, on a neighborhood of a non-degenerate singular point, the two *directional fields* (also called *projective vector fields*, see [84, §15]) of the principal directions are determined. Thus we can show the following corollary shown by Kossowski.

Corollary 5.3.4 ([48]). *Let $\mathcal{M}^2$ be a compact 2-manifold, and $f : \mathcal{M}^2 \to \mathbf{R}^3$ a front which admits only non-degenerate singular points. If its Gaussian curvature on the set of regular points is negative, then the Euler number of $\mathcal{M}^2$ vanishes.*

In Chapter 6 (Corollary 6.9.12), we give an alternative proof of this corollary, as a consequence of the Gauss–Bonnet type formula for fronts.

Proof. Since the Gaussian curvature is negative on the set of regular points of f, there are no umbilics, and the two principal direction fields V_1 and V_2, which are perpendicular each other, are determined on the set of regular points (see Section 15 in [84] for the definition of direction fields). We let $p \in \mathcal{M}^2$ be a singular point of f. By Corollary 5.3.3, there exist a neighborhood U and a real number $t \in \mathbf{R}$ such that by choosing a suitable parallel surface f^t, $f^t|_U$ is an immersion without umbilics. Since the principal directions of f coincide with those of f^t, the two principal direction fields can be extended on U. In particular, two fields can be extended smoothly to all of $\mathcal{M}^2$. Then by the Poincaré–Hopf index theorem for direction fields ([84]), the Euler number of $\mathcal{M}^2$ vanishes. □

Curves on surfaces whose tangent vectors give the principal directions are called *curvature lines*. Considering a principal curvature function along a curvature line, it may diverge at a singular point. As in the case of planar curves, one can define principal curvature maps along the curvature lines. See [59] for details.

5.4. Singular Curvature

In this section, we define the singular curvature along the set of singular points of the first kind, and study its properties.

For a frontal $f : U \to \mathbf{R}^3$ defined on a domain U in the uv-plane $\mathbf{R}^2$, let p be a singular point of the first kind, and let $\gamma(t)$ be the singular curve with $\gamma(0) = p$. Then the zero set of the function

$$\lambda := \det(f_u, f_v, \nu)$$

coincides with the image of $\gamma(t)$ on U. Let $\eta(t)$ be a null vector field. Since p is a singular point of the first kind, the null direction $\eta(t)$ and the singular direction $\gamma'(t)$ are linearly independent. Noticing that the singular curve is a level set of λ, the directional derivative

$$\lambda_\eta := d\lambda(\eta)$$

of the function λ with respect to the vector η does not vanish. Thus its sign

$$\operatorname{sgn}(\lambda_\eta) \in \{1, -1\}$$

can be defined. On the other hand, since $\gamma'(t)$ and $\eta(t)$ are linearly independent as vectors in $\mathbf{R}^2$, the determinant $\det(\gamma'(t), \eta(t))$ resulting by inserting these two vectors into a 2×2 matrix is defined. Setting $\hat{\gamma} := f \circ \gamma$, the value $\kappa_s(t)$ defined by

$$\kappa_s(t) := \varepsilon_\gamma(t) \frac{\det(\hat{\gamma}'(t), \hat{\gamma}''(t), \hat{\nu}(t))}{|\hat{\gamma}'(t)|^3},$$

$$\varepsilon_\gamma(t) = \operatorname{sgn}\Big(\det(\gamma'(t), \eta(t)) \lambda_\eta \Big), \tag{5.17}$$

is called the *singular curvature* of $\hat{\gamma}(t)$ consisting of singular points of the first kind, where $\hat{\nu}(t) = \nu \circ \gamma(t)$.

The singular curvature can be regarded as the "limit of the geodesic curvature adjusted by an appropriate sign $\varepsilon_\gamma(t)$" (see also (4.34) in Chapter 4). If $\hat{\gamma}(t)$ is a regular curve on a regular surface, then the definition of the singular curvature coincides with the geodesic curvature of $\hat{\gamma}(t)$ up to $\pm$ ambiguity (see (2.15) in Chapter 2). In fact, the sign of the geodesic curvature of a curve on a regular surface depends on the choice of the conormal vector $\boldsymbol{n} := \pm \nu \times \gamma'$. The singular curvature κ_s can be interpreted as the "geodesic curvature" of the singular curve when the vector $\boldsymbol{n}$ points toward where the image of the surface exists.

The sign $\varepsilon_\gamma(t)$ may change when $\gamma(t)$ passes through admissible singular points of the second kind, such as swallowtails. On the other hand, if γ consists only of singular points of the first kind, then $\{\gamma'(t), \eta(t)\}$ can be taken so that, for each t, the pair $\{\gamma'(t), \eta(t)\}$ agrees as a positively oriented basis of $\boldsymbol{R}^2$ with the orientation of the uv-plane. Then since $\det(\gamma', \eta) > 0$,

$$\kappa_s = \operatorname{sgn}(\lambda_\eta) \frac{\det(\hat{\gamma}', \hat{\gamma}'', \hat{\nu})}{|\hat{\gamma}'|^3} \tag{5.18}$$

holds. We note that

$$\operatorname{sgn}(\lambda_\eta) = \begin{cases} 1 & \text{(if the left hand side of } \gamma \text{ is } \lambda > 0\text{)}, \\ -1 & \text{(if the left hand side of } \gamma \text{ is } \lambda < 0\text{)}. \end{cases} \tag{5.19}$$

Recall that $\kappa_s(0)$ at the value $t = 0$ is called the "singular curvature" of the singular point p of the first kind. This value can be regarded as an invariant of singular points of the first kind.

Proposition 5.4.1 ([69]). *The singular curvature κ_s of a singular point of the first kind does not depend on the choice of the parametrization of the singular curve, nor the orientation of the surface, nor the choice of a unit normal vector field, nor the choice of a null vector field, nor the choice of the direction of the singular curve. Furthermore, it does not change under translation, rotation or reflection of $\boldsymbol{R}^3$.*

Proof. The independence of κ_s from the choice of parameter for the singular curve can be easily verified by the same method as we used when defining the geodesic curvature of regular surfaces. We show the remaining parts of the claim. By Lemma 5.2.10, we can take a coordinate system (u, v) such that the u-axis is the singular curve. Changing the coordinate system from (u, v) to $(u, -v)$, the sign changes as

$$\lambda \to -\lambda, \qquad \det(\gamma', \eta) \to -\det(\gamma', \eta),$$

and this does not affect the singular curvature. Changing the unit normal vector field ν to $-\nu$ then λ changes to $-\lambda$, and this also does not affect the singular curvature. On the other hand, changing the curve $\gamma(t)$ to $\gamma(-t)$ produces the sign changes

$$\gamma' \to -\gamma', \qquad \eta \to -\eta,$$

and again this does not affect the singular curvature. Furthermore, changing η to $-\eta$ creates the sign changes

$$\operatorname{sgn}(\lambda_\eta) \to -\operatorname{sgn}(\lambda_\eta), \qquad \det(\gamma', \eta) \to -\det(\gamma', \eta)$$

which also do not affect the singular curvature.

It is clear from the definition that the singular curvature does not change under translations of $\boldsymbol{R}^3$. So it is sufficient to show that the singular curvature does not change under orthogonal linear transformations T of $\boldsymbol{R}^3$. We identify T with an orthogonal matrix, and we note that $\det(T) = \pm 1$. If we replace f and ν by Tf and $T\nu$, respectively, then λ_η changes to $\det(T)\lambda_\eta$. Moreover, if we replace $\hat{\gamma}(t)$ by $T\hat{\gamma}(t)$, we have

$$|T\hat{\gamma}'(t)| = |\hat{\gamma}'(t)|, \qquad \det(T\hat{\gamma}', T\hat{\gamma}'', T\hat{\nu}) = \det(T)\det(\hat{\gamma}', \hat{\gamma}'', \hat{\nu}).$$

Thus, Tf and f have the same singular curvature for each singular point. □

Together with the limiting normal curvature, described in Section 5.2 as the oriented length component of $\hat{\gamma}''(s)$ normal to $\hat{\nu}$, we have the following proposition.

Proposition 5.4.2. *The curvature $\kappa(t)$ of a space curve $\hat{\gamma}(t)$ that parametrizes the image of a set of singular points of the first kind satisfies*

$$\kappa(t) = \sqrt{\kappa_s(t)^2 + \kappa_\nu(t)^2}.$$

Proof. Let $e(t)$ be the unit tangent vector of $\gamma(t)$, and take a unit vector $b(t)$ so that $\{e(t), \hat{\nu}(t), b(t)\}$ is an orthonormal basis. Without loss of generality, we may assume t is an arc-length parameter. By the definitions of the singular curvature and the limiting normal curvature, we have

$$|\kappa_s(t)| = |\hat{\gamma}''(t) \cdot b(t)|, \quad |\kappa_\nu(t)| = |\hat{\gamma}''(t) \cdot \hat{\nu}(t)|.$$

Thus, we have

$$\kappa(t) = |\hat{\gamma}''(t)| = \sqrt{\kappa_s(t)^2 + \kappa_\nu(t)^2}. \qquad (5.20)$$

□

Let $\varepsilon > 0$ be a real number. Let $\gamma_w : [a, b] \to U$ be a family of regular curves determined by each w ($|w| < \varepsilon$) satisfying $\gamma_0(t) = \gamma(t)$. We assume $\gamma(t)$ is the singular curve consisting of singular points of the first kind, and the correspondence

$$[a, b] \times (-\varepsilon, \varepsilon) \ni (t, w) \mapsto \gamma_w(t) \in U$$

is a restriction to U of an orientation preserving diffeomorphism from a domain of $\mathbf{R}^2$. The existence of such a map can be shown using the coordinate system given Lemma 5.2.10. Then γ_w does not have any intersections with γ_0 other than $w = 0$. We then take the left-ward co-normal vector field $\mathbf{n}_L^w(t)$ along $\gamma_w(t)$ for $w \neq 0$. Then $\det(\hat{\gamma}_w(t), \hat{\mathbf{n}}_L^w(t), \hat{\nu}(t)) > 0$ holds whenever the signed area density function $\lambda := \det(f_t, f_w, \nu)$ is positive, where $\hat{\mathbf{n}}_L^w(t) := df(\mathbf{n}_L^w(t))$. The following proposition holds.

Proposition 5.4.3. *Suppose that the signed area density $\lambda(t, w)$ is positive on γ_w for $w > 0$ and negative for $w < 0$. Then the geodesic curvature*

$$\kappa_g^w(t) := \frac{\hat{\gamma}_w''(t) \cdot \hat{\mathbf{n}}_L^w(t)}{\hat{\gamma}_w'(t) \cdot \hat{\gamma}_w'(t)} \tag{5.21}$$

of γ_w with respect to the left-ward co-normal vector field $\mathbf{n}_L^w(t)$ satisfies

$$\lim_{w \to 0^+} \kappa_g^w(t) = \kappa_s(t), \qquad \lim_{w \to 0^-} \kappa_g^w(t) = -\kappa_s(t). \tag{5.22}$$

In particular (like for the geodesic curvature), the singular curvature κ_s is determined from merely the information provided by the first fundamental form.

Remark 5.4.4. Since the statement of this proposition is essentially the same as that of Proposition 4.6.3 in Chapter 4, the proof given here implies that a proof of Proposition 4.6.3 in Chapter 4 has now also been given. Later, we shall give an explicit formula for κ_s written in terms of the coefficients of the first fundamental form of f, see Theorem 5.5.4.

Proof. We can write $\mathbf{n}_L^w(t) = a(t, w)\partial_t + b(t, w)\partial_w$ for $w \neq 0$. Since $\mathbf{n}_L^w(t)$ points into the left-hand side of γ in the tw-plane, we have $b(t, w) > 0$. The signed area density $\lambda = \det(f_t, f_w, \nu)$ is positive (resp. negative) for $w > 0$ (resp. $w < 0$). Since (t, w) is a local coordinate system whose orientation coincides with that of (u, v), by setting $\hat{\nu}^w(t) := \nu(t, w)$, it holds that $\lambda_w(t, w) > 0$ and

$$0 < \text{sgn}(w)\lambda(t, w)$$
$$= \text{sgn}(w) \det(\hat{\gamma}_w'(t), f_w(t, w), \hat{\nu}^w(t))$$
$$= \frac{\text{sgn}(w)}{b(t, w)} \det(\hat{\gamma}_w'(t), \hat{\mathbf{n}}_L^w(t), \hat{\nu}^w(t))$$

whenever $w \neq 0$. Since $b(t, w) > 0$, we have

$$\hat{\mathbf{n}}_L^w(t) = \begin{cases} \hat{\nu}^w(t) \times \hat{\gamma}_w'(t)/|\gamma_w'(t)| & (\text{if } w > 0), \\ -\hat{\nu}^w(t) \times \hat{\gamma}_w'(t)/|\gamma_w'(t)| & (\text{if } w < 0). \end{cases}$$

On the other hand, the singular curvature function along $\gamma(=\gamma_0)$ is given by

$$\kappa_s(t) = \mathrm{sgn}(\lambda_w)\frac{\det(\hat{\gamma}_0'(t), \hat{\gamma}_0''(t), \hat{\nu}^0(t))}{|\gamma_0'(t)|^3} = \frac{(\hat{\nu}^0(t) \times \hat{\gamma}_0'(t)) \cdot \hat{\gamma}_0''(t)}{|\gamma_0'(t)|^3}.$$

Then, substituting (5.21) into the expression for $\hat{n}_L^w(t)$ and letting $w \to 0$, we obtain (5.22). $\qquad\Box$

In Section 4.4 of Chapter 4, we defined folds and Whitney cusps as singular points between two domains on $\boldsymbol{R}^2$. The concept of singular curvature can be canonically extended for the definitions of folds and Whitney cusps on C^∞-maps between two manifolds as follows.

Let $\mathcal{M}^2$ and $\mathcal{N}^2$ be two two-dimensional manifolds, and let $f : \mathcal{M}^2 \to \mathcal{N}^2$ be a C^∞-map. If an immersion $g : \mathcal{N}^2 \to \boldsymbol{R}^3$ is given, then the composition of two maps given by

$$h := g \circ f : \mathcal{M}^2 \to \boldsymbol{R}^3$$

is a frontal. The singular point of this map coincides with that of f. A typical example is the case that $\mathcal{N}^2$ is the plane $\boldsymbol{R}^2$. Consider a C^∞-map $f : \mathcal{M}^2 \to \boldsymbol{R}^2$ and

$$g : \boldsymbol{R}^2 \ni (x,y) \mapsto (x,y,0) \in \boldsymbol{R}^3$$

is the inclusion map of the xy-plane in $\boldsymbol{R}^3$, which is an embedding. Then $g \circ f : \mathcal{M}^2 \to \boldsymbol{R}^3$ gives a frontal, since the constant vector field

$$\nu := (0,0,1)$$

gives the unit normal vector field of the map $g \circ f$. We can show the following proposition.

Proposition 5.4.5. *If p is a fold of h, then it is a singular point of the first kind, and if p is a Whitney cusp of h, then it is an admissible singular point of the second kind.*

Proof. It is sufficient to verify this when h is the standard fold map (4.10), and when it is the standard Whitney cusp (4.12) in Chapter 4. We leave this to the reader. $\qquad\Box$

In particular, the singular curvature can be defined on folds as singular curvature of the corresponding singularities of h, and the following proposition holds.

Proposition 5.4.6. *Let a C^∞-map $f : U \to \boldsymbol{R}^2(\subset \boldsymbol{R}^3)$ have a fold at $p \in U$, and let $\gamma(t)$ ($|t| < \varepsilon$) be the singular curve defined on U satisfying*

$\gamma(0) = p$. We orient the image of this curve $\hat{\gamma} := f \circ \gamma$ such that $f(U)$ lies on the left-hand side of $\hat{\gamma}$. Then the singular curvature of f at p coincides with the curvature of $\hat{\gamma}$ at p as a planar curve.

Proof. We may assume that the image of f lies in the xy-plane in $\mathbf{R}^3$. There exists a coordinate system (which agrees with the orientation of $\mathbf{R}^2$) centered at p such that the u-axis is the singular curve, and $\eta := \partial_v$ is a null vector field (see Lemma 5.2.10). Since $\lambda_u(u, 0) = 0$ and $d\lambda \neq \mathbf{0}$, setting $e_3 = (0, 0, 1)$, we have (cf. (5.2))

$$0 \neq \lambda_v(u, 0) = \det(f_u(u, 0), f_{vv}(u, 0), e_3).$$

Since $f_v(u, 0) = \mathbf{0}$, we can write

$$f(u, v) = f(u, 0) + \frac{f_{vv}(u, 0)}{2} v^2 + \text{(the higher order term)}. \tag{5.23}$$

Thus, $f_{vv}(0, 0)$ indicates the direction of $f(U)$. Since $f_u(u, 0) = \hat{\gamma}'(u)$, the fact that $f(U)$ lies on the left-hand side of $\hat{\gamma}$ implies $\lambda_v(u, 0) > 0$.

On the other hand, the singular curvature $\kappa_s(u)$ along this singular curve is given by

$$\kappa_s(u) = \operatorname{sgn}(\lambda_v(u, 0)) \frac{\det(\hat{\gamma}'(u), \hat{\gamma}''(u), e_3)}{|\hat{\gamma}'(u)|^3}.$$

Noting that $\lambda_v > 0$, we have

$$\kappa_s(u) = \frac{\det(\hat{\gamma}'(u), \hat{\gamma}''(u), e_3)}{|\hat{\gamma}'(u)|^3}.$$

This implies that κ_s coincides with the curvature of $\hat{\gamma}$ as a planar curve. $\square$

Now using Proposition 5.4.6, we exhibit a geometric property of the singular curvature of a cuspidal edge.

Proposition 5.4.7. *Let $f : U \to \mathbf{R}^3$ be a front, and ν a unit normal vector field of f. Let $p \in U$ be a cuspidal edge, and let the curve $\gamma(t)$ ($|t| < \varepsilon$) be the singular curve, satisfying $\gamma(0) = p$. We set the image of this curve to be $\hat{\gamma} := f \circ \gamma$, and project f to the limiting tangent plane of f, giving the projected map $g : U \to \mathbf{R}^2$. Then g has folds on γ, and the singular curvature of f at p coincides with the singular curvature of g at p as a fold.*

Proof. The map g is given by

$$g := f - (f \cdot \nu(p))\nu(p).$$

We set $\sigma := g \circ \gamma$. Since $\sigma = \hat{\gamma} - \big(\hat{\gamma} \cdot \nu(p)\big)\nu(p)$, we have

$$\hat{\gamma}'(0) \cdot \nu(p) = \sigma'(0) \cdot \nu(p) = 0,$$

and so $|\sigma'(0)| = |\hat{\gamma}'(0)|$. In particular, we have

$$\frac{\det(\sigma'(0), \sigma''(0), \nu(p))}{|\sigma'(0)|^3}$$

$$= \frac{\det\big(\hat{\gamma}'(0) - (\hat{\gamma}'(0) \cdot \nu(p))\nu(p), \hat{\gamma}''(0) - (\hat{\gamma}''(0) \cdot \nu(p))\nu(p), \nu(p)\big)}{|\hat{\gamma}'(0)|^3}$$

$$= \frac{\det(\hat{\gamma}'(0), \hat{\gamma}''(0), \nu(p))}{|\hat{\gamma}'(0)|^3}.$$

On the other hand, taking a local coordinate system (u, v) centered at p so that its orientation agrees with the orientation of the plane, as given by Lemma 5.2.10, and calculating the Jacobian Λ of g,

$$\Lambda = \det\big(f_u - (f_u \cdot \nu(p))\nu(p), f_v - (f_v \cdot \nu(p))\nu(p), \nu(p)\big) = \det\big(f_u, f_v, \nu(p)\big)$$

holds as a function on U. Noticing that $f_v(p) = \mathbf{0}$, we have

$$\Lambda_v(p) = \det\big(f_u(p), f_{vv}(p), \nu(p)\big) = \lambda_v(p).$$

Thus, the singular curvature μ of g is

$$\mu(0) = \mathrm{sgn}(\Lambda_v(p)) \frac{\det(\sigma'(0), \sigma''(0), \nu(p))}{|\sigma'(0)|^3}$$

$$= \mathrm{sgn}(\lambda_v(p)) \frac{\det(\hat{\gamma}'(0), \hat{\gamma}''(0), \nu(p))}{|\hat{\gamma}'(0)|^3} = \kappa_s(0).$$

This proves the assertion. □

Here, we recall that the limiting tangent plane of a frontal f at $f(p)$ is the plane which is perpendicular to the normal direction at p (cf. Section 2.7 in Chapter 2). In Chapter 4 (see Fig. 4.10), we defined the singular curvature so that it is negative for a saddle-shaped cuspidal edge, and is positive for a bowl-shaped cuspidal edge. As a consequence of Propositions 5.4.6 and 5.4.7, we have the following theorem, which shows that this former definition coincides with the definition given here.

Theorem 5.4.8. *If the singular curvature of a singular point p of the first kind on a frontal $f : \mathcal{M}^2 \to \mathbf{R}^3$ is positive (respectively, negative), then the*

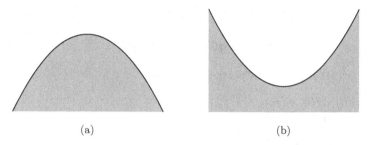

(a)   (b)

Fig. 5.2. Projection of cuspidal edges to the limiting tangent plane, (a) being the case
that the singular curvature is positive, and (b) being the case that it is negative.

*apparent contour (see Section 4.5 in Chapter 4) of the normal projection of
the surface is convex (respectively, concave) (see Fig. 5.2).*

By this theorem, the shape of a cuspidal edge looks like the convex side
of a rounded sword (pirates' sword) if $\kappa_s > 0$, and it looks like the concave
side of a rounded sword if $\kappa_s < 0$ (Fig. 4.10 in Chapter 4).

Proof. By Proposition 5.4.7, the singular curvature of f coincides with the
singular curvature of the apparent contour (see Section 4.5 in Chapter 4)
of the normal projection to the limiting tangent plane of the surface. Moreover, by Proposition 5.4.6, this curvature coincides with the curvature of
the apparent contour as a planar curve when giving it the orientation so
that the image of the surface lies on its left-hand side. This proves the
assertion.   □

Exercises 5.4

1 Show that the image of the singular curve of the tangent developable surface

$$f(t, v) := \hat{\gamma}(t) + v\hat{\gamma}'(t)$$

of the helix $\hat{\gamma}(t) := (a\cos t, a\sin t, bt)$ $(a, b > 0)$ is $\hat{\gamma}(t)$ itself. Calculate the
singular curvature of this curve and show it is negative.

5.5.   Further Properties of Singular Curvature

Firstly, we show that the singular curvature can be determined in terms of
the first fundamental form. The following lemma holds.

Lemma 5.5.1. *Let $f : \mathcal{M}^2 \to \mathbf{R}^3$ be a frontal map of a two-dimensional
manifold $\mathcal{M}^2$, and let $p \in \mathcal{M}^2$ be a singular point of the first kind. Then*

there exists a local coordinate neighborhood $(U; u, v)$ centered at p satisfying the following properties:

(1) *the u-axis parametrizes the singular set,*
(2) $\partial_v := \partial/\partial v$ *points in the null direction,*
(3) *we can write $f_v(u, v) = v\varphi(u, v)$ such that $\{f_u(u, 0), \varphi(u, 0), \nu(u, 0)\}$ gives an orthogonal basis for each point on the u-axis.*

Proof. By Lemma 5.2.10, we can take a local coordinate system $(V; u, v)$ satisfying (1) and (2). We denote by ds^2 the induced metric of f and set

$$ds^2 = E du^2 + 2F du dv + G dv^2.$$

We set $X := \partial_u / \sqrt{E}$ and

$$Y := \partial_v - ds^2(X, \partial_v)X.$$

Then it can be easily checked that $ds^2(X, Y)$ vanishes on V. Applying [84, B.5.4], there exists a local coordinate system (x, y) centered at p such that ∂_x (respectively ∂_y) is a scalar multiple of X (respectively Y). Then the x-axis corresponds to the u-axis, and $f_y(x, 0) = 0$ for each x. So $\varphi := f_y(x, y)/y$ can be smoothly extended to (cf. Appendix A) a vector field defined on a neighborhood of p. Since $ds^2(X, Y) = 0$, we have

$$y f_x \cdot \varphi = f_x \cdot f_y = ds^2(\partial_x, \partial_y) = 0, \quad y\varphi \cdot \nu = f_y \cdot \nu = 0,$$

where ν is the unit normal vector field of f. By the continuity of φ, we have

$$\varphi \cdot f_x = \varphi \cdot \nu = 0.$$

Since p is a non-degenerate singular point, we have

$$0 \neq \lambda_y(p) = \det(f_x(p), f_{yy}(p), \nu).$$

Since $\varphi(0, 0) = f_{yy}(0, 0)$, we may assume that the vector φ does not vanish at each point of V by choosing V to be sufficiently small. So $\{f_x, \varphi, \nu\}$ gives an orthogonal basis at each point of the x-axis. $\qquad \square$

Corollary 5.5.2. *Let $\gamma(t)$ be a space curve whose curvature function does not admit zeros. If γ parametrizes cuspidal edges on a surface S, then the singular curvature $\kappa_s(t)$ of the cuspidal edge is given by*

$$\kappa_s(t) = \kappa(t) \cos\theta(t)$$

where $\theta(t)$ is the cuspidal angle at $\gamma(t)$ (cf. (4.35) in Chapter 4).

Proof. We may assume that $U \subset \mathcal{M}^2$ and $f : (U, u, v) \to \mathbf{R}^3$ is a wave front such that the u-axis corresponds to a cuspidal edge. We denote by $\nu(u, v)$ be the unit normal vector field along f. We may assume that (u, v) satisfies the conditions (1), (2) and (3) of Lemma 5.5.1. Then the space curve

$$\gamma(u) := f(u, 0)$$

parametrizes a cuspidal edge of the surface. We may assume that (u, v) is a local coordinate which is compatible with the orientation of $\mathcal{M}^2$. Since ∂_v gives the null direction, $f_v(u, 0)$ vanishes identically, and so we have the expression

$$f(u, v) = f(u, 0) + \frac{f_{vv}(u, 0)}{2} v^2 + (\text{higher order term}).$$

By (3), $f_{vv}(u, 0)$ points in the cuspidal direction. Since $f_{vv}(u, 0)$ is orthogonal to $f_u(u, 0)$ and $\nu(u, 0)$, the two vectors

$$f_{vv}(u, 0), \qquad (\boldsymbol{w}(u) :=)\nu(u, 0) \times f_u(u, 0)$$

are linearly dependent. Since $f_v(u, 0) = \mathbf{0}$, we have

$$\begin{aligned}
\lambda_v(u, 0) &= \det(f_u(u, v), f_v(u, v), \nu(u, v))_v|_{v=0} \\
&= \det(f_u(u, 0), f_{vv}(u, 0), \nu(u, 0)) \\
&= f_{vv}(u, 0) \cdot (\nu(u, 0) \times f_u(u, 0)) = f_{vv}(u, 0) \cdot \boldsymbol{w}(u).
\end{aligned}$$

Thus,

$$\boldsymbol{v}(u) := \text{sgn}(\lambda_v(u, 0))\nu(u, 0) \times \frac{f_u(u, 0)}{|f_u(u, 0)|} = \text{sgn}(\lambda_v(u, 0))\nu(u, 0) \times e(u) \tag{5.24}$$

gives the unit vector field pointing in the cuspidal direction.

We denote by s the arc-length parameter of γ. Then we have $\gamma' = \dot{\gamma}s'$ and so $\gamma'' = \ddot{\gamma}(s')^2 + \dot{\gamma}s''$, where $\gamma' := d\gamma/du$ and $\dot{\gamma} := d\gamma/ds$. Since

$$s' = |\gamma'|, \quad \dot{\gamma} = e, \quad \ddot{\gamma} = \kappa n,$$

we have that

$$\det(\gamma', \gamma'', \nu) = \kappa|\gamma'|^3 \det(e, n, \nu) = \kappa|\gamma'|^3 n \cdot (\nu \times e) = \kappa|\gamma'|^3 \text{sgn}(\lambda_v) n \cdot v.$$

Since the cuspidal angle θ satisfies $\cos\theta = n \cdot v$, the formula (5.18) for κ_s is

$$\kappa_s = \text{sgn}(\lambda_\eta) \frac{\det(\hat{\gamma}', \hat{\gamma}'', \hat{\nu})}{|\hat{\gamma}'|^3}$$

and is equivalent to the relation $\kappa_s = \kappa \cos\theta$. □

We next prepare the following:

Lemma 5.5.3. *Let U be a domain on the uv-plane $\mathbf{R}^2$. Let $f : U \to \mathbf{R}^3$ be a smooth map with the first fundamental form ds^2 of f written as*

$$ds^2 = E du^2 + F dudv + G dv^2.$$

Then it holds that

$$f_{uu} \cdot f_{vv} - f_{uv} \cdot f_{uv} = F_{uv} - \frac{E_{vv} + G_{uu}}{2}. \tag{5.25}$$

Proof. It holds that

$$f_u \cdot f_{vv} = (f_u \cdot f_v)_v - f_{uv} \cdot f_v = F_v - \frac{(f_v \cdot f_v)_u}{2} = F_v - \frac{G_u}{2}.$$

Using this, we have that

$$
\begin{aligned}
f_{uu} \cdot f_{vv} &= (f_u \cdot f_{vv})_u - f_u \cdot f_{vvu} \\
&= \left(F_v - \frac{G_u}{2} \right)_u - (f_u \cdot f_{uv})_v + f_{uv} \cdot f_{uv} \\
&= F_{uv} - \frac{G_{uu}}{2} - \frac{E_{vv}}{2} + f_{uv} \cdot f_{uv},
\end{aligned}
$$

proving the assertion. $\square$

Using the above lemma, we prove the following:

Theorem 5.5.4. *Let $f : \mathcal{M}^2 \to \mathbf{R}^3$ be a frontal map of a differentiable manifold $\mathcal{M}^2$, and let $p \in \mathcal{M}^2$ be a singular point of the first kind. We let (u, v) be the local coordinate system as in Lemma 5.5.1. Then the singular curvature $\kappa_s(u)$ at $(u, 0)$ is given by*

$$\kappa_s(u) = \frac{-E_{vv}(u, 0)}{\sqrt{2} E(u, 0) \sqrt{G_{vv}(u, 0)}}. \tag{5.26}$$

In particular, the singular curvature is determined by the first fundamental form of f.

Proof. We set $ds^2 = E du^2 + 2F dudv + G dv^2$. Since (u, v) is the local coordinate system as in Lemma 5.5.1, $G = F = F_v = 0$ and $f_v = \mathbf{0}$ along the u-axis. So $F_{vu}(u, 0) = G_{uu}(u, 0) = 0$, $f_{vu}(u, 0) = \mathbf{0}$ and (5.25) reduces to

$$f_{uu}(u, 0) \cdot f_{vv}(u, 0) = -\frac{E_{vv}(u, 0)}{2}. \tag{5.27}$$

Since $\{f_u, f_{vv}, \nu\}$ is a basis of $\mathbf{R}^3$ along the u-axis, we can write

$$f_{uu}(u, 0) = A(u) f_u(u, 0) + B(u) f_{vv}(u, 0) + C(u) \nu(u, 0),$$

and

$$B(u,0) = \frac{f_{uu}(u,0) \cdot f_{vv}(u,0)}{|f_{vv}(u,0)|^2}. \tag{5.28}$$

So we have

$$\kappa_s = \mathrm{sgn}(\lambda_\eta) \frac{\det(\hat{\gamma}', \hat{\gamma}'', \hat{\nu})}{|\hat{\gamma}'|^3} = \mathrm{sgn}(\lambda_v) \frac{\det(f_u(u,0), f_{uu}(u,0), \nu(u,0))}{|f_u(u,0)|^3}$$
$$= \mathrm{sgn}(\lambda_v) B(u) \frac{\det(f_u(u,0), f_{vv}(u,0), \nu(u,0))}{|f_u(u,0)|^3}.$$

Here (5.24) yields

$$\mathrm{sgn}(\lambda_v) B(u) \det\left(\frac{f_u(u,0)}{|f_u(u,0)|}, f_{vv}(u,0), \nu(u,0) \right) = f_{vv}(u,0) \cdot v(u)$$
$$= |f_{vv}(u,0)|.$$

So we have

$$\kappa_s(u) = B(u) \frac{|f_{vv}(u,0)|}{|f_u(u,0)|^2} = B(u) \frac{|f_{vv}(u,0)|}{E(u,0)}.$$

By (5.27) and (5.28), we have

$$\kappa_s(u) = \frac{f_{uu}(u,0) \cdot f_{vv}(u,0)}{E(u,0)|f_{vv}(u,0)|} = -\frac{E_{vv}(u,0)}{2E(u,0)|f_{vv}(u,0)|}.$$

Since

$$f_{vv}(u,v) \cdot f_v(u,v) = \frac{G_v(u,v)}{2},$$

the fact that $f_v(u,0) = 0$ yields

$$f_{vv}(u,0) \cdot f_{vv}(u,0) = (f_{vv}(u,v) \cdot f_v(u,v))_v|_{v=0} = \frac{G_{vv}(u,0)}{2}.$$

So we obtain (5.26). □

Here, we study behaviors of the singular curvature and the Gaussian curvature near a singular point. Firstly, we show the following.

Theorem 5.5.5 ([69]). *At an admissible singular point of the second kind on a frontal, the singular curvature diverges to* $-\infty$. *In particular, if the singular point is a swallowtail, it diverges to* $-\infty$.

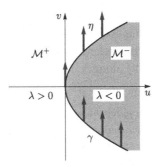

Fig. 5.3. Singular curve of a swallowtail.

Proof. Since p is non-degenerate, the set of singular points can be parametrized by a regular curve $\gamma(t)$ (singular curve). We can take a positive coordinate system (u, v) (i.e., a local coordinate system of $\mathcal{M}^2$ with compatible orientation), so that the v-direction is the null direction and the origin $(0, 0)$ corresponds to the singular point p of the second kind (see Lemma 4.2.2 in Chapter 4). Expressing γ as $\gamma(t) = (u(t), v(t))$ ($|t| < \varepsilon$), and we may assume $\gamma(0) = (0, 0)$ without loss of generality. We may assume that the left-hand side with respect to the direction of travel of γ is the domain $\lambda > 0$. We may assume $\gamma'(0) = (0, 1)$, and then we have $u'(0) = 0$ and $v'(0) = 1$. Moreover, we may also assume $\eta(t) = (0, 1)$ (see Fig 5.3). Since

$$\hat{\gamma}'(t) = u'(t) f_u(u(t), v(t)) + v'(t) f_v(u(t), v(t)) = u'(t) f_u(u(t), v(t)),$$

$$(5.29)$$

differentiating once again,

$$\hat{\gamma}'' = u'' f_u + (u')^2 f_{uu} + u'v' f_{uv}$$

holds. Moreover, since we oriented γ so that its left-hand side is the domain $\lambda > 0$ (see (5.17)), we have

$$(\varepsilon_\gamma(t) =) \operatorname{sgn}(\det(\gamma'(t), \eta(t))) \operatorname{sgn}(\lambda_\eta) = 1.$$

Thus,

$$\kappa_s = \frac{\det(\hat{\gamma}', \hat{\gamma}'', \hat{\nu})}{|\hat{\gamma}'|^3} = \operatorname{sgn}(u') \frac{\det(f_u, f_{uu}, \nu)}{|f_u|^3} + \frac{v' \det(f_u, f_{uv}, \nu)}{|u'||f_u|^3}. \quad (5.30)$$

Since the first term on the right-hand side is bounded, we draw our attention to the second term. Differentiating $\lambda(\gamma(t)) = 0$, we have

$$\lambda_u u' + \lambda_v v' = 0.$$

Since $u'(0) = 0$ and $v'(0) = 1$, it holds that $\lambda_v(0,0) = 0$. By non-degeneracy, $\lambda_u(0,0) \neq 0$. Then, the domain $\lambda > 0$ lies to the left of γ, and the right-hand side is where $\lambda < 0$. In particular, we have $\lambda_u(0,0) < 0$. Since $v' > 0$ and $(0 \neq)\lambda_u = \det(f_u, f_{uv}, \nu)$ along γ, the fact $\lim_{t \to 0} u'(t) = 0$ implies the conclusion. $\qquad\square$

Corollary 5.5.6. *On a neighborhood of an admissible singular point of the second kind, $\kappa_s ds$ gives a smooth 1-form on the singular curve, where $ds := |\dot{\hat{\gamma}}(t)| \, dt$. In particular, let the curve $\gamma : [a,b] \to \mathcal{M}^2$ be a regular curve parametrizing the singular curve such that $\gamma(a)$ is an admissible singular point of the second kind and the set $\gamma((a,b])$ consists of singular points of the first kind. Then the integral (cf. (4.37) in Chapter 4)*

$$\Theta_\gamma := \int_a^b \kappa_s ds$$

is well defined.

Proof. As in the proof of Theorem 5.5.5, the arc-length of $\hat{\gamma}$ is

$$s(t) = \int_0^t |u'(w)| |f_u(w,0)| \, dw$$

by (5.29), and we have $ds = |u'(t)| |f_u(u(t),0)| \, dt$. Thus, by (5.30),

$$\kappa_s \, ds = (u' \det(f_u, f_{uu}, \nu) + v' \det(f_u, f_{uv}, \nu)) \frac{dt}{|f_u|^2}.$$

Since f_u does not vanish near the origin, $\kappa_s ds$ is smooth. $\qquad\square$

Here, we give an example of a swallowtail, with unbounded Gaussian curvature.

Example 5.5.7. The origin of the C^∞-map

$$f^\pm(u,v) := \frac{1}{12}\left(3u^4 - 12u^2v \pm (6u^2 - 12v)^2, 8u^3 - 24uv, 6u^2 - 12v\right)$$

is a swallowtail, and the u-axis is the singular curve. At the origin, the limiting normal curvature κ_ν does not vanish. The Gaussian curvature of f^+ on the domain where $v > 0$ is negative, and the Gaussian curvature of f^+ on the domain where $v < 0$ is positive. Approaching $v = 0$, the Gaussian curvature diverges to $-\infty$ and $+\infty$, respectively. Similarly, the Gaussian curvature of f^- on the domain where $v > 0$ is positive, and the Gaussian curvature of f^- on the domain where $v < 0$ is negative, and in this case, approaching $v = 0$, the Gaussian curvature diverges to $+\infty$ and $-\infty$, respectively. We leave the verification of these facts to the reader as an exercise (Exercise **2** in this section).

To continue our study of the behavior of surfaces near cuspidal edges, we show the following:

Proposition 5.5.8 ([53]). *Let* $f : U \to \mathbf{R}^3$ *be a front, and* $p \in U$ *a cuspidal edge. Then we can take a local coordinate system* (u, v) *centered at* p, *and applying a rotation and translation of* $\mathbf{R}^3$, *we may assume that* f *has the form*

$$f(u, v) = (u, u^2 a(u) + v^2, u^2 b(u) + uv^2 r(u) + v^3 q(u, v)). \tag{5.31}$$

Furthermore, in this expression,

(1) $2a(0)$ *coincides with the singular curvature at* p,
(2) $2|b(0)|$ *is the absolute value of the limiting normal curvature at* p, *and*
(3) $q(0, 0) \neq 0$.

Proof. Taking a local coordinate system (u, v) as in Lemma 5.2.10 centered at p, we write $f = f(u, v)$. By applying a translation of $\mathbf{R}^3$ if necessary, we may set $f(0, 0) = \mathbf{0}$, and by applying a rotation in $\mathbf{R}^3$ if necessary, we may assume

$$f_u(0, 0) = (c_0, 0, 0) \qquad (c_0 > 0),$$

without loss of the generality. Let us set $f(u, v) = (x(u, v), y(u, v), z(u, v))$. Since

$$x_u(0, 0) = c_0 \neq 0,$$

we can take (x, v) as a new local coordinate system of f. This coordinate change still leaves the set of singular points at $v = 0$, and the null direction is the v-direction. Namely, we may assume

$$x(u, v) = u.$$

from the beginning. By the division lemma (Corollary A.2 in Appendix A), we may assume

$$y(u, v) = u^2 a(u) + vA(u, v), \qquad z(u, v) = u^2 b(u) + vB(u, v).$$

Since the set of singular points is the u-axis, and $f_v(u, 0) = \mathbf{0}$, we have that

$$A(u, 0) = B(u, 0) = 0.$$

Thus, again by the division lemma, A and B can be written as

$$A(u, v) = v\alpha(u, v), \qquad B(u, v) = v\beta(u, v).$$

By the assumption that we have a cuspidal edge, at least one of $\alpha(0,0)$ or $\beta(0,0)$ does not vanish. By a rotation of $\boldsymbol{R}^3$ fixing the x-axis, we may assume

$$(\alpha(0,0), \beta(0,0)) = (\alpha_0, 0) \qquad (\alpha_0 > 0). \tag{5.32}$$

The coordinate change $v \mapsto v\sqrt{\alpha(u,v)}$ does not change the fact that the set of singular points is $\{v = 0\}$, nor the fact that the null direction is the v-direction. We have

$$y(u,v) = u^2 a(u) + v^2, \qquad y(u,v) = u^2 b(u) + v^2 \hat{\beta}(u,v),$$

where $\hat{\beta}(u,v)$ is a smooth function defined on a neighborhood of $(0,0)$. Since $\hat{\beta}(0,0) = 0$ (cf. (5.32)), we also have

$$\hat{\beta}(u,v) = ur(u) + vq(u,v).$$

This shows the first assertion of the proposition. With this expression, since the set of singular point of f is still the u-axis,

$$\hat{\gamma}(u) := f(u,0) = (u, u^2 a(u), u^2 b(u))$$

holds. Since

$$f_u(0,0) = (1,0,0), \qquad f_{vv}(0,0) = \lim_{v \to 0} \frac{f_v(0,v)}{v} = (0,2,0),$$

$\nu_0 := (0,0,1)$ is a unit normal vector of f at the origin. Thus we have

$$\kappa_\nu(0) = \frac{\hat{\gamma}''(0) \cdot \nu_0}{|\hat{\gamma}'(0)|^2} = 2b(0).$$

Noting that there is a plus-minus ambiguity in the choice of unit normal vector field, (2) is shown. Since the signed area density $\lambda = \det(f_u, f_v, \nu)$ satisfies

$$\lambda_v(0,0) = \det(f_u(0,v), f_v(0,v), \nu_0)_v|_{v=0}$$
$$= \det(f_u(0,0), f_{vv}(0,0), \nu_0) = 2(> 0),$$

the singular curvature is

$$\kappa_s(0) = \frac{\det(\hat{\gamma}'(0), \hat{\gamma}''(0), \nu_0)}{|\hat{\gamma}'(0)|^3} = 2a(0).$$

This shows (1). Since $f_v(0,v) = \boldsymbol{0}$, f is a front if and only if $\nu_v(0,v) \neq \boldsymbol{0}$, and since

$$(\tilde{\nu}(v) :=) f_u(0,v) \times f_v(0,v)$$
$$= v(-2v^2 \left(vq_u(0,v) + r(0)\right), -v \left(3q(0,v) + vq_v(0,v)\right), 2)$$

and $\nu(0,v) = \tilde{\nu}(v)/|\tilde{\nu}(v)|$, one can verify by direct calculation that the condition $\nu_v(0,0) \neq \boldsymbol{0}$ is equivalent to $q(0,0) \neq 0$. $\qquad\square$

The non-zero real number $\kappa_c(p) := 3q(0,0)/\sqrt{2}$ is an invariant of a cuspidal edge called the *cuspidal curvature*. The following fact holds.

Fact 5.5.9 ([54]). For a cuspidal edge, the product $|\kappa_\nu \kappa_c|$ of the limiting normal curvature and the cuspidal curvature depends only on the first fundamental form.

Since $\kappa_c \neq 0$ for a cuspidal edge, this fact implies that the property of whether κ_ν is zero or not is determined by the first fundamental form. On the other hand, if the Gaussian curvature is bounded in a neighborhood, then κ_ν vanishes along the cuspidal edge, and the Gaussian curvature can take either positive or negative values. In fact, there are rotationally symmetric surfaces with cuspidal edges whose Gaussian curvatures are either positively or negatively constant (see [84], Appendices B-7 and B-8).

Definition 5.5.10. For a given cuspidal edge p, if there exist a positive (respectively, non-negative) number δ and a neighborhood $V(\subset U)$ of p such that the Gaussian curvature K satisfies

$$K(q) \geq \delta \qquad (q \in V \setminus \Sigma),$$

where Σ stands for the set of singular points, then we say that the Gaussian curvature is *positive* (respectively, *non-negative*) at p.

The following holds.

Theorem 5.5.11 ([69]). *If the Gaussian curvature is positive (respectively, non-negative) at a cuspidal edge, then the singular curvature is negative (respectively, non-positive).*

A generalization of this theorem for the behavior of the Gaussian curvature at cuspidal edges in arbitrary Riemannian 3-manifolds is also given in [69]. The reader can understand the meaning of the theorem by taking a look at the surface whose Gaussian curvature is a positive constant, and at the flat front (the tangent developable surface of a helix) in Fig. 5.4.

Proof. Using (5.31), let us calculate the first fundamental coefficients E, F, G. To avoid complexity of calculation, we calculate setting $u = 0$ and regarding them as functions of v. Since

$$f_u(0, v) = (1, 0, r(0)v^2 + v^3 q_u(0, v)),$$
$$f_v(0, v) = (0, 2v, 3v^2 q(0, v) + v^3 q_v(0, v)),$$

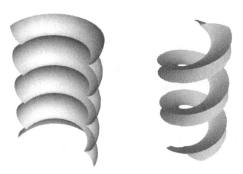

Fig. 5.4.   Evolution surface with $K = 1$ and the tangent developable surface of a helix.

we have

$$E = 1 + v^4 \left(r(0) + vq_u(0, v) \right)^2,$$

$$F = v^4 \left(3q(0, v) + vq_v(0, v) \right) \left(r(0) + vq_u(0, v) \right),$$

$$G = 4v^2 + v^4 \left(3q(0, v) + vq_v(0, v) \right)^2.$$

In particular, $EG - F^2$ can be written as

$$EG - F^2 = (\text{a positive function}) \cdot v^2. \qquad (5.33)$$

The properties that the Gaussian curvature is positive or negative or divergent do not change when multiplying K by a positive function. Thus, to show the theorem, it is enough to show that the sign of the limits $v \to 0^+$ and $v \to 0^-$ are different for the function $\varphi(v) := (LN - M^2)/v^2$. Since

$$\tilde{\nu}(v) := \frac{f_u(0, v) \times f_v(0, v)}{v}$$

$$= (-2v^2 r(0) - 2v^3 q_u(0, v), -3vq(0, v) - v^2 q_v(0, v), 2),$$

the unit normal vector field $\nu = \tilde{\nu}/\sqrt{\tilde{\nu} \cdot \tilde{\nu}}$ is defined even at $v = 0$. The inner products of f_{uu}, f_{uv}, f_{vv} and ν are the coefficients of the second fundamental form. We instead consider

$$\tilde{L} := f_{uu} \cdot \tilde{\nu}, \quad \tilde{M} := f_{uv} \cdot \tilde{\nu}, \quad \tilde{N} := f_{vv} \cdot \tilde{\nu}.$$

Since these three functions differ from L, M, N only by multiplication by a positive function, the sign of

$$\psi(v) := \frac{\tilde{L}\tilde{N} - \tilde{M}^2}{v^2} \qquad (5.34)$$

is equal to the sign of the Gaussian curvature K. Setting

$$c_1(v) := 2\Big(b(0) + v^2 r_u(0) \Big) + v^3 q_{uu}(0, v),$$

$$c_2(v) := v(2r(0) + 3vq_u(0, v) + v^2 q_{uv}(0, v)),$$

$$c_3(v) := v(6q(0, v) + 6vq_v(0, v) + v^2 q_{vv}(0, v)),$$

it holds that

$$f_{uu}(0,v) = (0, 2a(0), c_1(v)), \quad f_{uv}(0,v) = (0, 0, c_2(v)),$$
$$f_{vv}(0,v) = (0, 2, c_3(v)).$$

Thus,

$$\tilde{L}(0,v) = 2(2b(0) - a(0)v\,(3q(0,v) + vq_v(0,v)) + 2v^2r_u(0) + v^3q_{uu}(0,v)),$$
$$\tilde{M}(0,v) = 2v(2r(0) + 3vq_u(0,v) + v^2q_{uv}(0,v)),$$
$$\tilde{N}(0,v) = 2v(3q(0,v) + v\,(5q_v(0,v) + vq_{vv}(0,v)))$$

hold. In particular,

$$\tilde{L}(0,v) = 4b(0) - 6a(0)q(0,0)v + \text{(higher order terms)}, \qquad (5.35)$$
$$\tilde{M}(0,v) = 4r(0)v + \text{(higher order terms)}, \qquad (5.36)$$
$$\tilde{N}(0,v) = 6q(0,0)v + \text{(higher order terms)}. \qquad (5.37)$$

Then the cuspidal edge at $(0,0)$ is not generic, and in particular $b(0) = 0$. Thus, by (5.35), (5.36) and (5.37),

$$\psi(0) = -9a(0)q(0,0)^2 - 4r(0)^2$$

holds. If K is positive (respectively, non-negative), then $\psi(0) > 0$ (respectively, $\psi(0) \geq 0$). Then $a(0) < 0$ (respectively, $a(0) \leq 0$) holds. Noting that $2a(0) = \kappa_s(0)$, we have the conclusion. $\qquad \square$

Similar discussions like as in this section for cross caps are given in Appendix F. Further geometric properties at singular points are discussed in [33, 63, 53, 54, 72] and also [36, 21, 37, 38], and especially [37] is a book related to the subjects in this text.

Exercises 5.5

1 Consider the map

$$f(u,v) = \left(v + \frac{u^2}{2} - \frac{b^2u^2v}{2} - \frac{b^2u^4}{8},\ \frac{bu^3}{3} + buv,\ \frac{cv^2}{2} \right),$$

where b, c are positive numbers.

(1) By calculating $f_u \times f_v$, find a unit normal vector field of f.

(2) Show that the set of singular points of f is the u-axis and an identifier of singularities can be taken as $\Lambda = u$. Moreover, show that $\partial_u - u\partial_v$ gives an extended null vector field. Using them, show that the origin is a swallowtail.

(3) Find the Gaussian curvature K of f.

(4) Calculate the limiting normal curvature κ_ν and the singular curvature κ_s of f.

2 Show that the map

$$f^{\pm}(u,v) = \frac{1}{12}\left(3u^4 - 12u^2v \pm (6u^2 - 12v)^2, 8u^3 - 24uv, 6u^2 - 12v\right)$$

in Example 5.5.7 has a swallowtail at the origin, and show that the limiting normal curvature does not vanish there. Moreover, show that the Gaussian curvature changes sign across the set of singular points.

Chapter 6

Gauss–Bonnet Type Formulas and Applications

In this chapter, we give proofs for the Gauss–Bonnet-type formulas for wave fronts and frontals, which were described briefly at the end of Chapter 4. In the latter half of the chapter, applications of the formulas are introduced.

6.1. Triangulation of Two-Dimensional Manifolds

In this chapter, the two-dimensional manifold $\mathcal{M}^2$ is assumed to be a Hausdorff space satisfying the second countability axiom.

Take three distinct points A, B, C in the plane $\mathbf{R}^2$ in general position. We call the closed domain bounded by the line segments AB, BC and CA a *linear triangle* or a *linear triangular domain*, and denote it by $\triangle$ABC.

Definition 6.1.1. A closed domain T on a two-dimensional manifold $\mathcal{M}^2$ is called a *triangle* if there exists a local coordinate system $\varphi : U \to \mathbf{R}^2$ (i.e., a diffeomorphism from U to $\varphi(U)$) containing T such that $\varphi(T)$ is a linear triangle in $\mathbf{R}^2$.

In the setting of Definition 6.1.1, the interior of T is mapped to the interior $\varphi(T)$. The *vertices* (respectively, *edges*) of T are the points corresponding to the vertices (respectively, edges) of the linear triangle $\varphi(T)$. Identifying the vertices A, B, C of the linear triangle and those of T, we write[1]

$$T = \triangle\text{ABC}.$$

---

[1] In this chapter, we denote points on the manifold $\mathcal{M}^2$ by upright-font upper case letters (e.g., P) instead of lower case letters (e.g., p).

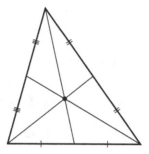

Fig. 6.1.   Barycentric refinement of a triangle.

A *Riemannian manifold* is a manifold endowed with a *Riemannian metric*, that is, a positive definite symmetric covariant 2-tensor on $\mathcal{M}^2$. When a 2-manifold is realized in $\boldsymbol{R}^3$ by an immersion, the first fundamental form is a Riemannian metric. Since a triangle on a Riemannian manifold is identified with a linear triangle by an appropriate local coordinate system, we have the following proposition.

Proposition 6.1.2. *Each interior angle of a triangle on a two-dimensional Riemannian manifold is positive and less than π.*

From now on, we assume that the manifold $\mathcal{M}^2$ is two-dimensional without boundary, which is compact, orientable and can assign a given Riemannian metric ds^2 if necessary.

The *barycentric refinement of a linear triangle* is a subdivision of a triangle into six triangles by line segments meeting the barycenter and the midpoints of the edges (cf. Fig. 6.1). Using local coordinate systems, one can then divide a triangle on $\mathcal{M}^2$ into six triangles.

A regular closed curve on a 2-manifold is said to be *simple* if it has no self-intersections. Let D be a domain in $\mathcal{M}^2$ whose closure $\bar{D}$ is compact, so that its boundary ∂D is the disjoint union of a finite number of simple regular curves (when ∂D is not empty).

Definition 6.1.3. A collection of a finite number of triangles $\{T_i\}_{i=1,...,N}$ on $\mathcal{M}^2$ is a *triangulation* of $\bar{D}$ if it satisfies the following:

(1) $\bar{D} = T_1 \cup \cdots \cup T_N$ and ∂D consists of edges and vertices of the triangles, when it is non-empty, and
(2) the intersection $T_i \cap T_j$ ($i \neq j$) of each pair of two distinct triangles T_i and T_j consists of a common edge or a common vertex of T_i and T_j, whenever it is not empty.

The existence of the triangulation is stated as follows.

Fact 6.1.4. Let D be a domain with compact closure such that either $\bar{D} = \mathcal{M}^2$ or the boundary ∂D consists of the disjoint union of a finite number of simple closed curves. Then, for any given finite number of points in ∂D, there exists a triangulation of $\bar{D}$ such that these given points lie in the collection of vertices of the triangles.

The proof for the case $\partial D = \emptyset$ can be found in [55, Chapter 1]. For the case that $\partial D \neq \emptyset$, it is given in [60, Theorem 10.6].

We fix a triangulation of $\bar{D}$ as in Fact 6.1.4. Since we have assumed that the manifold $\mathcal{M}^2$ is oriented, each coordinate system can be chosen so that it is compatible with the orientation of the manifold. Using these coordinate systems, the boundary of each triangle can be oriented such that the interior of the triangle is on the left-hand side of its boundary. When the boundaries of all triangles are oriented in this manner, common edges of two adjacent triangles have opposite orientations. We call such an orientation the *natural orientation* of the triangulation. We will assume the natural orientation of triangulations in this chapter.

6.2. Gauss–Bonnet Theorem for Two-Dimensional Manifolds

Let $(\mathcal{M}^2, ds^2)$ be an oriented Riemannian 2-manifold, and consider a triangle $\triangle ABC$ with vertices A, B, C. We assume that the triangle lies on a coordinate neighborhood $(U; u, v)$ compatible with the orientation of $\mathcal{M}^2$, and that the boundary of each triangle is oriented such that the interior of the triangle is on the left-hand side of its bounding edges. Let

$$\gamma_i(t) \qquad (0 \leq t \leq 1) \tag{6.1}$$

$(i = 1, 2, 3)$ be regular curves on U such that γ_1, γ_2 and γ_3 parametrize the edges AB, BC and CA, respectively. In addition, take the unit left-ward normal vector fields on the uv-plane

$$\boldsymbol{n}_i^L(t) \qquad (i = 1, 2, 3)$$

along each γ_i, pointing into the interior of the triangle so that $ds^2(\boldsymbol{n}_i^L(t), \boldsymbol{n}_i^L) = 1$ and $ds^2(\boldsymbol{n}_i^L(t), \gamma_i'(t)) = 0$ $(t \in [0, 1])$ for each $i = 1, 2, 3$. We call $\boldsymbol{n}_i^L(t)$ the *left-ward co-normal vector field* along γ_i. Denote the geodesic curvature function of γ_i by (cf. (4.29))

$$\kappa_i(t) := \frac{ds^2(\gamma_i''(t), \boldsymbol{n}_i^L(t))}{ds^2(\gamma_i'(t), \gamma_i'(t))} \qquad (i = 1, 2, 3) \tag{6.2}$$

(see [84, Equation (13.22)]). Here, we have denoted by $\gamma_i''(t)$ the acceleration vector field $\nabla_{\gamma_i'(t)}\gamma_i'(t)$ of $\gamma_i(t)$ with respect to the metric. (∇ is the Levi-Civita connection of the Riemannian metric ds^2.)

Remark 6.2.1. Let $f : \mathcal{M}^2 \to \mathbf{R}^3$ be an immersion and ds^2 its induced metric (the first fundamental form). Choose the unit normal vector field ν appropriately and set $\hat{\gamma}_i = f \circ \gamma_i$. Then, for a triangle $\triangle ABC$ in $\mathcal{M}^2$, it holds that

$$df\left(\boldsymbol{n}_i^L(t)\right) = \varepsilon \nu\left(\gamma_i(t)\right) \times \frac{\hat{\gamma}_i'(t)}{|\hat{\gamma}_i'(t)|}, \qquad \varepsilon := \mathrm{sgn}\left(\det(\hat{\gamma}_i', df(\boldsymbol{n}_i^L), \nu \circ \gamma_i)\right),$$

and hence

$$
\begin{aligned}
\kappa_i(t) &= \frac{ds^2(\gamma_i''(t), \boldsymbol{n}_i^L(t))}{ds^2(\gamma_i'(t), \gamma_i'(t))} = \frac{\hat{\gamma}_i''(t) \cdot df(\boldsymbol{n}_i^L(t))}{|\hat{\gamma}_i'(t)|^2} \\
&= \frac{\varepsilon}{|\hat{\gamma}_i'(t)|^2}\hat{\gamma}_i''(t) \cdot \left(\nu(\gamma_i(t)) \times \frac{\hat{\gamma}_i'(t)}{|\hat{\gamma}_i'(t)|}\right) \\
&= \frac{\varepsilon \det\left(\hat{\gamma}_i'(t), \hat{\gamma}_i''(t), \nu(\gamma_i(t))\right)}{|\hat{\gamma}_i'(t)|^3} \qquad (i = 1, 2, 3).
\end{aligned}
$$

That is, (6.2) coincides with the geodesic curvature defined in (2.15) of Chapter 2 for regular surfaces up to a sign.

When s is an arc-length parameter of $\hat{\gamma}$, we set

$$\int_{\partial \triangle ABC} \kappa_g \, ds := \sum_{i=1}^{3} \int_0^1 \kappa_i(t)|\gamma_i'(t)| \, dt,$$

noting that $ds = |\hat{\gamma}_i'(t)| \, dt$. This is the sum of the integrals of the geodesic curvature functions of the three edges. So, using the notation in Section 4.6 in Chapter 4, we can write

$$\int_{\partial \triangle ABC} \kappa_g \, ds = \Theta_{AB} + \Theta_{BC} + \Theta_{CA}. \tag{6.3}$$

We have the following:

Fact 6.2.2 ([84, Proposition 13.4]). Let $\triangle ABC$ be a triangle of a Riemannian 2-manifold $\mathcal{M}^2$, and denote by $\angle A$, $\angle B$ and $\angle C$ the interior angles of the domain at A, B and C, respectively. Then

$$\angle A + \angle B + \angle C - \pi = \int_{\partial \triangle ABC} \kappa_g \, ds + \int_{\triangle ABC} K \, dA. \tag{6.4}$$

6.3. Angles and the Area Element for Frontals

Let

$$f : \mathcal{M}^2 \longrightarrow \boldsymbol{R}^3$$

be a frontal whose singular points are all non-degenerate and either of the first kind or second kind with admissibility (cf. Section 5.2). The first fundamental form ds^2, by definition, satisfies

$$ds^2(\boldsymbol{v}, \boldsymbol{w}) = df(\boldsymbol{v}) \cdot df(\boldsymbol{w}) \qquad (6.5)$$

for any tangent vectors $\boldsymbol{v}$, $\boldsymbol{w}$ at each given point $P \in \mathcal{M}^2$. In this chapter, we assume there exists a globally defined unit normal vector field ν of the frontal f. Regarding ν as a unit vector based at the point O (the origin) by a translation, it can be identified with the Gauss map

$$\nu : \mathcal{M}^2 \longrightarrow S^2 \subset \boldsymbol{R}^3. \qquad (6.6)$$

Take two non-zero vectors $\boldsymbol{a}, \boldsymbol{b} \in T_P \mathcal{M}^2$ at $P \in \mathcal{M}^2$. When P is a singular point of f, we also assume that $\boldsymbol{a}, \boldsymbol{b}$ are not proportional to (i.e., do not point in) the null direction. We define the angle between these two vectors by

$$\angle(\boldsymbol{a}, \boldsymbol{b}) := \arccos\left(\frac{df(\boldsymbol{a}) \cdot df(\boldsymbol{b})}{|df(\boldsymbol{a})||df(\boldsymbol{b})|}\right) \in [0, \pi]. \qquad (6.7)$$

This definition is compatible with the definition of the angle in the Gauss–Bonnet type formulas in Section 4.6. We have the following lemma, immediately from the definition.

Lemma 6.3.1 (Continuity of angles). *Let $\gamma(t)$ be a continuous curve in the uv-plane defined on an interval $[0, c]$, and let $\boldsymbol{a}, \boldsymbol{b} \colon [0, c] \to \boldsymbol{R}^2$ be continuous vector fields along γ which do not vanish everywhere. If $\boldsymbol{a}(t), \boldsymbol{b}(t)$ are not proportional to the null direction for each t, then*

$$\lim_{t \to 0} \angle\big(\boldsymbol{a}(t), \boldsymbol{b}(t)\big) = \angle\big(\boldsymbol{a}(0), \boldsymbol{b}(0)\big).$$

Definition 6.3.2. Let P be a singular point of f. When there exists a regular curve in the uv-plane such that the tangent vector of the curve is proportional to the null direction (i.e. a non-zero vector belonging to the kernel of df) at P, we call it a *null curve* at P.

By definition, there exists a null curve at any non-degenerate singular point.

Lemma 6.3.3. *Let $P \in U$ be a singular point, and take linearly independent vectors $\boldsymbol{a}$ and $\boldsymbol{b}$ at P which are not proportional to the null direction at P.*

Then the angle $\angle(\boldsymbol{a}, \boldsymbol{b})$ of these two vectors is π (respectively, 0) if these vectors are on opposite sides (respectively, the same side) of a null curve at P.

Proof. Since the metric degenerates at P, the angle is either 0 or π. If there are no null curves between the two vectors, the continuous deformation

$$\boldsymbol{c}_t := (1 - t)\boldsymbol{a} + t\boldsymbol{b} \qquad (0 \le t \le 1)$$

of the two vectors is not proportional to the null direction for each t. Hence the angle $\angle(\boldsymbol{a}, \boldsymbol{c}_t)$ is continuous in t. Hence, by continuity of the angle (Lemma 6.3.1), we can conclude that $\angle(\boldsymbol{a}, \boldsymbol{b}) = 0$. On the other hand, when the two vectors are on opposite sides of a null curve, there are no null curves between $\boldsymbol{a}$ and $-\boldsymbol{b}$. Therefore, $\boldsymbol{a}$ can be deformed to $-\boldsymbol{b}$, and the conclusion follows. $\qquad\square$

Now we review the two area elements in Section 5.1 of Chapter 5. For a local coordinate system $(U; u, v)$ compatible with the orientation of $\mathcal{M}^2$, we called

$$d\hat{A} := \det(f_u, f_v, \nu)\, du \wedge dv$$

the "signed area element", and

$$dA := |\det(f_u, f_v, \nu)|\, du \wedge dv$$

the "area element". These are both differential forms on $\mathcal{M}^2$, and in particular, $d\hat{A}$ is of class C^∞ on $\mathcal{M}^2$, by Proposition 5.1.2. On the other hand, dA is of class C^0 and might not be differentiable at a singular point. We set

$$\mathcal{M}^+ := \{P \in \mathcal{M}^2 \,;\, dA_P = d\hat{A}_P\}, \quad \mathcal{M}^- := \{P \in \mathcal{M}^2 \,;\, dA_P = -d\hat{A}_P\}.$$
(6.8)

Then $\mathcal{M}^2 = \mathcal{M}^+ \cup \mathcal{M}^-$ and the singular set of f is

$$\Sigma(f) := \mathcal{M}^+ \cap \mathcal{M}^-.$$

As we are assuming that the manifold $\mathcal{M}^2$ is compact without boundary, the fact that the singular points of f are all non-degenerate implies that $\Sigma(f)$ is a union of regular simple closed curves on $\mathcal{M}^2$. We denote by K the Gaussian curvature of f defined on $\mathcal{R}^2 := \mathcal{M}^2 \setminus \Sigma(f)$. As shown in Proposition 5.1.3, $\hat{\Omega} := K\, d\hat{A}$ can be smoothly extended on $\mathcal{M}^2$, which is called the (signed) "Euler form" of f. On the other hand, the 2-form given by

$$\Omega := K\, dA$$

defined on $\mathcal{R}^2$ is called the "unsigned Euler form" of f.

6.4. Admissible Triangulations

As in the previous section, let $f : \mathcal{M}^2 \to \mathbf{R}^3$ be a co-orientable frontal (i.e. a C^∞-map with unit normal vector field on $\mathcal{M}^2$) defined on a compact oriented 2-manifold $\mathcal{M}^2$, whose singularities are non-degenerate, each of them being either of the first kind or an admissible singularity of the second kind (cf. Section 5.2 in Chapter 5). The compactness of $\mathcal{M}^2$ implies that the singular set $\Sigma(f)$ of f consists of closed curves embedded in $\mathcal{M}^2$. The closed subsets $\mathcal{M}^+$ and $\mathcal{M}^-$ are two-dimensional manifolds with boundary; hence, they can be triangulated, by Fact 6.1.4. We also have $\Sigma(f) = \mathcal{M}^+ \cap \mathcal{M}^-$, telling us that $\Sigma(f)$ is the union of a finite number of edges. Since admissible singularities of the second kind are isolated amongst singularities of the second kind, compactness of $\mathcal{M}^2$ yields that

- The number of admissible singular points of f of the second kind is finite.

Hence we may assume that singularities of the second kind are vertices of the triangulation. Using barycentric refinement if necessary, we may assume that such a triangulation has the following properties:

(a) The singular points are on edges of triangles. In particular, the interior of each triangle consists of regular points, and all points of an edge consists of a singular curve if it bounds two singular vertices or the edge contains any singular point at all.

(b) Each singular point of the second kind is a vertex of a triangle.

Applying barycentric refinements, we can obtain a triangulation such that each triangle is one of the following types (Fig. 6.2):

- A triangle whose vertices are all regular points: In this case, both the boundary edges and the interior consist of regular points. We call such a triangle a *regular triangle*

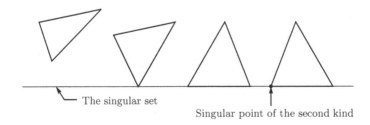

The singular set

Singular point of the second kind

Fig. 6.2.   A regular triangle, a quasi regular triangle, a singular triangle of type I, and a singular triangle of type II.

- A triangle having only one singular point at one vertex: In this case, all points except that one singular vertex are regular points. We call such a triangle a *quasi regular triangle*. The singular vertex can be a singular point of the second kind.
- A triangle having only one regular vertex amongst the three vertices: In this case, the edge joining the other two vertices must be a singular curve. We call such an edge a *singular edge*. The points other than these on the singular edge are regular points. We call such a triangle a *singular triangle*.

We fix a triangle $\triangle ABC$ such that the vertex A is a singular point, and assume the edge AB is not a singular edge. When the tangent vector at A of the edge AB coincides with the null direction, one can perturb the tangent vector of AB at A while preserving the ends A and B. Since the number of triangles is finite, we can apply this operation for all edges whose endpoints are singular. Thus, we may assume, without loss of generality:

(c) For each quasi regular or singular triangle, the tangent vector of any non-singular edge at a singular endpoint does not coincide with the null direction.

Moreover, we can assume that:

(d) Each triangle has at most one singular edge.
(e) All singular points of the second kind are vertices of triangles, and all edges have at most one singular point of the second kind.

In fact, if an admissible singular point of the second kind appears on an edge at a point that is not a vertex, by inserting an edge to join this singular point with the third vertex of the triangle not lying on the singular edge, we can arrive at a triangulation satisfying (d) and (e).

Thus, singular triangles are classified into the following two classes:

- An edge consists of singular points, and all vertices are not of the second kind. We call such an triangle a *singular triangle of type I*.
- An edge consists of singular points, and one vertex is of the second kind. We call such an triangle a *singular triangle of type II*.

Summing up, a triangulation consists of triangles of the following four kinds, see Fig. 6.2:

regular triangle,   quasi regular triangle,
singular triangle of type I,   singular triangle of type II.

6.5. Local Gauss–Bonnet Type Formula I

In this section, we consider a triangle $\triangle ABC$ in the triangulation of $\mathcal{M}^+$. If $\triangle ABC$ is a regular triangle, there are no singular points of f on its edges and vertices. Then the local Gauss–Bonnet formula (6.4) holds.

We now give an interpretation of the angles

$$\angle A, \qquad \angle B, \qquad \angle C$$

even for a triangle which is not necessarily regular. These values, the interior angles of the triangle, must coincide with the angles of the triangular region realized on the surface in $\mathbf{R}^3$. The meaning of the interior angle is evident when A is a regular point. So, we assume that A is either

- a singular vertex of a quasi regular triangle,
- a singular vertex of a singular triangle of type I, or
- a singular vertex of a singular triangle of type II, which is not a singular point of the second kind.

We may assume the two edges CA and AB are parametrized as regular curves $\gamma_3(t)$, $\gamma_1(t)$ $(0 \le t \le 1)$, respectively, with $A = \gamma_3(1) = \gamma_1(0)$. The vectors

$$\boldsymbol{c} := -\gamma_3'(1), \qquad \boldsymbol{b} := \gamma_1'(0)$$

do not vanish at A. Hence, by (6.7), $\angle A$ can be represented as

$$\angle A = \angle(\boldsymbol{c}, \boldsymbol{a}). \tag{6.9}$$

Let $\gamma(t)$ $(0 \le t \le 1)$ be a parametrization (on the uv-plane) of the edge AB in $\triangle ABC$. When the edge AB is not a singular edge, we define the *total geodesic curvature* of the edge by

$$\Theta_{AB} := \int_0^1 \kappa_g(t)ds \qquad (ds := |\hat{\gamma}'(t)|\, dt),$$

where $\kappa_g(t)$ is the geodesic curvature with respect to the left-ward co-normal vector field along γ in $\mathcal{M}^2$. This quantity does not depend on the choice of parameter t. On the other hand, consider the case that AB is a singular edge, on which the singular curvature $\kappa_s(t)$ is defined though the geodesic curvature cannot be defined. Then we define the *total singular curvature* by (cf. Corollary 5.5.6)

$$\Theta_{AB} := \int_0^1 \kappa_s(t)\, ds \qquad (ds := |\hat{\gamma}'(t)|\, dt).$$

Similarly, the total curvatures Θ_{CA} and Θ_{BC} of the edges CA and BC are defined, respectively. Our goal in this section is the following theorem:

Theorem 6.5.1 (Local Gauss–Bonnet theorem for frontals [68]). *Let $\triangle ABC$ be a triangle in a triangulation of $\mathcal{M}^+$, and assume that it is a regular triangle, a quasi regular triangle, or a singular triangle of type I (i.e., it is not a singular triangle of type II). Then*

$$\angle A + \angle B + \angle C - \pi = \Theta_{AB} + \Theta_{BC} + \Theta_{CA} + \int_{\triangle ABC} \Omega. \tag{6.10}$$

Here, Ω is the unsigned Euler form as defined in Definition 5.1.6 (see also Section 6.3). When the triangle $\triangle ABC$ is regular, (6.10) coincides with (6.4), that is, the usual "local Gauss–Bonnet theorem". Hence we shall prove (6.10) as an extension of (6.4) for non-regular triangles.

We remark that, since, at this moment, we have not defined the angle at a vertex that is a singular point of the second kind, we exclude the case of singular triangles of type II in the statement of Theorem 6.5.1. In the following section, we will extend the definition of interior angles and show that the same formula holds for singular triangles of type II as well (Proposition 6.6.3).

First we show the following lemma:

Lemma 6.5.2. *For a triangle $\triangle ABC$ in $\mathcal{M}^2$, there exists a family of triangles $\{\triangle A_s BC\}_{s\in[0,1)}$ satisfying the following properties (see Fig. 6.3 (a)).*

(1) *Each triangle $\triangle A_s BC$ is a subset of $\triangle ABC$, and $\triangle A_0 BC$ coincides with $\triangle ABC$.*

(2) *For each s, the edge CA_s is a part of the edge CA, and $\lim_{s\to 0} A_s = A$ holds.*

(3) *There exists a C^∞-embedding $\Gamma : [0,1] \times [0,1] \to \triangle ABC$ such that $t \mapsto \Gamma(t,s)$ parametrizes the edge $A_s B$ for each s.*

Proof. By definition, the triangle $\triangle ABC$ lies in a coordinate neighborhood U compatible with the orientation of $\mathcal{M}^2$. Moreover, there exists a domain

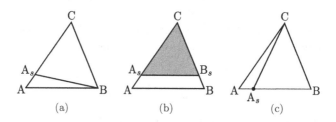

Fig. 6.3.   Deformations of triangles.

$D \subset \boldsymbol{R}^2$, a linear triangle $T \subset D$, and a diffeomorphism $\varphi : D \to U$ such that $\varphi(T) = \triangle ABC$. We may assume that T is a triangle with vertices

$$\boldsymbol{a}_0 := (0,0), \quad \boldsymbol{b}_0 := (0,1), \quad \boldsymbol{c}_0 := (1,0),$$

and

$$\varphi(\boldsymbol{a}_0) = A, \quad \varphi(\boldsymbol{b}_0) = B, \quad \varphi(\boldsymbol{c}_0) = C,$$

without loss of generality. Fixing $s \in [0,1]$, let T_s be a linear triangle in D with vertices

$$\boldsymbol{a}_s := (s,0), \quad \boldsymbol{b}_0 = (0,1), \quad \boldsymbol{c}_0 = (1,0).$$

Then T_s is a subset of $T = T_0$, and letting $A_s := \varphi(\boldsymbol{a}_s)$, $\varphi(T_s) = \triangle A_s BC$ ($s \in [0,1]$) gives a family of triangles on $\mathcal{M}^2$ satisfying the desired properties. In fact,

$$\Gamma(t,s) := \varphi((1-t)\boldsymbol{a}_s + t\boldsymbol{b}_0) \qquad (0 \le t, s \le 1)$$

is the map in (3). $\qquad \square$

Next, we prepare another perturbation of a triangle.

Lemma 6.5.3. *For a triangle $\triangle ABC$ in $\mathcal{M}^2$, there exists a family of triangles $\{\triangle A_s B_s C\}_{s \in [0,1)}$ satisfying the following properties (see Fig. 6.3 (b)).*

(1) *For each s, the triangle $\triangle A_s B_s C$ is a subset of $\triangle ABC$ and $\triangle A_0 B_0 C$ coincides with $\triangle ABC$.*
(2) *The edge CA_s is a part of the edge CA for each s, and $\lim_{s \to 0} A_s = A$ holds.*
(3) *The edge CB_s is a part of CB for each s, and $\lim_{s \to 0} B_s = B$ holds.*
(4) *There exists a C^∞-embedding $\Gamma : [0,1] \times [0,1] \to \triangle ABC$ such that $t \mapsto \Gamma(t,s)$ gives a parametrization of the edge $A_s B_s$ for each s.*

Proof. Instead of the three points $\boldsymbol{a}_s, \boldsymbol{b}_0, \boldsymbol{c}_0$ as in the proof of Lemma 6.5.2, we rather consider $\boldsymbol{a}_s, \boldsymbol{b}_s, \boldsymbol{c}_0$, where

$$\boldsymbol{b}_s =: (1-s, s)$$

for each $s \in [0,1]$, and let T_s be the linear triangle in D with vertices $\boldsymbol{a}_s$, $\boldsymbol{b}_s$ and $\boldsymbol{c}_0$. Setting

$$A_s := \varphi(0,s), \qquad B_s := \varphi(1-s,s),$$

$\varphi(T_s) = \triangle A_s BC$ ($s \in [0,1)$) gives a family satisfying the desired properties. In fact,

$$\Gamma(t,s) := \varphi((1-t)(0,s) + t(1-s,s)) \qquad (0 \le t, s \le 1)$$

is a map satisfying the property in (4). $\qquad \square$

Proposition 6.5.4 ([68]). *The local Gauss–Bonnet formula* (6.10) *holds for a quasi regular triangle* $\triangle ABC (\subset \mathcal{M}^+)$ *such that* A *is the singular vertex.*

Proof. By condition (c), the tangent vector of each edge at each vertex does not coincide with the null direction. Hence the geodesic curvature of the two edges starting at the singular vertex A is well defined. We take a family $\{\triangle A_s BC\}_{s \in [0,\delta)}$ of triangles as in Lemma 6.5.2. Since each triangle $\triangle A_s BC$ is regular,

$$\angle A_s + \angle B + \angle C - \pi = \Theta_{A_s B} + \Theta_{BC} + \Theta_{CA_s} + \int_{\triangle A_s BC} \Omega \qquad (6.11)$$

holds, by Fact 4.6.1. The total geodesic curvature and the integral of the Gaussian curvature are integrals of continuous functions on bounded closed sets, so they vary continuously as s varies. Here, the interior angle $\angle A_s$ of $\triangle A_s BC$ is the angle in the usual sense, and by the continuity of angles (Lemma 6.3.1) and Lemma 6.3.3, $\angle A_s$ converges to $\angle A$ as $s \to 0$. Hence, by letting $s \to 0$, we have (6.10).   □

Proposition 6.5.5 ([68]). *The local Gauss–Bonnet formula* (6.10) *holds for singular triangles of type I in* $\mathcal{M}^+$.

Proof. Let AB be the singular edge of a singular triangle ABC of type I. In a similar way as for Proposition 6.5.4, but applying Lemma 6.5.3 instead of Lemma 6.5.2, we can take a family of triangles $\{\triangle A_s B_s C\}_{s \in [0,1)}$. Since each triangle in this family is regular,

$$\angle A_s + \angle B_s + \angle C - \pi = \Theta_{A_s B_s} + \Theta_{B_s C} + \Theta_{CA_s} + \int_{\triangle A_s B_s C} \Omega \qquad (6.12)$$

holds for all non-zero s. Letting $s \to 0$, the total geodesic curvature $\Theta_{A_s B_s}$ of $A_s B_s$ converges to the total geodesic curvature Θ_{AB} of the edge AB, because of (5.22) and the assumption that the triangles lie on $\mathcal{M}^+$. Since the null direction is not a tangential direction of the edge AB, the property that the null line passes through or does not pass through the interior of the triangle is preserved as $s \to 0$. Hence, by the same reasoning as in the proof of Proposition 6.5.4, (6.10) is obtained by letting $s \to 0$ in (6.12).   □

6.6.   Local Gauss–Bonnet Type Formula II

In this section, we consider a singular triangle $\triangle ABC (\subset \mathcal{M}^+)$ of type II. First, we show the existence of a local coordinate system at a singular point of the second kind.

Lemma 6.6.1 ([68]). *Let* $f : \mathcal{M}^2 \to \mathbf{R}^3$ *be a frontal and* P *a singular point of type II. Then there exists a local coordinate system* $(U; u, v)$ *of* $\mathcal{M}^2$ *around* P *compatible with the orientation of* $\mathcal{M}^2$ *satisfying the following property.*

- *The point* P *corresponds to the origin of the uv-plane, and the null direction on the singular curve passing through* P *coincides with the direction* $\partial_u := \partial/\partial u$.

Proof. Applying Lemma B-5-6 in [84] for $\{\xi, \eta\}$, where η is the extended null vector field and ξ is a vector field linearly independent to η, we have a local coordinate system (u, v) such that ∂_u and ∂_v are proportional to η and ξ, respectively. This coordinate system is the desired one.   □

On the coordinate plane in Lemma 6.6.1, we may assume that the singular curve $\gamma(t)$ passing through P satisfies

$$\gamma(0) = (0, 0), \qquad \gamma'(0) = (1, 0), \tag{6.13}$$

without loss of generality. We have the following corollary.

Corollary 6.6.2. *Let* P *be an admissible singular point of the second kind, and take a local coordinate system* $(U; u, v)$ *as in Lemma 6.6.1, and assume the singular curve* γ *satisfies* (6.13). *Then* γ *satisfies the following properties in the uv-plane.*

(1) *The tangent vector* $\gamma'(t)$ *of* $\gamma(t)$ *is not horizontal except at* $t = 0$.

(2) $\gamma(t)$ *does not intersect the u-axis for sufficiently small t. In particular, the points on* γ *are all included in either the upper half-plane, or all in the lower half-plane.*

Proof. The property (1) is a conclusion coming from the fact that all singular points in a neighborhood of P are of the first kind. Since γ is tangent to the u-axis at P, it is the graph of a function, that is, the singular curve can be parametrized as $\gamma(u) = (u, \varphi(u))$ in a neighborhood of P. Assume that γ intersects the u-axis, that is, there exists a value u_0 such that $\varphi(0) = \varphi(u_0) = 0$. Then, by Rolle's theorem, there exists a value u such that $\varphi'(u) = 0$, a contradiction to (1).   □

We have already defined the interior angles, except at singular vertices of the second kind on singular triangles of type II. Let $\triangle ABC (\subset \mathcal{M}^+)$ be a singular triangle of type II such that A is a singular point of the second kind and AC is a singular edge. We define the interior angle of A as follows,

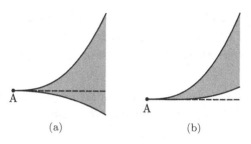

(a)   (b)

Fig. 6.4.   The interior angle of a singular vertex of the second kind. (a) Shows the case that the interior angle is π, and (b) shows the case of angle 0.

using the local coordinate system in Lemma 6.6.1 (cf. Fig. 6.4) so that ∂_u points in the null direction along the singular curve:

$$\angle A := \begin{cases} \pi, & \text{when the positive half of the } u\text{-axis} \\ & \text{lies in the interior of the triangle} \\ & \text{in a neighborhood of A,} \\ 0, & \text{otherwise.} \end{cases} \tag{6.14}$$

We shall prove, later, that this definition does not depend on the choice of local coordinate system satisfying Corollary 6.6.2. But before proving that, we shall prove the following proposition.

Proposition 6.6.3 ([68]). *Let* $\triangle ABC (\subset \mathcal{M}^+)$ *be a singular triangle of type II. Then the local Gauss–Bonnet formula* (6.10) *holds.*

Proof. Let A be a singular point of the second kind, and let AB be the singular edge. Applying Lemma 6.5.2, exchanging the roles of B and C, we have a family $\{\triangle A_s BC\}_{s \in [0,1)}$ of triangles satisfying the following properties (see Fig. 6.3 (c)).

(1) For each s, the triangle $\triangle A_s BC$ is a singular triangle of type I, and is a subset of $\triangle ABC$, and $\triangle A_0 BC$ coincides with $\triangle ABC$.

(2) The edge $A_s B$ is a part of the singular edge AB for each s, and $\lim_{s \to 0} A_s = A$ holds.

(3) There exists a C^∞-map $\Gamma : [0,1] \times [0,1] \to \triangle ABC$ such that $t \mapsto \Gamma(t,s)$ is a regular parametrization of CA_s for each s.

Since $\triangle A_s BC$ is a singular triangle of type I, (6.11) holds by Proposition 6.5.5. We shall prove the formula (6.10) as $s \to 0$. By Proposition 5.1.2 and Corollary 5.5.6, Ω and $\kappa_s \, ds$ are continuous on $\triangle ABC$. Then the convergence of the integral of Ω and the total geodesic curvature are obvious.

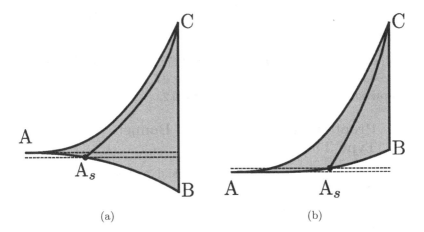

(a)   (b)

Fig. 6.5.   Proof of Lemma 6.6.4.

On the other hand, the angle $\angle C_s$ converges to $\angle C$, because C and C_s are regular points. Hence (6.10) is obtained by the next lemma.   □

Lemma 6.6.4. *The angle* $\angle A_s$ *converges to* $\angle A$ *defined in* (6.14) *as* $s \to 0$.

Proof. The null direction at A_s is parallel to the u-axis in the coordinate system, as in Lemma 6.6.1. By Corollary 6.6.2, this null direction cannot be the tangential direction of the edge A_sB. If $\angle A = \pi$, the right-half of the u-axis locally lies in the interior of the triangle on a neighborhood of A. In fact, Fig. 6.5(a) is this case. By (2) of Corollary 6.6.2, translating the dotted half-line downward slightly, the intersection of the line and lower edge is the point A_s. Since the translated half-line emanating from A_s lies in the interior of the triangle, $\angle A_s = \pi$ holds for sufficiently small $s(> 0)$ (cf. Lemma 6.3.3) and $\lim_{s \to 0} \angle A_s = \pi$ by Lemma 6.3.1.

On the other hand, consider the case $\angle A = 0$, that is, the case in Fig. 6.5(b). In this case, the dotted half-line lies in the exterior of the triangle. In the figure, translating the dotted line upward slightly, then the half-line meets the two edges AB and AC, respectively. For each fixed A_s for sufficiently small s, we can adjust the intersection of the line and the lower edge AB so that it consists of exactly one point A_s. Since the half-line emanating from A_s lies in the exterior of the triangle, $\angle A_s = 0$ holds for sufficiently small $s(> 0)$ (cf. Lemma 6.3.3) and $\lim_{s \to 0} \angle A_s = 0$ by Lemma 6.3.1.   □

Since the right-hand side of Eq. (6.14) is independent of coordinates, we have the next corollary.

Corollary 6.6.5. *The definition* (6.14) *of the interior angle of vertices of the second kind for singular triangles of type II does not depend on choice of coordinate system satisfying Corollary 6.6.2.*

6.7. Proof of the Global Gauss–Bonnet Type Theorem

Let $\mathcal{M}^2$ be a compact and oriented 2-manifold, and consider a frontal $f :$ $\mathcal{M}^2 \to \mathbf{R}^3$ whose singular points are either of the first kind or admissible singularities of the second kind. Then the two closed subsets $\mathcal{M}^+$ and $\mathcal{M}^-$ of $\mathcal{M}^2$ are defined. As pointed out in the beginning of Section 6.4, the number of admissible singular points of the second kind is finite. We classify admissible singular points of the second kind in the following three classes.

Proposition 6.7.1. *At an admissible singular point of the second kind, the interior angle on the $\mathcal{M}^+$ side is either 2π, π or 0.*

An admissible singular point of the second kind $\mathrm{P} \in \mathcal{M}^2$ is said to be *positive*, *zero* or *negative* when the interior angle on the $\mathcal{M}^+$ side is 2π, π or 0, respectively. We set

$$\mathrm{sgn}(\mathrm{P}) := \begin{cases} 1 & \text{when P is positive,} \\ 0 & \text{when P is zero,} \\ -1 & \text{when P is negative.} \end{cases} \qquad (6.15)$$

Lemma 6.7.2. *Let $\mathrm{P} \in \mathcal{M}^2$ be an admissible singular point of the second kind of a frontal f. If P is a swallowtail or a Whitney cusp on an embedded surface in $\mathbf{R}^3$ (cf. Proposition 5.4.5), then P is either positive or negative.*

Proof. Since the property that the interior angle is 2π, π or 0 is preserved by right-left equivalence, it is sufficient to check the assertion using the canonical forms of the swallowtail (cf. Example 2.5.3) and the Whitney cusp (cf. (4.13)). In fact, for example, let us consider the case of the standard swallowtail. Then, the left-hand side of the singular curve $\gamma(t) = (6t^2, t)$ is $\mathcal{M}^+$ as in Fig. 5.3. We set $\boldsymbol{b} := (-1, 0)$. Then $\angle(\gamma'(t), \boldsymbol{b})$ at $\gamma(t)$ $(t > 0)$ and $\angle(-\gamma'(t), \boldsymbol{b})$ at $\gamma(t)$ $(t < 0)$ are both π, by Lemma 6.3.3. By Lemma 6.3.1, the interior angle of $\mathcal{M}^+$ at P is equal to

$$\lim_{t \to 0+0} \angle(\gamma'(t), \boldsymbol{b}) + \lim_{t \to 0-0} \angle(-\gamma'(t), \boldsymbol{b}) = 2\pi.$$

$\square$

As a conclusion, we have the following:

Proposition 6.7.3. *Let* $f : \mathcal{M}^2 \to \mathbf{R}^3$ *be a frontal whose singular points are of the first kind or admissible singularities of the second kind, and let* ν *be a globally-defined unit normal vector field of* f*, and let* $\hat{\Omega}$ *be the Euler form of* f *(which depends on the choice of* ν*). We denote by* $\{P_1, \ldots, P_n\}$ *the set of singular points of the second kind in* $\Sigma(f)$*. It then holds that*

$$\int_{\mathcal{M}^+} \hat{\Omega} + \int_{\Sigma(f)} \kappa_s \, ds = 2\pi\chi(\mathcal{M}^+) + \pi \sum_{i=1}^{n} \mathrm{sgn}(P_i), \qquad (6.16)$$

where ds *is the line element of the singular set* $\Sigma(f)$ *(see (6.8) for the definition of* $\mathcal{M}^+$*). Here, each connected component of* $\Sigma(f)$ *is oriented so that* $\mathcal{M}^+$ *is on the left side (in each local coordinate system compatible with the orientation of* $\mathcal{M}^2$*).*

Proof. Denote by $\angle P_i$ ($i = 1, \ldots, n$) the interior angle with respect to $\mathcal{M}^+$ at a singular point of the second kind P_i. These points are vertices of our triangulation of $\mathcal{M}^+$, and the singular set $\Sigma(f) = \partial\mathcal{M}^+$ is a collection of edges. Let $\{V_1, \ldots, V_N\}$ be the set of vertices of the triangulation, and assume that m_j triangles $T_{j,1}, \ldots, T_{j,m_j}$ share the jth vertex V_j.

Each triangle $\triangle ABC$ is a subset of a coordinate neighborhood $(U; u, v)$ compatible with the orientation of the manifold. Noticing that $\hat{\Omega}$ coincides with the usual area element on $\mathcal{M}^+$, summing up the equality (6.10) over all triangles, we have

$$\int_{\mathcal{M}^+} \Omega = - \int_{\Sigma(f)} \kappa_s ds + 2\pi\chi(\mathcal{M}^+) - \sum_{i=1}^{n} (\pi - \angle P_i),$$

in the same way as in Proposition 4.6.2 in Chapter 4, because the total geodesic curvature of a non-singular edge cancels out with that of the adjacent triangle. Note that the total geodesic curvature on singular edges remains, because the union of such edges is $\partial\mathcal{M}^+$. Thus, in Eq. (4.6.2) of Chapter 4, the term

$$\int_{\Sigma(f)} \kappa_s \, ds$$

remains. Here, the interior angle $\angle P_i$ with respect to the $\mathcal{M}^+$ side at P_i is 2π, π or 0 when $\mathrm{sgn}(P_j)$ is positive, zero or negative, respectively. Since $\Omega = \hat{\Omega}$ holds on $\mathcal{M}^+$, we obtain the assertion. $\square$

Corollary 6.7.4. *Under the same situation as in Proposition* 6.7.3*, it holds that*

$$-\int_{\mathcal{M}^-} \hat{\Omega} + \int_{\Sigma(f)} \kappa_s \, ds = 2\pi\chi(\mathcal{M}^-) - \pi \sum_{i=1}^{n} \mathrm{sgn}(P_i). \qquad (6.17)$$

Here, $\Sigma(f)$ *is the set of singular curves oriented so that* $\mathcal{M}^-$ *is to the left-hand side, with respect to a coordinate system compatible with the orientation of* $\mathcal{M}^2$*, and* $\{P_1, \ldots, P_n\}$ *is the set of singular points of the second kind in* $\Sigma(f)$.

Proof. If we reverse the orientation of $\mathcal{M}^2$, then the roles of $\mathcal{M}^+$ and $\mathcal{M}^-$ are interchanged. Then we can apply the proof of Proposition 6.7.3 for $\mathcal{M}^-$. Since the sign of P_i is reversed and the singular curvature κ_s does not change by reversing the orientation of $\partial\mathcal{M}^-$, we have

$$\int_{\mathcal{M}^-} \Omega + \int_{\Sigma(f)} \kappa_s ds = 2\pi\chi(\mathcal{M}^-) - \pi\sum_{i=1}^{n} \mathrm{sgn}(P_i).$$

Since $\int_{\mathcal{M}^-} \Omega = -\int_{\mathcal{M}^-} \hat{\Omega}$, we obtain (6.17).   □

Adding (6.16) and (6.17), and subtracting (6.17) from (6.16), we obtain the following theorem (the first equality of (6.19) follows from (5.9)).

Theorem 6.7.5 ([69]). *Let* $f : \mathcal{M}^2 \to \mathbf{R}^3$ *be a frontal whose singular points are all non-degenerate, and either of the first kind or of the second kind and admissible. Let* ν *be a globally-defined unit normal vector field of* f*, and let* $d\hat{A}$ *(respectively, dA) be the signed (respectively, unsigned) area element associated with* ν*. Then*

$$\int_{\mathcal{M}^2} \Omega + 2\int_{\Sigma(f)} \kappa_s \, ds = 2\pi\chi(\mathcal{M}^2), \tag{6.18}$$

$$2\deg(\nu) = \frac{1}{2\pi}\int_{\mathcal{M}^2} \hat{\Omega} = \chi(\mathcal{M}^+) - \chi(\mathcal{M}^-) + \sum_{i=1}^{n} \mathrm{sgn}(P_i) \tag{6.19}$$

hold, where $\{P_1, \ldots, P_n\}$ *is the set of singular points of the second kind in* $\Sigma(f)$ *(cf. Proposition 6.7.1), and*[2] *(cf. Proposition 5.1.9)*

$$\int_{\mathcal{M}^2} \Omega := \int_{\mathcal{M}^+} \hat{\Omega} - \int_{\mathcal{M}^-} \hat{\Omega}. \tag{6.20}$$

The equalities (6.18) and (6.19) are essentially the same as those in Corollary 4.6.5 of Chapter 4. (By (5.7), $d\hat{A}_\nu = K \, d\hat{A}$ holds, and so the integral $\frac{1}{4\pi}\int_{\mathcal{M}^2} \hat{\Omega}$ is just equal to $\deg(\nu)$.)

---

[2]Even though Ω might be discontinuous on $\mathcal{M}^2$ (cf. Remark 5.1.8), $\hat{\Omega}$ is smooth on $\mathcal{M}^2$. So $\int_{\mathcal{M}^2} \Omega$ is well defined.

6.8. Applications of the Global Gauss–Bonnet Type Theorem I

Let $\mathcal{N}^2$ be an oriented 2-manifold. Here we assume that there exists an embedding

$$\iota : \mathcal{N}^2 \to \boldsymbol{R}^3.$$

(For example, if $\mathcal{N}^2$ is compact or $\mathcal{N}^2 = \boldsymbol{R}^2$, then such an embedding can be easily constructed.) By this embedding ι, we may regard the 2-manifold $\mathcal{N}^2$ as a surface embedded in $\boldsymbol{R}^3$. The induced metric (i.e., the first fundamental form) ds^2 under ι is a Riemannian metric on $\mathcal{N}^2$.

Let $\mathcal{M}^2$ be an oriented 2-manifold and $f : \mathcal{M}^2 \to \mathcal{N}^2$ a C^∞-map whose singular points are folds and Whitney cusps (cf. Section 4.4 of Chapter 4). Then the composition $\iota \circ f$ is a frontal in $\boldsymbol{R}^3$, and a fold singular point (respectively, a Whitney cusp) corresponds to a singular point of the first kind (respectively, an admissible singular point of the second kind) of $\iota \circ f$. Applying Theorem 6.7.5, we have the following.

Theorem 6.8.1 (Levine's formula [50]). *Let $\mathcal{M}^2$ be a compact, orientable 2-manifold, and let $f : \mathcal{M}^2 \to \boldsymbol{R}^2$ be a C^∞-map whose singular point set $\Sigma(f)$ consists of fold singular points and Whitney cusps. Then $\Sigma(f)$ is the union of a finite number of simple closed regular curves $C_1, C_2, \ldots, C_k$ in $\mathcal{M}^2$. Defining an orientation on the images*

$$f(C_1), \ f(C_2), \ \ldots, \ f(C_k)$$

of singular curves such that $f(\mathcal{M}^2)$ lies to the left side, the rotation indices $I(C_i)$, with $f(C_i)$ regarded as closed wave fronts $(i = 1, \ldots, k)$, are defined. (Note that $I(C_i)$ takes values in the set of half integers, since $f(C_i)$ may have cusps.) It then holds that

$$\frac{\chi(\mathcal{M}^2)}{2} = I(C_1) + \cdots + I(C_k).$$

Proof. By (6.18) in Theorem 6.7.5 in the previous section, we have

$$2 \int_{\Sigma(f)} \kappa_s \, ds = 2\pi \chi(\mathcal{M}^2), \tag{6.21}$$

since the Gaussian curvature of the plane is 0. Here, under the orientation of $f(C_i)$ as given in the statement, the curvature κ of $f(C_i)$ as a plane curve coincides with the singular curvature κ_s, because of Proposition 5.4.6.

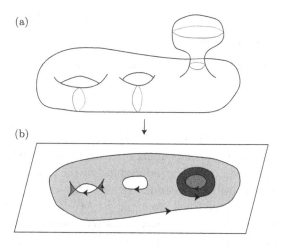

Fig. 6.6. Example 6.8.2.

Hence we have

$$\int_{\Sigma(f)} \kappa_s \, ds = \sum_{i=1}^{k} \int_{f(C_i)} \kappa ds = 2\pi \sum_{i=1}^{k} I(C_i), \qquad (6.22)$$

and by (6.21) and (6.22), the conclusion follows. □

The Euler number of an orientable closed surface is expressed as $2 - 2g$ ($g = 0, 1, 2, \dots$) using its genus g.

Example 6.8.2. Let S be a closed surface in $\mathbf{R}^3$ of genus 2 without self-intersections, and let $f : S \to \mathbf{R}^2$ be a projection map as in Fig. 6.6. Then each singular value set is oriented as in Fig. 6.6(b), and the outermost circle is of rotation index 1, the leftmost (respectively, middle) inner closed front is of rotation index -1 (respectively, -1), and the rightmost two circles have rotation indices $+1$ (outside) and -1 (inside). The total sum of the rotation indices is -1, which is half of the Euler number of S.

Theorem 6.8.3 (Quine's formula [66]). *Let $\mathcal{M}^2$ and $\mathcal{N}^2$ be compact orientable 2-manifolds, and let $f \colon \mathcal{M}^2 \to \mathcal{N}^2$ be a C^∞-map whose singular point set consists of folds and Whitney cusps. Then the mapping degree* $\deg(f)$ *of f satisfies*

$$\deg(f)\chi(\mathcal{N}^2) = \chi(\mathcal{M}^+) - \chi(\mathcal{M}^-) + \#C_f^+ - \#C_f^-.$$

Here, $\mathcal{M}^+$ and $\mathcal{M}^-$ are the subsets of $\mathcal{M}^2$ where the Jacobian of f is non-negative and non-positive, respectively, and $\#C_f^+$ and $\#C_f^-$ are the number of positive and negative Whitney cusps (cf. Lemma 6.7.2), respectively.

Proof. As mentioned in the beginning of this section, we may assume that $\mathcal{N}^2$ is realized as a surface in $\mathbf{R}^3$ by an embedding $\iota : \mathcal{N}^2 \to \mathbf{R}^3$, without loss of generality. Then the Riemann metric induced on $\mathcal{N}^2$ is the first fundamental form of ι, and the degree of the Gauss map $\nu : \mathcal{N}^2 \to S^2$ of ι satisfies (see [42, Note 20])

$$\deg(\nu) = \frac{\chi(\mathcal{N}^2)}{2}. \qquad (6.23)$$

On the other hand, applying (6.19) in Theorem 6.7.5 of the previous section to f, we have

$$2\deg(\nu \circ f) = \frac{1}{2\pi}\int_{\mathcal{M}^2}\hat{\Omega} = \chi(\mathcal{M}^+) - \chi(\mathcal{M}^-) + \#C_f^+ - \#C_f^-. \qquad (6.24)$$

Then properties of the mapping degree and (6.23) yield

$$\deg(\nu \circ f) = \deg(f)\deg(\nu) = \frac{\deg(\nu)\chi(\mathcal{N}^2)}{2}.$$

Substituting this into (6.24), the conclusion follows. $\qquad\square$

Example 6.8.4. Projecting a torus of revolution $\mathcal{M}^2$ to a plane obliquely, and mapping it to the unit sphere S^2 centered at the origin by the inverse of stereographic projection, we have a map from $\mathcal{M}^2$ to the sphere as in Fig. 6.7. There are four Whitney cusps in this map, and the degree of the map is zero because it is not surjective. Thus, by Theorem 6.8.3, we have

$$0 = \chi(\mathcal{M}^+) - \chi(\mathcal{M}^-) + \#C_f^+ - \#C_f^-.$$

There are two Whitney cusps in the part surrounded by the dotted loop in of Fig. 6.7(b). Since the upper one is visible and the lower one is hidden, the signs of the two cusps are different. So, the numbers of positive and negative Whitney cusps are the same, and hence we have $\chi(\mathcal{M}^+) = \chi(\mathcal{M}^-)$.

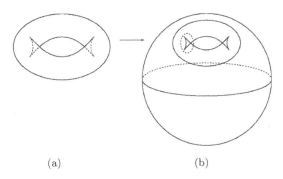

(a)   (b)

Fig. 6.7. Example 6.8.4.

Next, we apply Theorem 6.8.3 to Gauss maps of surfaces. Let $\mathcal{M}^2$ be a compact, connected and oriented 2-manifold, and let $f : \mathcal{M}^2 \to \mathbf{R}^3$ be an immersion. Assume that the singular points of the Gauss map ν of f consists of fold singular points and Whitney cusps. A point where the Gaussian curvature is positive (respectively, negative) is a point where the Gauss map preserves (respectively, reverses) the orientation [84]. Recall that a point where the Gauss map has a positive (respectively, negative) Whitney cusp is called a positive (respectively, negative) godron point (cf. Section 4.4). Since the Euler number of the sphere is 2, Theorem 6.8.3 yields

$$2 \deg \nu = \chi(\mathcal{M}^+) - \chi(\mathcal{M}^-) + \#G^+ - \#G^-. \tag{6.25}$$

Here, $\mathcal{M}^+$ (respectively, $\mathcal{M}^-$) is the set of points in $\mathcal{M}^2$ where the Gaussian curvature is non-negative (respectively, non-positive), and $\#G^{\pm}$ denotes the number of positive/negative godron points.

6.9. Applications of the Global Gauss–Bonnet Type Theorem II

Let $\mathcal{M}^2$ be a compact and orientable 2-manifold, and $f : \mathcal{M}^2 \to \mathbf{R}^3$ a wave front whose singular point set consists of cuspidal edges and swallowtails. We fix a unit normal vector field ν of f defined on $\mathcal{M}^2$. Then the signed area element associated with ν is expressed as

$$d\hat{A} = \det(f_u, f_v, \nu)du \wedge dv \tag{6.26}$$

for each local coordinate system. Then the two equalities in Theorem 6.7.5 turn out to be

$$\int_{\mathcal{M}^2} \Omega + 2 \int_{\Sigma(f)} \kappa_s \, ds = 2\pi\chi(\mathcal{M}^2), \tag{6.27}$$

$$\frac{1}{2\pi} \int_{\mathcal{M}^2} \hat{\Omega} = \chi(\mathcal{M}_f^+) - \chi(\mathcal{M}_f^-) + \#S_f^+ - \#S_f^-, \tag{6.28}$$

where $\hat{\Omega}$ and Ω are the Euler form and the unsigned Euler form, respectively, and $\#S_f^+$ (respectively, $\#S_f^-$) is the number of positive (respectively, negative) swallowtails. Here, since Ω is not continuous on $\mathcal{M}^2$ in general, the left-hand side of (6.27) must be interpreted as in (6.20). On the other hand, the Gauss map $\nu : \mathcal{M}^2 \to \mathbf{R}^3$ of f is a frontal and its signed area element $d\hat{A}_\nu$ and area element dA_ν are expressed as (cf. (5.6))

$$d\hat{A}_\nu = K \, d\hat{A}(= \hat{\Omega}), \qquad dA_\nu = |K| \, dA, \tag{6.29}$$

where $d\hat{A}$, dA, and K are the signed area element, the area element, and the Gaussian curvature of f (cf. the proof of Proposition 5.1.3). Since the image of ν lies in the unit sphere, the Gaussian curvature of ν is identically 1 on the set of regular points. We assume that ν (as the Gauss map) admits only fold singular points and Whitney cusps as singularities. Applying Theorem 6.7.5 to the Gauss map ν as a frontal (see also (6.33)),we have

$$\int_{\mathcal{M}^2} |K| dA + 2 \int_{\Sigma(\nu)} \kappa_\# ds = 2\pi \chi(\mathcal{M}^2), \qquad (6.30)$$

$$\frac{1}{2\pi} \int_{\mathcal{M}^2} \hat{\Omega} = \chi(\mathcal{M}_\nu^+) - \chi(\mathcal{M}_\nu^-) + \#C_\nu^+ - \#C_\nu^-. \qquad (6.31)$$

Here, as defined in Proposition 6.7.3, $\mathcal{M}_\nu^+$ (respectively, $\mathcal{M}_\nu^-$) denotes the set of points where the area element and the signed element of ν coincide (respectively, having different signs), $\#C_\nu^+$ (respectively, $\#C_\nu^-$) is the number of positive (respectively, negative) Whitney cusps of ν, and $\kappa_\#$ is the singular curvature along the singular curve of ν.

Remark 6.9.1. The equality (6.31) is equivalent to (6.25). Note that $|K| dA (= dA_\nu)$ is continuous on $\mathcal{M}^2$ while KdA is discontinuous along the singular set $\Sigma(f)$, because of (6.29) and Proposition 5.1.13.

Summing up, there are four Gauss–Bonnet type formulas (6.27), (6.28), (6.30) and (6.31) for frontals in Euclidean 3-space. We introduce some applications of these formulas.

Theorem 6.9.2 (Bleeker-Wilson's formula [4]). *Let $\mathcal{M}^2$ be a compact oriented 2-manifold, and $f : \mathcal{M}^2 \to \mathbf{R}^3$ an immersion. Assume that the singular point set of the Gauss map $\nu : \mathcal{M}^2 \to S^2$ of f consists of fold singular points and Whitney cusps. If ν is chosen so that $\det(f_u, f_v, \nu) > 0$ for a positive[3] local coordinate system (u, v), then*

$$2\chi(\{K < 0\}) = \#C_\nu^+ - \#C_\nu^-,$$

where K denotes the Gaussian curvature of f, and $\chi(\{K < 0\})$ is the Euler number of the set of points in $\mathcal{M}^2$ where K is negative. In addition, $\#C_\nu^+$ and $\#C_\nu^-$ are the numbers of positive and negative Whitney cusps (here, the Gaussian curvature of f is positive (respectively, negative) on the outside of each positive (respectively, negative) cusp), respectively.

---

[3]Here "positive" means the coordinate system is compatible with respect to the orientation of $\mathcal{M}^2$.

Proof. Since f is an immersion, (6.28) yields

$$\frac{1}{2\pi} \int_{\mathcal{M}^2} \hat{\Omega} = \chi(\mathcal{M}^2). \tag{6.32}$$

On the other hand, in a positively oriented local coordinate system (u, v) of $\mathcal{M}^2$, the signed area density of ν is expressed as (cf. Proposition 2.1.5)

$$\det(\nu_u, \nu_v, \nu) = K \det(f_u, f_v, \nu), \tag{6.33}$$

by (5.6). By our choice of ν, we have $\det(f_u, f_v, \nu) > 0$,

$$\chi(\mathcal{M}_\nu^+) = \chi(\{K \geq 0\}), \qquad \chi(\mathcal{M}_\nu^-) = \chi(\{K \leq 0\}), \tag{6.34}$$

and

$$\chi(\mathcal{M}^2) = \chi(\mathcal{M}_\nu^+) + \chi(\mathcal{M}_\nu^-). \tag{6.35}$$

So, (6.28) can be rewritten as

$$\frac{1}{2\pi} \int_{\mathcal{M}^2} \hat{\Omega} = \chi(\{K \geq 0\}) - \chi(\{K \leq 0\}) + \#C_\nu^+ - \#C_\nu^-. \tag{6.36}$$

Subtracting (6.36) from (6.32), we have the conclusion, by (6.35). □

Example 6.9.3. Consider a deformation of a torus of revolution

$$f^a(u, v) = (\cos v(2 + \varepsilon(v) \cos u), \sin v(2 + \varepsilon(v) \cos u), \varepsilon(v) \sin u),$$

where $\varepsilon(v) := 1 + a \cos v$, and the parameter a runs over $0 \leq a \leq 4/5$. When $a = 0$, $f^0(u, v)$ represents the torus of revolution as shown in Fig. 6.8(a) and $f^{4/5}(u, v)$ is the "heart-shaped" torus as in Fig. 6.8(b).

In the case of $a = 0$, the part with negative Gaussian curvature is an annulus, whose Euler number is 0. On the other hand, in the case of $a = 4/5$, the part that is everywhere locally convex has positive Gaussian

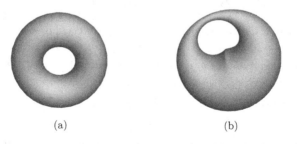

(a)   (b)

Fig. 6.8.   A torus of revolution and its deformation.

curvature, and then the part with negative Gaussian curvature is homeomorphic to a disc, whose Euler number is 1. By the Bleeker–Wilson formula (Theorem 6.9.2), it holds that

$$2 = 2\chi(\{K < 0\}) = \#C_\nu^+ - \#C_\nu^-.$$

Since the Gauss map of $f^{4/5}$ has two singular points of the second kind, $\#C_\nu^+$ must be 2. Hence these Whitney cusps of ν are positive.

Interchanging the roles of f and ν in Theorem 6.9.2, we have

Theorem 6.9.4 (Dual of the Bleeker–Wilson formula [73]). *Let* $f :$ $S^2 \to \mathbf{R}^3$ *be a wave front whose singular point set consists of cuspidal edges and swallowtails, and whose Gauss map is a diffeomorphism. We choose the unit normal vector field* ν *so that* $\det(\nu_u, \nu_v, \nu) > 0$ *for a positive local coordinate system* (u, v). *Then*

$$2\chi(\{K < 0\}) = \#S_f^+ - \#S_f^-$$

holds, where $\#S_f^+$ *(resp.* $\#S_f^-$*) is the number of positive (resp. negative) swallowtails.*

Proof. Let $\mathcal{M}^2$ be the 2-sphere S^2. Since ν is a diffeomorphism, (6.31) yields

$$\frac{1}{2\pi} \int_{\mathcal{M}^2} \hat{\Omega} = \chi(\mathcal{M}^2). \qquad (6.37)$$

By (5.3), the signed area density of ν is computed as

$$0 < \det(\nu_u, \nu_v, \nu) = K \det(f_u, f_v, \nu). \qquad (6.38)$$

Thus, the sign of $\det(f_u, f_v, \nu)$ coincides with that of K. Namely, $\chi(\mathcal{M}_f^+) = \chi(\{K \geq 0\})$ and $\chi(\mathcal{M}_f^-) = \chi(\{K \leq 0\})$ hold. Then (6.28) can be rewritten as

$$\frac{1}{2\pi} \int_{\mathcal{M}^2} \hat{\Omega} = \chi(\{K \geq 0\}) - \chi(\{K \leq 0\}) + \#S_f^+ - \#S_f^-. \qquad (6.39)$$

Subtracting (6.39) from (6.37), $\chi(\mathcal{M}^2) = \chi(\mathcal{M}_f^+) + \chi(\mathcal{M}_f^-)$ yields the conclusion. □

A connected regular surface whose Gauss map is a diffeomorphism is a convex closed surface, which is homeomorphic to the sphere. Parallel surfaces of such a surface may have singularities, and satisfy the assumption of Theorem 6.9.4. We observe this for an ellipsoid as follows.

Example 6.9.5. Let $f : S^2 \to \mathbf{R}^3$ be an embedding whose image is the ellipsoid $x^2/25 + y^2/16 + z^2 = 1$, and take the unit normal vector field ν

Table 6.1.   Parallel surfaces of an ellipsoid.

Value of t stage	$t = 0$ initial	$t = 0.3$ 1st	$t = 5.5$ 2nd	$t = 8$ 3rd	$t = 14$ 4th
Parallel surfaces					
Topological type of $\mathcal{M}_t^-$	$\emptyset$				
$\chi(\mathcal{M}_t^-)$	0	2	-2	0	0
$(\#S_f^+, \#S_f^-)$	(0,0)	(4,0)	(0,4)	(4,4)	(4,4)

as the inward normal. Then the parallel surface f^t of f with distance t varies as in Table 6.1. The stage of $f^t(S^2)$ is an open interval I of t such that the singular set of f^t consists of cuspidal edges and swallowtails for all $t \in I$, and if $t \in \partial I$, f^t has a singular point different from cuspidal edges and swallowtails. We denote by $\mathcal{M}_t^-$ the subset of S^2 where the Gaussian curvature of f^t is non-positive. Table 6.1 shows the topological types of $\mathcal{M}_t^-$ and numbers of swallowtails in f^t. That is, $\mathcal{M}_t^-$ is homeomorphic to the union of two discs when $t = 0.3$, to a disc with three discs excluded when $t = 5.5$, and to two annuli when $t = 8$ and 14. Black (respectively, white) circles in these figures indicate positive (respectively, negative) swallowtails.

Next, we consider applications of the formulas with singular curvatures.

Theorem 6.9.6 ([73]). *Let $\mathcal{M}^2$ be a compact orientable 2-manifold, and $f : \mathcal{M}^2 \to \mathbf{R}^3$ an immersion. Assume that the Gauss map $\nu : \mathcal{M}^2 \to S^2$ of f admits fold singular points and Whitney cusps as singularities. If ν is chosen so that $\mathcal{M}^2 = \mathcal{M}^+$, then*

$$\int_{\Sigma(\nu)} \kappa_\# \, ds = \int_{\mathcal{M}^2} K^- \, dA,$$

where K is the Gaussian curvature f and $K^- = \min\{0, K\}$, $\kappa_\#$ is the singular curvature of $\Sigma(\nu)$ and $ds = |\hat{\nu}'(t)| \, dt$.

Proof. Since f is an immersion,

$$\int_{\mathcal{M}^2} K^+ \, dA + \int_{\mathcal{M}^2} K^- \, dA = \int_{\mathcal{M}^2} \Omega = 2\pi \chi(\mathcal{M}^2) \qquad (6.40)$$

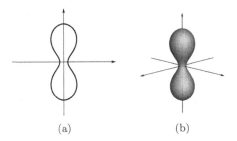

(a)   (b)

Fig. 6.9.   A non-convex simple closed curve (a) and its associated surface of revolution (b).

holds, where $K^+ = \max\{0, K\}$. On the other hand, since

$$\int_{\mathcal{M}^2} K^+ \, dA - \int_{\mathcal{M}^2} K^- \, dA + 2 \int_{\Sigma(\nu)} \kappa_\# ds = 2\pi\chi(\mathcal{M}^2) \qquad (6.41)$$

holds, because of (6.30), subtracting (6.41) from (6.40), we have the conclusion. □

Example 6.9.7. Consider a surface of revolution whose generating curve is a simple closed planar curve as in Fig. 6.9(a), which is symmetric with respect to the vertical axis, see Fig. 6.9(b). Then the singular set of its Gauss map consists of two small circles on the unit sphere, with constant curvature. Let r (> 0) be the radius of the small circles. Then the geodesic curvature of the circles with respect to the orientation such that the image of the surface is to the left side is $\sqrt{1 - r^2}/r$. Hence the total geodesic curvature is $2\pi\sqrt{1 - r^2}$, and then the absolute value of the total curvature of the part where the Gaussian curvature is negative is $4\pi\sqrt{1 - r^2}$, because of Theorem 6.9.6.

Interchanging the roles of f and ν in Theorem 6.9.6, we have the following theorem. The proof is left as an exercise for readers.

Theorem 6.9.8 ([73]). *Let $f : S^2 \to \mathbf{R}^3$ be a wave front defined on the 2-sphere and assume that the Gauss map ν is a diffeomorphism. If we choose ν so that $\det(\nu_u, \nu_v, \nu) > 0$, then*

$$\int_{\Sigma(f)} \kappa_s \, ds = \int_{\mathcal{M}^2} K^- \, dA,$$

where K is the Gaussian curvature of f and $K^- = \min\{0, K\}$.

On surfaces with constant Gaussian curvature in $\boldsymbol{R}^3$, cuspidal edges and swallowtails appear frequently. Since the limiting normal curvature of such cuspidal edges vanishes identically, they have the following special property.

Proposition 6.9.9 ([73]). *Let $\mathcal{M}^2$ be an orientable 2-manifold, and let $f : \mathcal{M}^2 \to \boldsymbol{R}^3$ be a wave front such that $\log|K|$ is bounded on the set of regular points, where K is the Gaussian curvature. Then the singular set of ν coincides with that of f.*

Proof. Since $\log|K|$ is bounded, K does not have any zeros. Regarding $K(= \det W)$, (5.6) yields that zeros of the signed area density function of ν coincide with zeros of the signed area density function of f.   □

Comparing this with Theorem 5.2.7 in Chapter 5, we have, in particular, the following:

Proposition 6.9.10 ([73]). *Let $\mathcal{M}^2$ be an orientable 2-manifold, and let $f : \mathcal{M}^2 \to \boldsymbol{R}^3$ be a wave front such that $\log|K|$ is bounded on the set of regular points, where K is the Gaussian curvature. Then*

$$\#S_f^+ - \#S_f^- = \operatorname{sgn}(K)(\#C_\nu^+ - \#C_\nu^-)$$

holds, where $\#S_f^+$, $\#S_f^-$ are the numbers of positive and negative swallowtails of f, and $\#C_\nu^+$, $\#C_\nu^-$ are the numbers of positive and negative Whitney cusps of ν, respectively. In particular, the signed sum of the Whitney cusps of the Gauss map coincides with the signed sum of the swallowtails on the surface, up to sign.

Proof. Applying the Gauss–Bonnet type formula to f and ν, we obtain (6.28) and (6.31), respectively. Then we have

$$\frac{1}{2\pi} \int_{\mathcal{M}^2} \hat{\Omega} = \chi(\mathcal{M}_f^+) - \chi(\mathcal{M}_f^-) + \#S_f^+ - \#S_f^-, \tag{6.42}$$

$$\frac{1}{2\pi} \int_{\mathcal{M}^2} \hat{\Omega} = \chi(\mathcal{M}_\nu^+) - \chi(\mathcal{M}_\nu^-) + \#C_\nu^+ - \#C_\nu^-. \tag{6.43}$$

Here

$$\det(\nu_u, \nu_v, \nu) = K \det(f_u, f_v, \nu) \tag{6.44}$$

holds, by (5.6). As pointed out in the proof of Proposition 6.9.9, K does not change sign and the singular set of ν coincides with that of f. When K is positive, (6.44) implies that the sign of $\det(\nu_u, \nu_v, \nu)$ coincides with that of $\det(f_u, f_v, \nu)$. So we have

$$\mathcal{M}_f^+ = \mathcal{M}_\nu^+, \quad \mathcal{M}_f^- = \mathcal{M}_\nu^-.$$

Then (6.43) can be rewritten as

$$\frac{1}{2\pi} \int_{\mathcal{M}^2} \hat{\Omega} = \chi(\mathcal{M}_f^+) - \chi(\mathcal{M}_f^-) + \#C_\nu^+ - \#C_\nu^-.$$

Hence we have the conclusion for $K > 0$, by (6.42).

Similarly, when K is negative, $\det(\nu_u, \nu_v, \nu)$ and $\det(f_u, f_v, \nu)$ have different signs, because of (6.44). Hence

$$\mathcal{M}_f^+ = \mathcal{M}_\nu^-, \quad \mathcal{M}_f^- = \mathcal{M}_\nu^+$$

hold, proving the conclusion for $K < 0$. $\qquad\square$

The following is another important property of wave fronts with bounded $\log |K|$.

Theorem 6.9.11 ([73]). *Let $\mathcal{M}^2$ be an orientable 2-manifold, and $f : \mathcal{M}^2 \to \boldsymbol{R}^3$ be a wave front such that $\log |K|$ is bounded on the set of regular points, where K is the Gaussian curvature. Let $\gamma(t)$ $(a \le t \le b)$ be a singular curve on $\mathcal{M}^2$ consisting of cuspidal edges of f and fold singular points of ν at the same time. We denote by $\kappa_s(t)$ (respectively, $\kappa_\#(t)$) the singular curvature of $\gamma(t)$ with respect to f (respectively, ν). Then*

$$\kappa_s(t)|\hat{\gamma}'(t)| = \mathrm{sgn}(K)\kappa_\#(t)|\hat{\nu}'(t)| \qquad (t \in [a,b])$$

holds, where $\mathrm{sgn}(K)$ denotes the sign of the Gaussian curvature of f, and

$$\hat{\gamma}(t) := f \circ \gamma(t), \quad \hat{\nu}(t) := \nu \circ \gamma(t) \qquad (a \le t \le b).$$

Proof. Let the singular curve $\gamma(t)$ be oriented such that $\mathcal{M}^+$ is on the left side of the curve on a positive local coordinate neighborhood of $\mathcal{M}^2$. Since $\gamma(t)$ consists of cuspidal edges, $\hat{\gamma}(t)$ is a regular curve in $\boldsymbol{R}^3$, so we may assume that t is an arc-length parameter of $\hat{\gamma}(t)$, without loss of generality. Then the singular curvature $\kappa_s(t)$ is given by

$$\kappa_s(t) = \det(\hat{\gamma}'(t), \hat{\gamma}''(t), \hat{\nu}(t)). \tag{6.45}$$

Let $\boldsymbol{n}(t) := -\hat{\gamma}'(t) \times \hat{\nu}(t)$. Then $(\hat{\gamma}', \boldsymbol{n}, \hat{\nu})$ gives a positive frame of $\boldsymbol{R}^3$. By Theorem 5.2.7, the limiting normal curvature vanishes, because $\log |K|$ is bounded. Thus, by (6.45), the derivative of $\boldsymbol{e}(t) := \hat{\gamma}'(t)$ satisfies

$$\boldsymbol{e}'(t)(= \hat{\gamma}''(t)) = \kappa_s(t)\boldsymbol{n}(t).$$

Here, since $\{e(t), n(t), \hat{\nu}(t)\}$ is an orthonormal frame along the curve, in the same way as we derive the Frenet–Serret formula for space curves (cf. [84, §5]), we can write

$$\frac{d}{dt}(e, n, \hat{\nu}) = (e, n, \hat{\nu})\begin{pmatrix} 0 & -\kappa_s & 0 \\ \kappa_s & 0 & -\mu \\ 0 & \mu & 0 \end{pmatrix} \tag{6.46}$$

for some C^∞-function $\mu(t)$ $(a \le t \le b)$. Since f has cuspidal edges along γ, it can be easily checked that $\mu(t)$ never vanishes. By (6.44), we have

$$\kappa_\# |\nu'| = \mathrm{sgn}(K)\frac{\det(\hat{\nu}', \hat{\nu}'', \hat{\nu})}{|\hat{\nu}'(t)|^2} = \mathrm{sgn}(K)\frac{\det(\hat{\nu}', \hat{\nu}'', \hat{\nu})}{\mu^2}.$$

Since $\hat{\nu}' = -\mu n$ and $\hat{\nu}'' = -\mu'n - \mu n'$, we have $\det(\hat{\nu}', \hat{\nu}'', \hat{\nu}) = \mu^2\kappa_s$, by (6.46). Hence we have

$$\kappa_\# |\hat{\nu}'| = \mathrm{sgn}(K)\kappa_s = \mathrm{sgn}(K)\kappa_s|\hat{\gamma}'|,$$

proving this theorem. □

As an application, we have:

Corollary 6.9.12 ([48]). *Let $\mathcal{M}^2$ be an orientable 2-manifold, and let $f : \mathcal{M}^2 \to \mathbf{R}^3$ be a wave front such that $\log|K|$ is bounded on the set of regular points, where K is the Gaussian curvature of f. If K is negative, the Euler number of $\mathcal{M}^2$ is zero.*

This was already proved in Corollary 5.3.4 in Chapter 5.

Proof. Since the Gaussian curvature is negative, Theorem 6.9.11 implies that the total singular curvature of f (i.e., the integral of the singular curvature over the singular set) and the total singular curvature of ν have the same absolute values but opposite signs. Since $K\,dA = -|K|\,dA$, adding (6.27) and (6.30), we obtain $\chi(\mathcal{M}^2) = 0$. □

There exist closed wave fronts with negative constant Gaussian curvature homeomorphic to tori, see [56]. Since the Euler number of a torus is zero, this is an example for Corollary 6.9.12.

Corollary 6.9.12 can be generalized as follows: Let $(\mathcal{N}^3, g)$ be a Riemannian 3-manifold of constant curvature 0, and let $f : \mathcal{M}^2 \to \mathcal{N}^3$ be a closed wave front. Then f can be lifted to a wave front in $\mathbf{R}^3$, which is the universal covering of $\mathcal{N}^3$. Though such a lift is a singly, doubly or triply periodic surface in $\mathbf{R}^3$, its Gauss map $\nu : \mathcal{M}^2 \to S^2$ is well defined on $\mathcal{M}^2$. Then the proof of Corollary 6.9.12 is valid, that is, if the Gaussian curvature K is negative and $\log|K|$ is bounded, then $\chi(\mathcal{M}^2) = 0$.

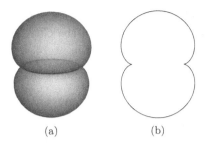

(a)   (b)

Fig. 6.10. A closed wave front with positive Gaussian curvature and its profile curve.

Next, we introduce a result for wave fronts with positive Gaussian curvature.

Corollary 6.9.13. *Let $f : S^2 \to \mathbf{R}^3$ be a wave front defined on the 2-sphere S^2 such that $\log |K|$ is bounded. If f has a singular point and K is positive, then (cf. (6.20))*

$$\int_{\mathcal{M}^2} \Omega > 4\pi. \tag{6.47}$$

Proof. Since $\chi(S^2) = 2$, the equality (6.27) and Theorem 5.5.11 in Chapter 5 yield the conclusion. □

As an example, a wave front of revolution with an epicycloid having two cusps as a generating curve (cf. Fig. 6.10) has positive Gaussian curvature.

Corollary 6.9.13 also holds for singly, doubly or triply periodic wave fronts. In fact, (6.47) holds for such a wave front with positive Gaussian curvature K such that $\log K$ is bounded.

Intrinsic Gauss–Bonnet theorem with singularities. The Gauss–Bonnet theorem holds not only for closed surfaces in the Euclidean 3-space, but also for compact Riemannian 2-manifolds. Since singularities are defined only under existence of an ambient space, they cannot be treated intrinsically without an ambient space present, even though the classical Gauss–Bonnet theorem can be treated intrinsically. However, this can be rectified for 2-manifolds with certain vector bundles of rank 2, called *coherent tangent bundles*, as in Chapter 9. As an application, theorems in this section can be generalized to surfaces in spaces of constant curvature, such as the 3-sphere and hyperbolic 3-space. See [73] and also [10] for an interesting application. On the other hand, cross caps are the singular points most frequently appearing on surfaces in $\mathbf{R}^3$, and their induced metrics generalize to a certain class of semi-positive definite metrics called *Whitney* *metrics.* For 2-manifolds having such metrics, a Gauss–Bonnet type theorem holds, see [28]. A generalization of the Gauss–Bonnet type formula (6.39) for higher dimensions is discussed in [9, 74].

Exercises 6.9

1 Prove Theorem 6.9.8.

2 Let $\gamma(t) := (\cos t, \sin t) - \frac{1}{3}(\cos 3t, \sin 3t)$, which is a closed wave front as a plane curve with two cusps looking something like a snowman, and consider a closed surface of revolution with profile curve γ (Fig. 6.10). Verify that the Gaussian curvature K of such a front is positive and $\log|K|$ is bounded.

Chapter 7

Flat Surfaces in R^3

In this chapter, we will discuss flat surfaces (i.e., surfaces whose Gaussian curvature vanishes). Caustics of regular surfaces are the union of singular sets appearing in the family of parallel surfaces. When the original surface is flat, its caustic is also a flat surface. We next define completeness of wave front surfaces and discuss the properties of complete flat surfaces.

7.1. Singularities of Caustics

We denote by λ_1, λ_2 the principal curvature functions of a regular surface $f : U \to R^3$ defined on a region $U(\subset R^2)$ which has no umbilics (see Fact 2.1.13 in Chapter 2). We suppose that λ_i $(i = 1, 2)$ does not vanish on U. We set

$$C_i = f + \frac{1}{\lambda_i}\nu, \tag{7.1}$$

and this is called a *caustic* (or the *focal surface*), where ν is the unit normal vector of f, a concept that is analogous to the caustic of plane curves (cf. (1.11) in Chapter 1).

Caustics of surfaces may have singular points in general. We will examine the criteria for cuspidal edges and swallowtails (cf. Theorem 4.2.3 in Chapter 4) of caustics. Without loss of generality, we may consider the caustic $C := C_1$ for the first principal curvature. Suppose that λ_1 has no zeros on U. Then C is defined by (7.1) on U. Since we assumed f has no umbilics, we may assume that (u, v) is a curvature line coordinate system on U (cf. Theorem 2.1.19), and the u-directions point in the principal direction corresponding to λ_1. Then Weingarten's formula (cf. Proposition 2.1.5) yields

$$\nu_u = -\lambda_1 f_u, \quad \nu_v = -\lambda_2 f_v, \tag{7.2}$$

and we have

$$C_u = f_u - \frac{(\lambda_1)_u}{(\lambda_1)^2}\nu + \frac{1}{\lambda_1}\nu_u$$

$$= f_u - \frac{(\lambda_1)_u}{(\lambda_1)^2}\nu - \frac{1}{\lambda_1}\lambda_1 f_u = -\frac{(\lambda_1)_u}{(\lambda_1)^2}\nu. \qquad (7.3)$$

Similarly, we get

$$C_v = f_v - \frac{(\lambda_1)_v}{(\lambda_1)^2}\nu + \frac{1}{\lambda_1}\nu_v = \left(1 - \frac{\lambda_2}{\lambda_1}\right)f_v - \frac{(\lambda_1)_v}{(\lambda_1)^2}\nu, \qquad (7.4)$$

which never vanishes since $\lambda_1 \neq \lambda_2$. Since f_u and f_v are mutually orthogonal (Proposition 2.1.17),

$$\boldsymbol{n} := \frac{f_u}{|f_u|} \qquad (7.5)$$

gives the unit normal vector field of C. This implies C is a frontal. Since f_v and ν are linearly independent, the singular points of C are characterized by the equation $(\lambda_1)_u = 0$. In particular, $C_u = \boldsymbol{0}$ holds. Since $\boldsymbol{n}$ is perpendicular to ν, (7.2) and (7.5) yield

$$\boldsymbol{n}_u \cdot \nu = (\boldsymbol{n} \cdot \nu)_u - \boldsymbol{n} \cdot \nu_u = \frac{\lambda_1}{|f_u|}f_u \cdot f_u = \lambda_1|f_u| \neq 0.$$

In particular, $\boldsymbol{n}_u \neq \boldsymbol{0}$. Thus, the facts $C_u = \boldsymbol{0}$ and $C_v \neq \boldsymbol{0}$ at the singular points of C implies that C is a wave front on U. Moreover, we can prove the following assertion:

Theorem 7.1.1. *Let $f : U \to \boldsymbol{R}^3$ be a regular surface without umbilic points, and suppose one of the principal curvature functions λ_1 does not have zeros on U. Then the caustic $C : U \to \boldsymbol{R}^3$ with respect to the principal curvature function λ_1 is a wave front. Moreover, if (u, v) is a curvature line coordinate system of U such that the u-direction corresponds to λ_1, then $p \in U$ is a singular point of $C(:= C_1)$ if and only if $(\lambda_1)_u(p) = 0$. Moreover,*

(1) *p is a cuspidal edge of C if and only if $(\lambda_1)_{uu}(p) \neq 0$,*
(2) *p is a swallowtail of C if and only if*

$$(\lambda_1)_{uu}(p) = 0, \quad (\lambda_1)_{uuu}(p) \neq 0, \quad (\lambda_1)_{uv}(p) \neq 0.$$

Proof. We have already proved that C is a wave front. So, it is sufficient to show the criteria for cuspidal edges and for swallowtails: Since we have assumed $\lambda_1 \neq 0$, the area density function of C is given by

$$\det(C_u, C_v, \boldsymbol{n}) = -\left(1 - \frac{\lambda_2}{\lambda_1}\right)\frac{(\lambda_1)_u}{(\lambda_1)^2}.$$

Since f is umbilic free, $\lambda_1 \neq \lambda_2$ holds on U. So, we can choose $\Lambda_C := (\lambda_1)_u$ as an identifier of singularities. By (7.3), ∂_u points in the null direction. Thus, the assertions (1) and (2) are obtained by applying Theorem 4.2.3 to Λ_C (if $(\Lambda_C)_u = (\lambda_1)_{uu}$ vanishes, $d\Lambda_C \neq \mathbf{0}$ is equivalent to $(\lambda_1)_{uv} \neq 0$). $\square$

We note that (1) of Theorem 7.1.1 together with (2) of Theorem 4.3.1 yields the following corollary.

Corollary 7.1.2. *Let $f : U \to \mathbf{R}^3$ be a regular surface without umbilic points, and suppose λ_1 does not have zeros on U. If we set $t_0 := 1/\lambda_1$, then a point where the parallel surface f^{t_0} (cf. (2.23) of Chapter 2) has a swallowtail singular point is a cuspidal edge singular point of the caustic of f whenever $d\lambda_1 \neq \mathbf{0}$.*

7.2. Parabolic Points of Caustics

We denote by $\hat{\Sigma}(f^t)$ the image of the singular sets of the parallel surfaces f^t ($t \in \mathbf{R}$) of a regular surface $f : U \to \mathbf{R}^3$. Take the union of the singular sets

$$\mathcal{C}_f := \bigcup_{t \in \mathbf{R}} \hat{\Sigma}(f^t),$$

and consider the relationship between $\mathcal{C}_f$ and f. By equation (2.28) given in the proof of Theorem 2.2.1, $\hat{\Sigma}(f^t)$ can be expressed as

$$\hat{\Sigma}(f^t) := \left\{ f^t(p); \ \frac{1}{\lambda_1(p)} = t \text{ or } \frac{1}{\lambda_2(p)} = t, \ p \in U \right\}, \tag{7.6}$$

where λ_j ($j = 1, 2$) are the principal curvatures of f. (Here we interpret $1/\lambda_j$ as ∞ at the points where $\lambda_j = 0$.) In particular, we can write

$$\mathcal{C}_f := C_1(U) \cup C_2(U).$$

Points on a regular surface at which the Gaussian curvature vanishes are called *parabolic points*. We are interested in the points on a surface f which correspond to the parabolic points of $\mathcal{C}_f$.

Definition 7.2.1. We let $\mathbf{v}_1, \mathbf{v}_2$ be the tangent vectors at p which give the principal directions with respect to the principal curvature functions λ_1, λ_2, respectively. If the derivative of the function λ_1 with respect to the direction $\mathbf{v}_2$ (resp. λ_2 with respect to the direction $\mathbf{v}_1$) vanishes, that is,

$$d\lambda_1(\mathbf{v}_2) = 0 \quad (\text{resp. } d\lambda_2(\mathbf{v}_1) = 0)$$

holds, then the point p is called a *sub-parabolic point* with respect to the principal curvature function λ_1 (resp. λ_2).

We prove the following assertion.

Theorem 7.2.2. *Let $f : U \to \mathbf{R}^3$ be a regular surface which has no umbilics, and let λ_1, λ_2 be its principal curvature functions. If λ_1 has no zeros on U, then the following two assertions are equivalent:*

(1) p *is a parabolic point of* C_1.

(2) p *is a sub-parabolic point of f with respect to the principal curvature* λ_2.

Proof. We take a curvature line coordinate system (u, v) centered at p so that $\mathbf{v}_1$ gives the u-direction and $\mathbf{v}_2$ gives the v-direction. Since the unit normal vector field of C_1 is given by (cf. (7.5))

$$\mathbf{n} := f_u / |f_u|, \tag{7.7}$$

the differential of $\mathbf{n}$ with respect to the direction $(\cos\theta, \sin\theta)$ $(\theta \in \mathbf{R})$ at p is given by

$$\mathbf{n}_\theta := (\cos\theta)\mathbf{n}_u + (\sin\theta)\mathbf{n}_v,$$

which is perpendicular to $\mathbf{n}$ (that is f_u).

Since (u, v) are curvature line coordinates, f_u is perpendicular to f_v. In particular, $\mathbf{n}_\theta(p)$ is expressed as a linear combination of $f_v(p)$ and $\nu(p)$. The necessary and sufficient condition that p is a parabolic point of C_1 gives that the unit normal vector field of C_1 as a map $\mathbf{n} : U \to S^2$ has a singular point at p, that is, $\mathbf{n}_u$, $\mathbf{n}_v$ are linearly dependent. This is equivalent to the condition that there exists an angle $\theta \in [0, 2\pi)$ such that $\mathbf{n}_\theta(p) = \mathbf{0}$. Since $\mathbf{n}_\theta$ is a linear combination of f_v and ν, the vector $\mathbf{n}_\theta(p)$ vanishes if and only if

$$\mathbf{n}_\theta(p) \cdot f_v(p) = \mathbf{n}_\theta(p) \cdot \nu(p) = 0. \tag{7.8}$$

On the other hand, we have (cf. (7.7))

$$\mathbf{n}_u = \frac{f_{uu}}{|f_u|} + \left(\frac{1}{|f_u|}\right)_u f_u, \quad \mathbf{n}_v = \frac{f_{uv}}{|f_u|} + \left(\frac{1}{|f_u|}\right)_v f_u.$$

Since f_u is perpendicular to f_v,

$$\mathbf{n}_u(p) \cdot f_v(p) = \frac{f_{uu}(p) \cdot f_v(p)}{|f_u(p)|}, \quad \mathbf{n}_u(p) \cdot \nu(p) = \frac{f_{uu}(p) \cdot \nu(p)}{|f_u(p)|}$$

hold. Similarly, we have

$$\mathbf{n}_v(p) \cdot f_v(p) = \frac{f_{uv}(p) \cdot f_v(p)}{|f_u(p)|}, \quad \mathbf{n}_v(p) \cdot \nu(p) = \frac{f_{uv}(p) \cdot \nu(p)}{|f_u(p)|}.$$

In particular, (7.8) is equivalent to

$$(\cos\theta f_{uu} + \sin\theta f_{uv}) \cdot f_v = (\cos\theta f_{uu} + \sin\theta f_{uv}) \cdot \nu = 0 \tag{7.9}$$

holding at p.

Since (u, v) is a curvature line coordinate system,

$$(M =)f_{uv} \cdot \nu = 0 \tag{7.10}$$

holds by Proposition 2.1.17. Since $f_{uu} \cdot \nu = -f_u \cdot \nu_u$, (7.2) implies $\lambda_1 = (f_{uu} \cdot \nu)/(f_u \cdot f_u)$ and the second equation of (7.9) reduces to

$$0 = \cos\theta f_{uu}(p) \cdot \nu(p) = \lambda_1(p) \cos\theta \big(f_u(p) \cdot f_u(p)\big),$$

which implies $\theta = \pm\pi/2$, since $\lambda_1(p) \neq 0$. Since $\mathbf{n}_{\pi/2} = -\mathbf{n}_{-\pi/2}$, we may set $\theta = \pi/2$ without loss of generality. Then (7.9) is equivalent to

$$f_{uv}(p) \cdot f_v(p) = 0. \tag{7.11}$$

By (7.10),

$$0 = (f_{uv} \cdot \nu)_v = f_{uvv} \cdot \nu + f_{uv} \cdot \nu_v$$

holds. This fact, with $\lambda_2 = f_{vv} \cdot \nu/(f_v \cdot f_v)$, yields

$$
\begin{aligned}
(\lambda_2)_u &= \frac{f_{vv} \cdot \nu_u}{f_v \cdot f_v} + \frac{f_{vvu} \cdot \nu}{f_v \cdot f_v} - \frac{2(f_{vu} \cdot f_v)(f_{vv} \cdot \nu)}{(f_v \cdot f_v)^2} \\
&= \frac{f_{vv} \cdot \nu_u}{f_v \cdot f_v} - \frac{f_{uv} \cdot \nu_v}{f_v \cdot f_v} - 2\lambda_2 \frac{f_{uv} \cdot f_v}{f_v \cdot f_v}.
\end{aligned} \tag{7.12}
$$

Since $f_u \cdot f_v = 0$, we have

$$f_u \cdot f_{vv} = (f_u \cdot f_v)_v - f_{uv} \cdot f_v = -f_{uv} \cdot f_v.$$

Since $\nu_u = -\lambda_1 f_u$, we have $f_{vv} \cdot \nu_u = -\lambda_1 f_u \cdot f_{vv} = \lambda_1 f_v \cdot f_{uv}$ and

$$\text{The right-hand side of (7.12)} = -\lambda_1 \frac{f_{vv} \cdot f_v}{f_v \cdot f_v} + \lambda_2 \frac{f_{uv} \cdot f_v}{f_v \cdot f_v} - 2\lambda_2 \frac{f_{uv} \cdot f_v}{f_v \cdot f_v}$$

$$= (f_{uv} \cdot f_v)\frac{\lambda_1 - \lambda_2}{f_v \cdot f_v}.$$

In particular, $(\lambda_2)_u$ vanishes at p if and only if (7.11) holds. $\qquad\square$

By definition, the Gaussian curvature of a flat regular surface vanishes identically. In particular, one of its principal curvatures vanishes identically. If the regular surface has no umbilics, then we may assume that λ_1 is identically zero and λ_2 does not admit zero points. Then C_1 is the empty set, and the image of C_2 coincides with $\mathcal{C}_f$. We take curvature line coordinates (u, v) such that the u-direction corresponds to λ_1. Since $\lambda_1 = 0$, we have $(\lambda_1)_v = 0$, that is, all points are sub-parabolic points with respect to λ_1. Thus, C_2 must be a flat surface, giving us the following classical fact:

Corollary 7.2.3. *The caustic of a regular flat surface is a flat surface wherever it is regular.*

Exercise 7.2

1 Show that a parallel surface of the tangential developable f of a helix (given in Exercise 5.4 in Section 5.4) is congruent to f. Moreover, show that the caustic of f is a circular cylinder.

7.3. Completeness and Weak Completeness of Wave Fronts

It is classically known that a complete flat regular surface in the Euclidean space $\boldsymbol{R}^3$ must be a cylinder over a plane curve (Example 2.1.18), that is, there exists a regular curve $\gamma(t)$ $(t \in \boldsymbol{R})$ on the xy-plane such that the surface is congruent to the ruled surface (cf. [34])

$$\gamma(t) + v \begin{pmatrix} 0 \\ 0 \\ 1 \end{pmatrix}.$$

So, it is natural to discuss flat surfaces with singular points to see global properties of such surfaces. We defined flatness of a wave front from a two-dimensional manifold $\mathcal{M}^2$ to $\boldsymbol{R}^3$, in Definition 2.3.6. By this definition, flatness of regular surfaces is equivalent to that for wave fronts. We now define two notions of completeness of wave fronts as generalizations of completeness of Riemannian manifolds.

Definition 7.3.1. Let $\mathcal{M}^2$ be a 2-manifold and $f : \mathcal{M}^2 \to \boldsymbol{R}^3$ a wave front. The map f is called *complete* if the singular set $\Sigma(f)$ is contained in a compact subset A of $\mathcal{M}^2$ and there exists a complete Riemannian metric g on $\mathcal{M}^2$ such that g coincides with the first fundamental form of f on $\mathcal{M}^2 \setminus A$.

If f has no singular points, then this completeness coincides with the original notion of completeness of Riemannian manifolds. If f is complete, then it is an immersion on the exterior of a sufficiently large open subset of $\mathcal{M}^2$. This is convenient for analyzing ends of complete surfaces.

We next define another generalization of completeness: Let $f : \mathcal{M}^2 \to \boldsymbol{R}^3$ be a wave front, and ν a unit normal vector field. Then the sum

$$d\sigma^2 := df \cdot df + d\nu \cdot d\nu \tag{7.13}$$

of the first fundamental form $df \cdot df$ and the third fundamental form[1] $d\nu \cdot d\nu$ gives a Riemannian metric, since $(f, \nu) : \mathcal{M}^2 \to \mathbf{R}^3 \times S^2$ is an immersion.

Definition 7.3.2. The wave front f is called *weakly complete* if this $d\sigma^2$ is a complete Riemannian metric.

The following assertion holds:

Proposition 7.3.3. *If a front* $f : \mathcal{M}^2 \to \mathbf{R}^3$ *is complete, then it is weakly complete.*

Let g, h be quadratic forms on a real vector space V. We write $g \geq h$ if $g(\boldsymbol{v}, \boldsymbol{v}) \geq h(\boldsymbol{v}, \boldsymbol{v})$ holds for each $\boldsymbol{v} \in V$. Before proving the proposition, we prepare the following lemma:

Lemma 7.3.4. *Let* $\mathcal{M}^2$ *be a manifold, and let* g_1, g_2 *be two Riemannian metrics on* $\mathcal{M}^2$. *If* $g_2 \geq g_1$ *and* g_1 *is a complete Riemannian metric on* $\mathcal{M}^2$, *then so is* g_2.

Proof. Let d_1 and d_2 be the distance functions induced by g_1 and g_2, respectively. Then these are compatible with respect to the topology of $\mathcal{M}^2$. Since $g_2 \geq g_1$, it holds that

$$d_2(p, q) \geq d_1(p, q) \quad (p, q \in \mathcal{M}^2).$$

Since g_1 is complete, Hopf–Rinow's theorem (cf. [41]) yields that $(\mathcal{M}^2, d_1)$ is a complete metric space. We let $\{p_n\}_{n=1,2,3,\dots}$ be a Cauchy sequence of $(\mathcal{M}^2, d_2)$. Since $d_2 \geq d_1$, $\{p_n\}_{n=1,2,3,\dots}$ is also a Cauchy sequence with respect to d_1. Since d_1 is a complete metric, $\{p_n\}_{n=1,2,3,\dots}$ converges to a point in $\mathcal{M}^2$. Since d_1, d_2 induce the same topology, this means that $(\mathcal{M}^2, d_2)$ is also a complete metric space. Again by Hopf–Rinow's theorem, g_2 is a complete Riemannian metric on $\mathcal{M}^2$. □

Proof of Proposition 7.3.3. Since f is complete, the singular set $\Sigma(f)$ is compact, and there exist a compact subset A of $\mathcal{M}^2$ and a complete Riemannian metric g on $\mathcal{M}^2$ such that g coincides with the first fundamental form ds^2 of f on $\mathcal{M}^2 \setminus A$. Since A is compact, there exists a positive constant $c > 0$ such that $d\sigma^2 \geq cg$ on A, where $d\sigma^2$ given by (7.13). In fact, the set

$$S(T\mathcal{M}^2) := \{\boldsymbol{v} \in T\mathcal{M}^2 \, ; \, g(\boldsymbol{v}, \boldsymbol{v}) = 1\}$$

---

[1] We consider ν as a map into $S^2 (\subset \mathbf{R}^3)$. Then the first fundamental form of ν as a map into $\mathbf{R}^3$ is $d\nu \cdot d\nu$, called the *third fundamental form* of f.

is called the unit tangent bundle with respect to the Riemannian metric g, which can be considered as a submanifold of the tangent bundle $T\mathcal{M}^2$ having a canonical projection $\pi : S(T\mathcal{M}^2) \to \mathcal{M}^2$. Since

$$\varphi : S(T\mathcal{M}^2) \ni v \mapsto d\sigma^2(v, v) \in \boldsymbol{R}$$

is a continuous function and $\pi^{-1}(A)$ is compact, the positive lower bound c $(0 < c < 1)$ of the function φ on $\pi^{-1}(A)$ exists. This, together with the fact that $ds^2 = g$ on $\mathcal{M}^2 \setminus A$, implies that $d\sigma^2 > cg$ on $\mathcal{M}^2$.

Since g is a complete Riemann metric, so is cg, and then $d\sigma^2$ is also complete by Lemma 7.3.4.   □

Remark 7.3.5. The converse assertion of Proposition 7.3.3 is not true in general. In fact, the tangent developable along a helix is weakly complete but not complete (cf. Corollary 7.3.7). We also remark that completeness is not stable under taking covering spaces, whereas weak completeness is stable. In fact, if $\pi : \tilde{\mathcal{M}}^2 \to \mathcal{M}^2$ is a covering projection and $\pi^{-1}(p)$ $(p \in \mathcal{M}^2)$ is not a finite set, then the inverse image of singular set of $f : \mathcal{M}^2 \to \boldsymbol{R}^3$ is not compact, and $f \circ \pi$ cannot be complete.

We now investigate the weak completeness of the tangential developables defined in Example 2.3.8 in Chapter 2. We denote by κ and τ the curvature function and torsion of a regular curve $\hat{\gamma}\colon \boldsymbol{R} \to \boldsymbol{R}^3$ with non-vanishing curvature function. Without loss of generality we may assume that t is an arc-length parameter of γ. The flat surface given by

$$f(t, v) := \hat{\gamma}(t) + v\hat{\gamma}'(t)$$

is the tangential developable associated to $\gamma(t)$, and the sum of the first and third fundamental forms is estimated by

$$d\sigma^2 = df \cdot df + d\nu \cdot d\nu$$

$$= \{(1 + v^2\kappa^2)dt^2 + 2\,dt\,dv + dv^2\} + \tau^2\,dt^2$$

$$= (1 + v^2\kappa^2 + \tau^2)\,dt^2 + 2\,dt\,dv + dv^2$$

$$\geq (1 + \tau^2)\,dt^2 + 2\,dt\,dv + dv^2.$$

Using this, we prove the following assertion.

Proposition 7.3.6. *Let $\hat{\gamma} : \boldsymbol{R} \to \boldsymbol{R}^3$ be a regular space curve of positive curvature $\kappa(t)$ $(t \in \boldsymbol{R})$. If the absolute value of the torsion function $\tau(t)$ is greater than a positive constant $c > 0$, then the tangential developable associated to $\hat{\gamma}$ is a weakly complete flat wave front.*

Proof. The quadratic form $h := (1 + \tau^2)dt^2 + 2dtdv + dv^2$ is positive definite if $|\tau| > 0$. We suppose that $|\tau(t)| \geq c$ holds for each $t \in \mathbf{R}$ for some positive constant c. Then h gives a Riemannian metric on $\mathbf{R}^2$. We consider a symmetric matrix

$$A := \begin{pmatrix} 1 + \tau(t)^2 & 1 \\ 1 & 1 \end{pmatrix},$$

whose eigenvalues are

$$a_1 := \frac{1}{2}(\sqrt{\tau^4 + 4} + \tau^2 + 2), \quad a_2 := \frac{1}{2}(-\sqrt{\tau^4 + 4} + \tau^2 + 2).$$

Then it holds that

$$a_1 \geq a_2 \geq \delta, \quad \delta := \frac{1}{2}(-\sqrt{c^4 + 4} + c^2 + 2) > 0.$$

Since A is a symmetric matrix, there exists an orthogonal matrix P such that

$$P^T A P = \begin{pmatrix} a_1 & 0 \\ 0 & a_2 \end{pmatrix}.$$

Take a vector $\boldsymbol{v} = (x, y)$ and regard it is a column vector, then

$$\begin{pmatrix} X \\ Y \end{pmatrix} := P\boldsymbol{v}$$

satisfies $X^2 + Y^2 = x^2 + y^2$. Thus, we have

$$h(\boldsymbol{v}, \boldsymbol{v}) = \boldsymbol{v}^T A \boldsymbol{v} = a_1 X^2 + a_2 Y^2$$

$$\geq \delta(X^2 + Y^2) = \delta(x^2 + y^2).$$

In particular, $h \geq \delta(dt^2 + dv^2)$ holds. Since $dt^2 + dv^2$ is the standard complete Riemannian metric on $\mathbf{R}^2$, h is also a complete metric by Lemma 7.3.4. Thus, so is the metric $d\sigma^2 = df \cdot df + d\nu \cdot d\nu$, proving the assertion. $\qquad\square$

Corollary 7.3.7. *The tangential developable of the helix defined in Exercise 1 in Section 5.4 is weakly complete, but not complete.*

Proof. The completeness follows from Proposition 7.3.6. However, the map cannot be complete since the helix as the singular set is not compact. $\qquad\square$

Several concepts of completeness and weak completeness are discussed in [83].

7.4.   Complete Flat Wave Fronts in $\boldsymbol{R}^3$

In this section, we construct complete flat wave fronts, which have no self-intersections outside of a certain compact set.

We let $\xi(t)$ $(t \in [0,c])$ be a simple closed regular curve on the unit sphere $S^2 := \{\boldsymbol{v} \in \boldsymbol{R}^3 \,;\, |\boldsymbol{v}| = 1\}$ such that the geodesic curvature (cf. (2.13) of Section 2.1) is positive everywhere, where c is a positive constant. Such a closed curve is called a *convex curve* on the sphere. Without loss of generality, we may assume that t is proportional to an arc-length parameter and is extended to $\boldsymbol{R}$ so that $\xi(t + c) = \xi(t)$ for $t \in \boldsymbol{R}$, where $c > 0$. It is well known that convex curves on S^2 lie in an open hemisphere and meet each great circle at most twice (see [2, §20]). Thus the image of ξ lies in an open hemisphere. Then the surface defined by

$$f_0(t, v) := v\xi(t) \quad (0 \le t \le c, \ v \in \boldsymbol{R})$$

satisfies the definition of a cone given in Example 2.3.7, and gives a flat wave front. We set

$$\nu(t) := \frac{\xi(t) \times \xi'(t)}{|\xi(t) \times \xi'(t)|},$$

which gives a unit vector field along the curve ξ so that $\{\xi(t), \xi'(t), \nu(t)\}$ is an orthogonal basis of $\boldsymbol{R}^3$ for each t. Then $\nu(t)$ is a unit normal vector field on the cone $f_0(t, v)$. Since the geodesic curvature of the spherical curve ξ is positive, we have $\nu'(t) \ne \boldsymbol{0}$, and this fact implies f is a wave front. Moreover, the unit normal vector $\nu(t)$ does not depend on the parameter v, so f is a flat wave front.

The cone f_0 above is an example of a wave front whose image of the singular set is one point. We now construct a more general family of flat surfaces containing the cone as a special case, as follows.

We denote by $C^\infty(S^1)$ $(S^1 := \boldsymbol{R}/c\boldsymbol{Z})$ the set of real-valued c-periodic smooth functions on $\boldsymbol{R}$. Then the linear map defined by

$$\Phi : C^\infty(S^1) \ni a(t) \longmapsto \int_0^c a(t)\xi(t)dt \in \boldsymbol{R}^3$$

has the kernel

$$\mathrm{Ker}(\Phi) = \{a \in C^\infty(S^1) \,;\, (a, \xi_i) = 0, \ (i = 1, 2, 3)\},$$

where $\xi = (\xi_1, \xi_2, \xi_3)$ and $(,)$ is the L^2-inner product of $C^\infty(S^1)$ defined by

$$(a, b) := \int_0^c a(t)b(t)dt \quad (a, b \in C^\infty(S^1)).$$

In particular, $\mathrm{Ker}(\Phi)$ is an infinite-dimensional subspace and we can take infinitely many smooth functions $a \in C^\infty(S^1)$ satisfying

$$\int_0^c a(t)\xi(t)dt = \mathbf{0}. \tag{7.14}$$

So we fix such a function $a(t)$ arbitrarily, and set

$$\hat{\gamma}(t) := \int_0^t a(u)\xi(u)du. \tag{7.15}$$

By condition (7.14), $\hat{\gamma}(t)$ is a closed curve in $\mathbf{R}^3$. However, $\hat{\gamma}(t)$ may not be a regular curve. The zeros of the function $a(t)$ correspond to the singular points of the curve $\hat{\gamma}(t)$. By definition, $\hat{\gamma}'(t)$ is proportional to the vector $\xi(t) \in \mathbf{R}^3$. Like as in Example 2.3.8,

$$f(t,v) := \hat{\gamma}(t) + v\xi(t) \tag{7.16}$$

gives the tangential developable of the curve $\hat{\gamma}(t)$. (If $a(t)$ vanishes identically, then this surfaces coincides with the cone f_0 above.) Like as in the case of f_0,

$$\nu(t) := \frac{\xi(t) \times \xi'(t)}{|\xi(t) \times \xi'(t)|}$$

gives a unit normal vector field of f. Since the geodesic curvature of the spherical curve ξ is positive, we have $\nu'(t) \neq \mathbf{0}$, and this fact implies f is a flat wave front, since $\nu(t)$ does not depend on the parameter v. Moreover, it is easily checked that the image of the singular set of f coincides with the image of the space curve $\hat{\gamma}(t)$. It can also be easily checked that there are no other singular points of f other than $\hat{\gamma}(t)$. Moreover, it can be proved that f is a complete wave front (see [59] for this fact).

We now would like to discuss singular points of the surface f. The space curve $\hat{\gamma}(t)$ itself has singular points as a space curve, and we will see that these singular points of the curve give the singular points of f other than the cuspidal edges. Let δ be a positive constant, and consider a family of surfaces $f_\delta(t,v) := f(t,\delta v)/\delta$. Then $f_0 := \lim_{\delta \to 0} f_\delta$ coincides with the cone associated to the spherical curve ξ. This means that f_δ has no self-intersections outside of a sufficiently large compact subset of $\mathbf{R}^3$.

Example 7.4.1. Consider a circle

$$\xi(t) = (\cos\beta\cos t, \cos\beta\sin t, \sin\beta) \quad (0 \le t \le 2\pi) \tag{7.17}$$

on S^2, where $|\beta| < \pi/2$. For each integer $n \ge 2$, we set

$$a(t) := \cos nt. \tag{7.18}$$

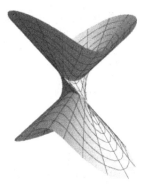

Fig. 7.1.   An example of a complete flat front which is not a cone.

Then this satisfies the period condition (7.14) and the resulting surface $f(t, v)$ given in (7.16) is a complete flat wave front. Figure 7.1 indicates the image of f for $n = 2$.

The following fact is known:

Fact 7.4.2 ([59]). If a complete flat wave front in $\mathbf{R}^3$ has non-empty singular set, and the surface does not have self-intersections outside of a ball of sufficiently large radius, then this surface is congruent to one of the surfaces given in (7.16).

We can prove the following assertion, which can be considered as a variant of the four vertex theorem for convex plane curves [84, §4].

Theorem 7.4.3 ([59]). *Let ξ be a convex curve in S^2, and let f be a flat front given in (7.16). Then, f has at least four singular points other than cuspidal edges.*

Since the swallowtail is the most generic singularity amongst the wave front singularities other than cuspidal edges, the above theorem asserts that typically there are at least four swallowtails on the surface. In fact, this estimate is the best possible because the complete flat wave front given in Figure 7.1 has exactly four swallowtails. The circular cone

$$f(u, v) := (u \cos v, u \sin v, u)$$

is a special case that has only one singular point as its image. However, in the domain of definition, $f^{-1}(0, 0, 0)$ has singular points consisting of a curve with infinitely many non-cuspidal-edge points, so the assertion of

Theorem 7.4.3 does not conflict with the fact that f has only one (image) cone point. As the reader will see in the proof of this theorem, this is related to the four vertex theorem for convex plane curves. An interesting example of flat surfaces are Möbius strips. Recently, Naokawa [61] gave a similar estimate for lower bounds on the number of singular points other than swallowtails on a given flat Möbius strip.

Proof of Theorem 7.4.3. We set

$$\xi = (\xi_1, \xi_2, \xi_3) : S^1 := \boldsymbol{R}/c\boldsymbol{Z} \to S^2 (\subset \boldsymbol{R}^3).$$

By definition, we have

$$\int_0^c a(t)\xi_i(t)dt = 0 \quad (i = 1, 2, 3). \tag{7.19}$$

Since ξ is a spherical convex curve, it lies in an open hemisphere of S^2. So, we may assume that $\xi_3(t) > 0$ without loss of generality. If $a(t)$ does not change sign, then $\int_0^c a(t)\xi_3(t)dt$ cannot be equal to zero, contradicting (7.19). Thus, the function $a(t)$ has at least one sign change on the interval $[0, c)$, whenever $a(t)$ vanishes identically. (A function $\varphi(t)$ is said to have a sign change at t if for any sufficiently small $\varepsilon > 0$, there exist $t_1, t_2 \in (t - \varepsilon, t + \varepsilon)$ such that $\varphi(t_1) < 0$ and $\varphi(t_2) > 0$, or vice versa.) We have already pointed out that the assertion is obvious when $a(t)$ vanishes identically. So, we may assume $a(t)$ changes sign.

It can be easily checked that the singular points of f are the set $\{v = 0\}$ in the tv-plane. So $\hat{\gamma}(t) := f(t, 0)$ gives the image of the singular set of f. We know that the restriction of the surface along the cuspidal edge singularities consists of regular space curves (cf. Theorem 4.2.3). (In fact, the null direction is different from the singular direction $\partial/\partial t$, and thus $df(\partial/\partial t) = \hat{\gamma}'(t)$ does not vanish.) Thus, the singular points of f which are not cuspidal edges are the points where $a(t)\xi(t) = \hat{\gamma}'(t) = \boldsymbol{0}$. Such points coincide with the zero set of the function $a(t)$. So, it is sufficient to show that $a(t)$ has at least four sign changes. We know that $a(t)$ must have at least one sign change. However, we also know that the number of sign changes is always even, whenever it is finite. So, we may assume that $a(t)$ has exactly two sign changes at $t = t_1, t_2$ ($0 < t_1 < t_2 < c$), and it is sufficient to find a contradiction. We prepare the following lemma. □

Lemma 7.4.4. *Let* $(\delta_1, \delta_2, \delta_3) \neq (0, 0, 0)$ *be a constant vector in* $\boldsymbol{R}^3$. *We denote by* $\xi(t) = (\xi_1(t), \xi_2(t), \xi_3(t))$ *the spherical convex curve* $\xi : S^1 \to S^2$. *Then the function*

$$\varphi(t) = \delta_1\xi_1(t) + \delta_2\xi_2(t) + \delta_3\xi_3(t)$$

has at most two distinct zeros. Moreover, if the function $\varphi(t)$ has exactly two zeros t_1, t_2, then this function changes sign at these two points.

Proof. The plane $\delta_1 x + \delta_2 y + \delta_3 z = 0$ in $\boldsymbol{R}^3$ passing through the origin meets the unit sphere S^2 at a great circle. We denote this circle by C. If the function $\varphi(t)$ has more than two zeros, then the spherical curve $\xi(t)$ must meet the great circle C at least three times. This contradicts the convexity of ξ in S^2. So we have obtained the first assertion.

We suppose that $\varphi(t)$ has exactly two distinct zeros at t_1, t_2, and suppose that $\varphi(t)$ does not change sign at $t = t_1$. Then the great circle C must be tangent to $\xi(t)$ at t_1 and also again meet $\xi(t)$ at $t = t_2$, which again contradicts the convexity of $\xi(t)$. $\qquad\square$

The proof of Theorem 7.4.3 continued. Let $\mathcal{A}$ be a finite-dimensional subspace of $C^\infty(S^1)$ defined by

$$\mathcal{A} := \{\delta_1 \xi_1(t) + \delta_2 \xi_2(t) + \delta_3 \xi_3(t) \in C^\infty(S^1); \delta_1, \delta_2, \delta_3 \in \boldsymbol{R}\}.$$

By Lemma 7.4.4, the linear map defined by

$$L : \mathcal{A} \ni \varphi \mapsto (\varphi(0), \varphi(t_1), \varphi(t_2)) \in \boldsymbol{R}^3$$

is an injection. Since the dimension of $\mathcal{A}$ is 3, the dimension formula of linear maps yields that L is a linear isomorphism, in particular L is a bijection. Since the great circle of S^2 passing through $\xi(t_1)$ and $\xi(t_2)$ is uniquely determined (we used the fact that ξ lies in an open hemisphere), there exists a function $\varphi \in \mathcal{A}$ such that

$$(\varphi(t_1), \varphi(t_2)) = (0, 0) \quad (0 < t_1 < t_1 < c).$$

Since $a(t)$ does not vanish identically, we may assume that $a(0) \neq 0$. Without loss of generality we may assume that $t_1, t_2 \neq 0$ and $a(0) \neq 0$. By Lemma 7.4.4, the function $\varphi(t)$ changes sign only at $t = t_1, t_2$. Since L is injective, such a function $\varphi(t)$ is determined up to a scalar multiplication, and so we may assume that $\varphi(0) = a(0)$. Since $a(t)$ also changes sign exactly at $t = t_1, t_2$, we have $a(t)\varphi(t) > 0$ for $t \neq t_1, t_2$. Thus,

$$\sum_{i=1}^{3} \delta_i \int_0^c a(t)\xi_i(t)dt = \int_0^c a(t)\varphi(t)dt > 0$$

holds, contradicting the period condition (7.19). So, we can conclude that $a(t)$ changes sign at least four times. $\qquad\square$

Chapter 8

Proof of the Criterion for Swallowtails

In this chapter, we prove the criteria for swallowtails using the notion of "versal unfolding". This method also enables us to give an alternative proof of the criterion for cuspidal edges given in Chapter 3.

8.1. Map-Germs

First, we introduce the notion "map-germ", in preparation for giving a proof of criteria for swallowtails (Theorems 2.6.3 in Chapter 2 and 4.2.3 in Chapter 4). Let U be a neighborhood of the origin of $\boldsymbol{R}^m$, and let

$$f : U \longrightarrow \boldsymbol{R}$$

be a C^∞-function. In the following discussion, we sometimes restrict a given neighborhood U of the origin as a domain of a function to a smaller neighborhood V of the origin. In this way, the particular choice of domain U is not essential, and we simply consider f to be defined in some neighborhood of the origin. Let g be another C^∞-function

$$g : U \longrightarrow \boldsymbol{R}.$$

If g coincides with f on a neighborhood $V \subset U$ of the origin, then we would like to identify them with each other: For C^∞-functions f and g, if

$$f(\boldsymbol{x}) = g(\boldsymbol{x}) \qquad (\boldsymbol{x} \in V)$$

holds on a neighborhood $V(\subset U)$ of the origin, then we say that f and g are *equivalent*, and this is denoted by $f \sim g$. Let m be a positive integer and

$$C_0^\infty(\boldsymbol{R}^m)$$

the quotient set of the set of C^∞-functions defined on a neighborhood of the origin by this equivalence relation $\sim$. Then the equivalence class of each function f belonging to $C_0^\infty(\boldsymbol{R}^m)$ is called a *function-germ* of $\boldsymbol{R}^m$ at the origin.

For two function-germs f, $g \in C_0^\infty(\boldsymbol{R}^m)$, one can see that the sum $f + g$ and the product fg can be defined on the new domain which is the intersection of the domains of f and g. For a positive integer n, we consider the set

$$C_0^\infty(\boldsymbol{R}^m, \boldsymbol{R}^n) = \{(f_1, \ldots, f_n)\,;\, f_1, \ldots, f_n \in C_0^\infty(\boldsymbol{R}^m)\}$$

of n-tuples of function-germs in $C_0^\infty(\boldsymbol{R}^m)$. This can be regarded as the quotient set of the set of C^∞ maps from a neighborhood of $\boldsymbol{R}^m$ to $\boldsymbol{R}^n$ by an equivalence relation which is defined as follows: two maps are *equivalent* if they coincide on a sufficiently small neighborhood of the origin. We call $F := (f_1, \ldots, f_n) \in C_0^\infty(\boldsymbol{R}^m, \boldsymbol{R}^n)$ a *map-germ*. We write

$$F\colon (\boldsymbol{R}^m, 0) \longrightarrow (\boldsymbol{R}^n, x)$$

when a map-germ F maps the origin $0 \in \boldsymbol{R}^m$ to $x \in \boldsymbol{R}^n$. If a map-germ F satisfies the following three conditions

- F maps the origin of $\boldsymbol{R}^m$ to the origin of $\boldsymbol{R}^n$, namely $F(0) = 0$,
- $m = n$,
- the Jacobian (i.e., the determinant of the Jacobi matrix) of F at the origin does not vanish,

then, by the implicit function theorem, there exists a neighborhood V of the origin such that F gives a diffeomorphism from V to $F(V)$. Regarding this, we say that a map-germ F satisfying the above three conditions is a *local diffeomorphism-germ* at $\boldsymbol{0}$.

Furthermore, for subsets containing the origin $\boldsymbol{0}$ of $\boldsymbol{R}^n$, we sometimes restrict our considerations to a neighborhood of $\boldsymbol{0}$. For this purpose, we introduce a notion of subset-germs.

Definition 8.1.1. Let $S_1, S_2 \subset \boldsymbol{R}^n$ be subsets containing $\boldsymbol{0}$. We say that S_1 and S_2 *determine the same subset-germ at the origin* if there exists a neighborhood U of $\boldsymbol{0}$ such that

$$S_1 \cap U = S_2 \cap U$$

holds. We call the equivalence class of a subset containing the origin of $\boldsymbol{R}^n$ by this equivalence relation a *subset-germ*.

8.2. Unfoldings and Discriminant Sets

Let r be a positive integer, and let

$$(t, \boldsymbol{x}) := (t, x_1, \ldots, x_r)$$

denote the coordinate system of $\boldsymbol{R} \times \boldsymbol{R}^r$, with $\boldsymbol{x} := (x_1, \ldots, x_r)$. Let $C^\infty_{(0,0)}(\boldsymbol{R} \times \boldsymbol{R}^r)$ be the set of function-germs at the origin $(0, \boldsymbol{0}) = (0, \ldots, 0) \in \boldsymbol{R}^{r+1}$ of $\boldsymbol{R} \times \boldsymbol{R}^r (= \boldsymbol{R}^{r+1})$. Although in essence this is the same as $C^\infty_0(\boldsymbol{R}^{r+1})$, we use the former notation, since we would like to distinguish the variables of the domain as in $(t, \boldsymbol{x})$.

Definition 8.2.1. For a positive integer k, the map-germ $\Phi \in C^\infty_{(0,0)}(\boldsymbol{R} \times \boldsymbol{R}^r)$ is said to have *regularity of order* $k + 1$ *in the variable* t if the function of one variable $\varphi(t) := \Phi(t, 0, \ldots, 0)$ satisfies

$$(\varphi(0) =) \, \varphi'(0) = \cdots = \varphi^{(k)}(0) = 0, \quad \varphi^{(k+1)}(0) \neq 0,$$

where $\varphi^{(k)}(t)$ stands for the kth derivative of $\varphi(t)$.

For instance, the function-germ $\Phi(t, x) := t^{k+1} + x : (\boldsymbol{R} \times \boldsymbol{R}, (0, 0)) \to (\boldsymbol{R}, 0)$ has regularity of order $k + 1$ in t. On the other hand, $\Phi(t, x) := tx$ and $\Phi(t, x) := x$ do not have regularity of order $l(\geq 2)$ in t.

From now on, we fix two integers $k(\geq 0)$ and $r(> 0)$.

The Malgrange preparation theorem. We state the Malgrange preparation theorem, which is useful for investigating singularities of maps.

Fact 8.2.2 (Malgrange preparation theorem). *Suppose that* $\Phi \in C^\infty_{(0,0)}(\boldsymbol{R} \times \boldsymbol{R}^r)$ *is a* C^∞ *function-germ which has regularity of order* $k + 1$ *$(k \geq 0)$ in the variable* t. *Then there exist a function-germ* $\rho \in C^\infty_0(\boldsymbol{R} \times \boldsymbol{R}^r)$ *and a map-germ*

$$\boldsymbol{H} = (H_1, \ldots, H_{k+1}) \in C^\infty_0(\boldsymbol{R}^r, \boldsymbol{R}^{k+1})$$

such that $\boldsymbol{H}(\boldsymbol{0}) = \boldsymbol{0}$, $\rho(0, \boldsymbol{0}) \neq 0$ *and*

$$\Phi(t, \boldsymbol{x}) = \rho(t, \boldsymbol{x}) \left(t^{k+1} + \sum_{i=1}^{k+1} H_i(\boldsymbol{x}) t^{k+1-i} \right) \tag{8.1}$$

holds, where $\boldsymbol{x} = (x_1, \ldots, x_r) \in \boldsymbol{R}^r$.

This theorem is a generalization of the Weierstrass preparation theorem for real analytic functions, and was proved by Malgrange in the 1960s. A full book was devoted to the first proof, and now simpler proofs are known. See [22, 52] for details.

The theorem means that a given function-germ can be identified with a polynomial

$$t^{k+1} + \sum_{i=1}^{k+1} H_i(x) t^{k+1-i}$$

of variable t up to multiplication by a nowhere-vanishing function.

Unfoldings and $P\mathcal{K}$-equivalence. We fix a C^∞-function-germ φ : $(\mathbf{R}, 0) \to (\mathbf{R}, 0)$ of one variable. A map-germ

$$\Phi : (\mathbf{R} \times \mathbf{R}^r, (0, \mathbf{0})) \to (\mathbf{R}, 0)$$

is an *r-dimensional unfolding* of φ if $\Phi(t, \mathbf{0}) = \varphi(t)$ holds.

Remark. In general, the notion of unfoldings can be generalized from one variable t to multi-variables $t_1, \ldots, t_s$, which is useful for investigating singularities of wave fronts in $\mathbf{R}^r$ whose rank of the Jacobi matrix is $r - s - 1$. The purpose of this chapter is to show the criteria for cuspidal edges and swallowtails which correspond to the case $r = 3$ and $s = 1$. Thus, we only deal with unfoldings Φ of functions $\varphi(t)$ of one variable. See [52] for general unfoldings of functions of general numbers of variables.

Definition 8.2.3. Let φ and ψ : $(\mathbf{R}, 0) \to (\mathbf{R}, 0)$ be two C^∞-function-germs of one variable, and let Φ and Ψ be r-dimensional unfoldings of φ and ψ, respectively. The unfoldings Φ and Ψ are *$P\mathcal{K}$-equivalent* if there exist a diffeomorphism-germ $\Xi : (\mathbf{R} \times \mathbf{R}^r, 0) \to (\mathbf{R} \times \mathbf{R}^r, 0)$ of the form

$$\Xi(t, \mathbf{x}) = (\vartheta(t, \mathbf{x}), \Theta(\mathbf{x})) \quad (\vartheta(t, \mathbf{x}) \in \mathbf{R}, \ \Theta(\mathbf{x}) \in \mathbf{R}^r), \qquad (8.2)$$

and a function-germ $\rho \in C_0^\infty(\mathbf{R} \times \mathbf{R}^r)$ satisfying $\rho(0, \mathbf{0}) \neq 0$ such that

$$\Phi(t, \mathbf{x}) = \rho(t, \mathbf{x}) \Psi(\vartheta(t, \mathbf{x}), \Theta(\mathbf{x})) \qquad (8.3)$$

holds.

By definition, it holds that

$$\frac{\partial \vartheta(t, \mathbf{x})}{\partial t} \neq 0 \qquad (8.4)$$

for sufficiently small $|t|$ and $|\mathbf{x}|$.

Remark 8.2.4. Two function-germs Φ, Ψ : $(\mathbf{R}^{r+1}, \mathbf{0}) \to (\mathbf{R}, 0)$ are *$\mathcal{K}$-equivalent* if there exist a function-germ $\rho \in C_0^\infty(\mathbf{R}^{r+1})$ satisfying $\rho(\mathbf{0}) \neq 0$ and a diffeomorphism-germ $\Xi : (\mathbf{R}^{r+1}, \mathbf{0}) \to (\mathbf{R}^{r+1}, \mathbf{0})$, which do not need to satisfy (8.2), so that $\Phi = \rho \Psi \circ \Xi$, that is, Φ coincides with the ρ multiple of $\Psi \circ \Xi$. In particular, $P\mathcal{K}$-equivalence means $\mathcal{K}$-equivalence[1] with Θ not depending on t.

By using $P\mathcal{K}$-equivalence, we can rewrite the Malgrange preparation theorem as follows.

Theorem 8.2.5 (Malgrange preparation theorem II). *If a C^∞-function-germ $\Phi : (\mathbf{R} \times \mathbf{R}^r, (0,0)) \to (\mathbf{R}, 0)$ has regularity of order $k+1$ in the variable t, then there exists an r-dimensional unfolding $\Psi : (\mathbf{R} \times \mathbf{R}^r, (0,\mathbf{0})) \to (\mathbf{R}, 0)$, which is $P\mathcal{K}$-equivalent to Φ, of the form*

$$\Psi(s, \boldsymbol{y}) = s^{k+1} + \sum_{i=1}^{k+1} H_i(\boldsymbol{y}) s^{k+1-i},$$

where $H_i \in C_0^\infty(\mathbf{R}^r, \mathbf{R})$ $(i = 1, \ldots, k+1)$ satisfy $H_i(\mathbf{0}) = 0$.

Proof. Setting $\Xi(t, \boldsymbol{x}) := (t, \boldsymbol{x})$ in Definition 8.2.3, we can show this theorem. Namely, an unfolding which has regularity of order $k+1$ in the variable t is $P\mathcal{K}$-equivalent to a polynomial in t of degree $k + 1$. So we obtain the conclusion. $\qquad \square$

As we remarked in Example 1.3.1 in Chapter 1, the polynomial

$$t^{k+1} + \sum_{i=1}^{k+1} h_i t^{k+1-i} \qquad (h_1, \ldots, h_{k+1} \in \mathbf{R})$$

can be modified into the form

$$P(s) := s^{k+1} + \sum_{i=2}^{k+1} h_i s^{k+1-i},$$

which has no kth degree term, by a transformation of the variable t to s. By this transformation, we can improve Theorem 8.2.5 as follows.

Theorem 8.2.6 (Malgrange preparation theorem III). *If an unfolding $\Phi : (\mathbf{R} \times \mathbf{R}^r, (0,\mathbf{0})) \to (\mathbf{R}, 0)$ of a C^∞-function-germ $\varphi(t) : (\mathbf{R}, 0) \to (\mathbf{R}, 0)$ has regularity of order $k + 1$ $(k \geq 1)$ in the variable t, then there exists an r-dimensional unfolding $\Psi : (\mathbf{R} \times \mathbf{R}^r, (0,\mathbf{0})) \to (\mathbf{R}, 0)$, which is $P\mathcal{K}$-equivalent to Φ, of the form*

$$\Psi(s, \boldsymbol{x}) = s^{k+1} + H_2(\boldsymbol{x}) s^{k-1} + \cdots + H_k(\boldsymbol{x}) s + H_{k+1}(\boldsymbol{x}),$$

where $H_i \in C_0^\infty(\mathbf{R}^r, \mathbf{R})$ $(i = 2, \ldots, k+1)$ satisfy $H_i(\mathbf{0}) = 0$.

---

[1] The "$\mathcal{K}$" in $\mathcal{K}$-equivalence comes from the word "Kontakt" (German), and the "P" of $P\mathcal{K}$-equivalence comes from the phrase "parameter preserving".

Proof. By Fact 8.2.2, we may assume Φ has the form (8.1), that is,

$$\Phi(t, \boldsymbol{x}) = \rho(t, \boldsymbol{x}) \left(t^{k+1} + \sum_{i=1}^{k+1} H_i(\boldsymbol{x}) t^{k+1-i} \right).$$

Since $H_1(\boldsymbol{0}) = 0$ in (8.1), a map-germ

$$\Xi(s, \boldsymbol{x}) := \left(s - \frac{H_1(\boldsymbol{x})}{k+1}, \boldsymbol{x} \right)$$

is a diffeomorphism-germ at the origin in $\boldsymbol{R}^{r+1}$. We set

$$\vartheta(s, \boldsymbol{x}) := s - \frac{H_1(\boldsymbol{x})}{k+1}, \qquad \Theta(\boldsymbol{x}) := \boldsymbol{x}.$$

Then this satisfies (8.2). We also set $s := \vartheta(t, \boldsymbol{x})$. Then we have the expression

$$\Phi \circ \Xi(s, \boldsymbol{x}) = \rho \circ \Xi(s, \boldsymbol{x}) \left(s^{k+1} + \sum_{i=2}^{k+1} \tilde{H}_i(\boldsymbol{x}) s^{k+1-i} \right),$$

where $\tilde{H}_i(\boldsymbol{x})$ are polynomials of $H_1(\boldsymbol{x}), \dots, H_{k+1}(\boldsymbol{x})$. In particular,

$$\Psi(s, \boldsymbol{x}) := s^{k+1} + \sum_{i=2}^{k+1} \tilde{H}_i(\boldsymbol{x}) s^{k+1-i}$$

is $P\mathcal{K}$-equivalent to Φ. By definition, $\tilde{H}_i(\boldsymbol{0}) = 0$ $(i = 2, \dots, k+1)$ hold. $\qquad\square$

Admissible unfolding and discriminant set

Definition 8.2.7. An unfolding $\Phi : (\boldsymbol{R} \times \boldsymbol{R}^r, (0, \boldsymbol{0})) \to (\boldsymbol{R}, 0)$ of a C^∞-function-germ $\varphi : (\boldsymbol{R}, 0) \to (\boldsymbol{R}, 0)$ is said to be *admissible* if

(1) $\Phi'(0, \boldsymbol{0}) = 0$, and
(2) the rank of the $(r+1) \times 2$ matrix

$$\left. \begin{pmatrix} \Phi' & \Phi_{x_1} & \cdots & \Phi_{x_r} \\ \Phi'' & \Phi'_{x_1} & \cdots & \Phi'_{x_r} \end{pmatrix} \right|_{(t, \boldsymbol{x}) = (0, \boldsymbol{0})}$$

is 2,

where $(t, \boldsymbol{x}) = (t, x_1, \dots, x_r)$ is the coordinate system of $\boldsymbol{R} \times \boldsymbol{R}^r$, and

$$\Phi' = \frac{\partial \Phi}{\partial t}, \quad \Phi_{x_j} = \frac{\partial \Phi}{\partial x_j}, \quad \Phi'' = \frac{\partial^2 \Phi}{\partial t^2}, \quad \Phi'_{x_j} = \frac{\partial^2 \Phi}{\partial x_j \partial t} \qquad (j = 1, \dots, r).$$

Setting a subset-germ

$$\Sigma_\Phi := \{(t, \boldsymbol{x}) \in \boldsymbol{R} \times \boldsymbol{R}^r \,;\, \Phi(t, \boldsymbol{x}) = \Phi'(t, \boldsymbol{x}) = 0\} \qquad (8.5)$$

at the origin of $\boldsymbol{R} \times \boldsymbol{R}^r$ for an admissible unfolding Φ, we have:

Proposition 8.2.8. *Let* $\Phi : (\boldsymbol{R} \times \boldsymbol{R}^r, (0, \boldsymbol{0})) \to (\boldsymbol{R}, 0)$ *be an admissible unfolding of* φ. *Then the subset-germ* Σ_Φ *given by* (8.5) *is an* $(r - 1)$-*dimensional submanifold in a neighborhood* U *of the origin of* $\boldsymbol{R} \times \boldsymbol{R}^r$.

Proof. For a map-germ $(\Phi, \Phi') \in C_0^\infty(\boldsymbol{R} \times \boldsymbol{R}^r, \boldsymbol{R}^2)$, $\Sigma_\Phi = (\Phi, \Phi')^{-1}(0, 0)$ holds. By (1) of Definition 8.2.7, we have $(0, \boldsymbol{0}) \in \Sigma_\Phi$, and so Σ_Φ is not empty. Furthermore, by (2) of Definition 8.2.7, restricting the map $\tilde{\Phi} :=$ (Φ, Φ') to a sufficiently small neighborhood U of the origin $(0, \boldsymbol{0})$, we can assume that the rank of the Jacobian matrix of $\tilde{\Phi}$ is always 2. Then by the implicit function theorem (cf. [22]), $\tilde{\Phi}^{-1}(0, 0)$ is an $(r - 1)$-dimensional (codimension 2) submanifold of $\boldsymbol{R} \times \boldsymbol{R}^r$. $\qquad\square$

We have the following proposition.

Proposition 8.2.9. *An unfolding which is* $P\mathcal{K}$-*equivalent to an admissible unfolding is also admissible.*

Proof. Let $\Phi(t, \boldsymbol{x})$ be an r-dimensional unfolding of $\varphi(t)$, which is $P\mathcal{K}$-equivalent to an admissible r-dimensional unfolding $\Psi(s, \boldsymbol{y})$ of $\psi(s)$. It is sufficient to show that Φ is also admissible. We may assume that two unfoldings satisfy (8.3) with (ϑ, Θ) and ρ as in Definition 8.2.3. Then the variables $(s, \boldsymbol{y})$ and $(t, \boldsymbol{x})$ satisfy

$$s = \vartheta(t, \boldsymbol{x}), \qquad \boldsymbol{y} = \Theta(\boldsymbol{x}).$$

Differentiating (8.3) by t, we have

$$\Phi'(t, \boldsymbol{x}) = \rho'(t, \boldsymbol{x})\Psi(\vartheta(t, \boldsymbol{x}), \Theta(\boldsymbol{x})) + \rho(t, \boldsymbol{x})\dot{\Psi}(\vartheta(t, \boldsymbol{x}), \Theta(\boldsymbol{x}))\vartheta'(t, \boldsymbol{x}).$$
$$(8.6)$$

Here, the prime $(')$ stands for the derivative with respect to the variable t, and the dot $(\dot{\ })$ stands for the derivative with respect to the variable s, namely,

$$\dot{\Psi}(s, \boldsymbol{y}) := \frac{\partial\Psi(s, \boldsymbol{y})}{\partial s}, \quad \Phi'(t, \boldsymbol{x}) := \frac{\partial\Phi(t, \boldsymbol{x})}{\partial t}, \quad \vartheta'(t, \boldsymbol{x}) := \frac{\partial\vartheta(t, \boldsymbol{x})}{\partial t}.$$

Since Ψ is an unfolding, $\Psi(0, \boldsymbol{0}) = 0$ holds, and by the admissibility of Ψ (cf. (1) of Definition 8.2.7), we have $\dot{\Psi}(0, \boldsymbol{0}) = 0$. Substituting $(t, \boldsymbol{x}) = (0, \boldsymbol{0})$

into (8.6), we have

$$(\Phi(0, \mathbf{0}) =) \, \Phi'(0, \mathbf{0}) = 0. \tag{8.7}$$

This shows Φ satisfies (1) of Definition 8.2.7.

Next, differentiating (8.3) by x_i, we have

$$\Phi_{x_i}(t, \mathbf{x}) = \rho_{x_i}(t, \mathbf{x}) \Psi(\vartheta(t, \mathbf{x}), \Theta(\mathbf{x})) \tag{8.8}$$
$$+ \rho(t, \mathbf{x}) \dot{\Psi}(\vartheta(t, \mathbf{x}), \Theta(\mathbf{x})) \vartheta_{x_i}(t, \mathbf{x})$$
$$+ \rho(t, \mathbf{x}) \sum_{j=1}^{r} \Psi_{y_j}(\vartheta(t, \mathbf{x}), \Theta(\mathbf{x})) \Theta_{x_i}^{j}(\mathbf{x})$$

for $i = 1, \ldots, r$, where

$$\Theta(\mathbf{x}) = (\Theta^1(\mathbf{x}), \ldots, \Theta^r(\mathbf{x})), \qquad \Theta_{x_i}^{j}(\mathbf{x}) := \frac{\partial \Theta^j(\mathbf{x})}{\partial x_i}.$$

Then by $\Psi(0, \mathbf{0}) = \dot{\Psi}(0, \mathbf{0}) = 0$, it holds that

$$\Phi_{x_i}(0, \mathbf{0}) = \rho(0, \mathbf{0}) \sum_{j=1}^{r} \Theta_{x_i}^{j}(\mathbf{0}) \Psi_{y_j}(0, \mathbf{0}) \qquad (i = 1, \ldots, r). \tag{8.9}$$

Here, we define the notation ∇ as follows: We set $\mathbf{x} = (x_1, \ldots, x_r)$. For a C^{∞}-function $\alpha(t, \mathbf{x})$ of the variables $(t, \mathbf{x})$ and for a C^{∞}-function $\alpha(\mathbf{x})$ of the variables $\mathbf{x}$, we set the gradient vectors $\nabla \alpha(t, \mathbf{x})$ and $\nabla \alpha(\mathbf{x})$ to be

$$\nabla \alpha(t, \mathbf{x}) := \begin{pmatrix} \alpha_{x_1}(t, \mathbf{x}) \\ \vdots \\ \alpha_{x_r}(t, \mathbf{x}) \end{pmatrix}, \qquad \nabla \alpha(\mathbf{x}) := \begin{pmatrix} \alpha_{x_1}(\mathbf{x}) \\ \vdots \\ \alpha_{x_r}(\mathbf{x}) \end{pmatrix}, \tag{8.10}$$

respectively. Usually, the gradient vector is defined by taking derivatives in all variables. However, here we define the notation ∇ taking derivatives only in the variables of $\mathbf{x}$ (ignoring the variable t). Using this notation, (8.9) can be written as

$$\nabla \Phi(0, \mathbf{0}) = \rho(0, \mathbf{0}) J(\mathbf{0}) \nabla \Psi(0, \mathbf{0}), \tag{8.11}$$

where

$$J(\mathbf{x}) := \begin{pmatrix} \Theta_{x_1}^{1}(\mathbf{x}) & \ldots & \Theta_{x_1}^{r}(\mathbf{x}) \\ \vdots & \ddots & \vdots \\ \Theta_{x_r}^{1}(\mathbf{x}) & \ldots & \Theta_{x_r}^{r}(\mathbf{x}) \end{pmatrix}. \tag{8.12}$$

Since Θ is a diffeomorphism-germ, J is a regular matrix. Noting that Ψ is admissible and $\Psi'(0, \mathbf{0}) = 0$, by (2) of Definition 8.2.7, it holds that

$\nabla\Psi(0,\mathbf{0}) \neq \mathbf{0}$. Thus, by (8.11), we have $\nabla\Phi(0,\mathbf{0}) \neq \mathbf{0}$. If $\Phi''(0,\mathbf{0}) \neq 0$, then by (8.7),

$$\begin{pmatrix} \Phi' & \Phi_{x_1} & \cdots & \Phi_{x_r} \\ \Phi'' & \Phi'_{x_1} & \cdots & \Phi'_{x_r} \end{pmatrix}^T \Bigg|_{(t,x)=(0,0)} = \begin{pmatrix} 0 & \Phi''(0,\mathbf{0}) \\ \nabla\Phi(0,\mathbf{0}) & \nabla\Phi'(0,\mathbf{0}) \end{pmatrix},$$

and the rank of this matrix is 2, and Φ also satisfies (2) of Definition 8.2.7. So we consider the case that $\Phi''(0,\mathbf{0}) = 0$. Differentiating (8.6) by t, we have

$$\Phi'' = \rho''\Psi + 2\rho'\dot{\Psi}\vartheta' + \rho\ddot{\Psi}(\vartheta')^2 + \rho\dot{\Psi}\vartheta''.$$

We substitute

$$\Psi(0,\mathbf{0}) = \dot{\Psi}(0,\mathbf{0}) = \Phi''(0,\mathbf{0}) = 0$$

into this equation. By (8.4), $\vartheta'(0,\mathbf{0}) \neq 0$ holds. So we have $\ddot{\Psi}(0,\mathbf{0}) = 0$. Next for $i = 1, \ldots, r$, differentiating (8.8) by t, we have

$$\Phi'_{x_i} = \rho'_{x_i}\Psi + \rho_{x_i}\dot{\Psi}\vartheta' + \rho'\dot{\Psi}\vartheta_{x_i} + \rho\ddot{\Psi}\vartheta_{x_i}\vartheta' + \rho\dot{\Psi}\vartheta'_{x_i}$$
$$+ \rho'\sum_{j=1}^{r}\Theta^j_{x_i}\Psi_{y_j} + \rho\sum_{j=1}^{r}\Theta^j_{x_i}\dot{\Psi}_{y_j}\vartheta'.$$

Substituting $\Psi(0,\mathbf{0}) = \dot{\Psi}(0,\mathbf{0}) = \ddot{\Psi}(0,\mathbf{0}) = 0$ into this equation, we have

$$\Phi'_{x_i}(0,\mathbf{0}) =$$
$$\rho'(0,\mathbf{0})\sum_{j=1}^{r}\Psi_{y_j}(0,\mathbf{0})\Theta^j_{x_i}(\mathbf{0}) + \rho(0,\mathbf{0})\vartheta'(0,\mathbf{0})\sum_{j=1}^{r}\dot{\Psi}_{y_j}(0,\mathbf{0})\Theta^j_{x_i}(\mathbf{0}). \quad (8.13)$$

Using the notation ∇ given in (8.10), (8.13) can be rewritten as

$$\nabla\Phi'(0,\mathbf{0}) = \rho'(0,\mathbf{0})J(\mathbf{0})\nabla\Psi(0,\mathbf{0}) + \vartheta'(0,\mathbf{0})\rho(0,\mathbf{0})J(\mathbf{0})\nabla\dot{\Psi}(0,\mathbf{0}). \quad (8.14)$$

Since $\rho \neq 0$,

$$\mathrm{rank}(\nabla\Phi, \nabla\Phi') = \mathrm{rank}(\rho J\nabla\Psi, \rho'J\nabla\Psi + \rho\vartheta'J\nabla\dot{\Psi})$$
$$= \mathrm{rank}(J\nabla\Psi, J\nabla\dot{\Psi})$$

holds at $(0,\mathbf{0})$, where $\mathrm{rank}\, M$ denotes the rank of a matrix M. Moreover, since J is regular,

$$\mathrm{rank}(\nabla\Phi(0,\mathbf{0}), \nabla\Phi'(0,\mathbf{0})) = \mathrm{rank}(\nabla\Psi(0,\mathbf{0}), \nabla\dot{\Psi}(0,\mathbf{0})) = 2.$$

Hence the admissibility of Φ is shown. $\qquad\square$

For a given admissible unfolding Φ, we set

$$D_\Phi := \pi(\Sigma_\Phi) \subset \boldsymbol{R}^r \qquad (8.15)$$

and call the subset-germ (Definition 8.1.1) at the origin of $\boldsymbol{R}^r$ the *discriminant set* of Φ, where $\pi : \boldsymbol{R} \times \boldsymbol{R}^r \ni (t, \boldsymbol{x}) \mapsto \boldsymbol{x} \in \boldsymbol{R}^r$. Namely,

$$D_\Phi = \{\boldsymbol{x} \in U \,;\, \text{there exists } t \text{ such that } \Phi(t, \boldsymbol{x}) = \Phi'(t, \boldsymbol{x}) = 0\}$$

for a neighborhood U of the origin of $\boldsymbol{R}^r$.

Example 8.2.10. Let $\gamma : (\boldsymbol{R}, 0) \to (\boldsymbol{R}^2, \boldsymbol{0})$ be a map-germ of a (wave) front, and $\gamma'(0) = \boldsymbol{0}$. For $\boldsymbol{x} \in \boldsymbol{R}^2$, we set

$$\Phi(t, \boldsymbol{x}) := (\boldsymbol{x} - \gamma(t)) \cdot \boldsymbol{n}(t), \qquad (8.16)$$

where $\boldsymbol{n}(t)$ is the unit normal vector field along $\gamma(t)$. Here "$\cdot$" stands for the Euclidean inner product of $\boldsymbol{R}^2$. Then we see that D_Φ and the image of γ coincide as set-germs at the origin of $\boldsymbol{R}^2$. In fact, $\Phi(t, \boldsymbol{x}) = 0$ is equivalent to

$$(\boldsymbol{x} - \gamma(t)) \cdot \boldsymbol{n}(t) = 0, \qquad (8.17)$$

and since $\gamma'(t) \cdot \boldsymbol{n}(t) = 0$, the condition $\Phi'(t, \boldsymbol{x}) = 0$ is equivalent to

$$(\boldsymbol{x} - \gamma(t)) \cdot \boldsymbol{n}'(t) = 0. \qquad (8.18)$$

Since γ is a front, $\boldsymbol{n}(0)$ and $\boldsymbol{n}'(0)$ are linearly independent near $t = 0$. In fact, $\boldsymbol{n}'$ and $\boldsymbol{n}$ are perpendicular, and since γ is a front, $\boldsymbol{n}'(0) \neq \boldsymbol{0}$. Thus by (8.17) and (8.18), $\boldsymbol{x} \in D_\Phi$ near $\boldsymbol{0}$ implies that the existence of t such that $\boldsymbol{x} - \gamma(t) = \boldsymbol{0}$. Thus D_Φ coincides the image of γ as subset-germs.

Discriminant sets have the following important property.

Theorem 8.2.11. *Let φ and $\psi : (\boldsymbol{R}, 0) \to (\boldsymbol{R}, 0)$ be two C^∞-function-germs of one variable, and let Φ and Ψ be r-dimensional unfoldings of φ and ψ, respectively. If Ψ is admissible, and Φ is $P\mathcal{K}$-equivalent to Ψ, then the two discriminant sets D_Φ and D_Ψ are diffeomorphic as set-germs at the origin of $\boldsymbol{R}^r$. Namely, for any sufficiently small neighborhood V of the origin in $\boldsymbol{R}^r$ there exists a diffeomorphism $\Theta : \Theta^{-1}(V) \to V$ such that*

$$D_\Psi \cap V = \Theta(D_\Phi) \cap V. \qquad (8.19)$$

Proof. Since Φ is $P\mathcal{K}$-equivalent to Ψ, the unfolding Φ is admissible, by Proposition 8.2.9. Moreover, there exist ρ, ϑ and a diffeomorphism $\Theta : U \to \Theta(U)$ on a neighborhood U of the origin in $\boldsymbol{R}^r$ such that (8.3) holds. Thus, for $\boldsymbol{x} \in U$, the condition $\Phi(t, \boldsymbol{x}) = 0$ is equivalent to $\Psi(\vartheta(t, \boldsymbol{x}), \Theta(\boldsymbol{x})) = 0$. By differentiating (8.3) with respect to t, we obtain (8.6). Since $\Phi'(t, \boldsymbol{x}) = 0$ under $\Phi(t, 0) = 0$, regarding (8.4), we have that

$$\dot{\Psi}(\vartheta(t, \boldsymbol{x}), \Theta(\boldsymbol{x})) = 0,$$

under the assumption that $\Psi(\vartheta(t,\boldsymbol{x}),\Theta(\boldsymbol{x})) = 0$. Thus, $\Phi(t,\boldsymbol{x}) = \Phi'(t,\boldsymbol{x}) = 0$ if and only if

$$\Psi\big(\vartheta(t,\boldsymbol{x}),\Theta(\boldsymbol{x})\big) = \dot{\Psi}\big(\vartheta(t,\boldsymbol{x}),\Theta(\boldsymbol{x})\big) = 0.$$

Hence for $\boldsymbol{x} \in U$, the conditions $\Theta(\boldsymbol{x}) \in D_\Psi \cap \Theta(U)$ and $\boldsymbol{x} \in D_\Phi \cap U$ are equivalent. Thus, we have

$$D_\Psi \cap \Theta(U) = \Theta(D_\Phi \cap U) = \Theta(D_\Phi) \cap \Theta(U).$$

We let V be a neighborhood of $\boldsymbol{0}$ contained in $\Theta(U)$. Then we have

$$D_\Psi \cap V = \Theta(D_\Phi) \cap V.$$

which is equivalent to (8.19). $\qquad\qquad\qquad\qquad\qquad\qquad\square$

8.3. Versal Unfoldings

Definition 8.3.1. Let $\varphi : (\boldsymbol{R},0) \to (\boldsymbol{R},0)$ be a C^∞-function-germ satisfying

$$(\varphi(0) =)\varphi'(0) = \cdots = \varphi^{(k)}(0) = 0, \quad \varphi^{(k+1)}(0) \neq 0 \qquad (8.20)$$

for a positive integer k. For $k \leq r$, an r-dimensional unfolding $\Phi : (\boldsymbol{R} \times \boldsymbol{R}^r, (0,\boldsymbol{0})) \to (\boldsymbol{R},0)$ of φ is a *versal unfolding*[2] (of φ) if

$$\mathrm{rank}(\nabla\Phi, \nabla\Phi', \ldots, \nabla\Phi^{(k-1)})|_{(t,\boldsymbol{x})=(0,\boldsymbol{0})} = k,$$

that is,

$$\mathrm{rank}\begin{pmatrix} \Phi_{x_1} & \Phi'_{x_1} & \Phi^{(2)}_{x_1} & \cdots & \Phi^{(k-1)}_{x_1} \\ \Phi_{x_2} & \Phi'_{x_2} & \Phi^{(2)}_{x_2} & \cdots & \Phi^{(k-1)}_{x_2} \\ \vdots & \vdots & \vdots & \ddots & \vdots \\ \Phi_{x_r} & \Phi'_{x_r} & \Phi^{(2)}_{x_r} & \cdots & \Phi^{(k-1)}_{x_r} \end{pmatrix}\Bigg|_{(t,\boldsymbol{x})=(0,\boldsymbol{0})} = k \qquad (8.21)$$

holds, where $(t,\boldsymbol{x})$ is the coordinate system of $\boldsymbol{R} \times \boldsymbol{R}^r$. Here, $' = \partial/\partial t$, and

$$\varphi^{(l)} = \frac{d^l\varphi}{dt^l}, \quad \Phi^{(l)} = \frac{\partial^l\Phi}{\partial t^l}, \quad \Phi^{(l)}_{x_j} = \frac{\partial^{l+1}\Phi}{\partial t^l \partial x_j} \quad (1 \leq l \leq k-1,\ 1 \leq j \leq r).$$

In Remark 8.3.12, we will mention where the word "versal" is derived from.

Remark 8.3.2. By definition, versal unfoldings for $k \geq 2$ are admissible in the sense of Definition 8.2.7.

---

[2]Strictly speaking, it is usually called $\mathcal{K}$-*versal unfolding*.

Example 8.3.3. Let $r \geq k$, and set

$$\Psi_{k,r}(t, \boldsymbol{x}) := t^{k+1} + x_1 t^{k-1} + x_2 t^{k-2} + \cdots + x_{k-1} t + x_k$$

$$(\boldsymbol{x} := (x_1, \ldots, x_r)). \quad (8.22)$$

Then this is a versal unfolding of the C^∞-function $\psi(t) = t^{k+1}$. The function $\Psi_{k,r}(t, \boldsymbol{x})$ contains only the first k variables of $\boldsymbol{x} = (x_1, \ldots, x_r)$. Thus only $\Psi_{r,r}(t, \boldsymbol{x})$ contains the rth variable x_r. The discriminant set

$$D_{k,r} := \{\boldsymbol{x} \in \boldsymbol{R}^r \,;\, \text{there exists } t \text{ such that } \Phi(t, \boldsymbol{x}) = \Phi'(t, \boldsymbol{x}) = 0\} \quad (8.23)$$

of $\Phi := \Psi_{k,r}$ coincides with the set of coefficients of the polynomial $\Phi(t, \boldsymbol{x})$ in t so that it has a real multiple root. In particular,

(i) as we saw in Example 1.3.1 of Chapter 1, $D_{2,2}$ coincides with the image of the plane curve $\gamma_C(t) := (-3t^2, 2t^3)$, which is right–left equivalent to the cusp $t \mapsto (t^2, t^3)$, as set-germs in $\boldsymbol{R}^2$ at the origin,

(ii) as we saw in Example 2.5.2 of Chapter 2, $D_{2,3}$ coincides with the image of the cuspidal edge $g_C(u, v) := (-3u^2, 2u^3, v)$ as set-germs,

(iii) as we saw in Example 2.5.3 of Chapter 2, $D_{3,3}$ coincides with the image of the swallowtail $g_S(u, v) := (u, -4v^3 - 2uv, 3v^4 + uv^2)$ as set-germs.

In general, we have:

Lemma 8.3.4. *Let k be a positive integer. If a map-germ $\varphi : (\boldsymbol{R}, 0) \to (\boldsymbol{R}, 0)$ satisfies*

$$(\varphi(0) =)\varphi'(0) = \cdots = \varphi^{(k)}(0) = 0, \quad \varphi^{(k+1)}(0) \neq 0,$$

then for any $r \geq k$, there exists an r-dimensional versal unfolding of φ.

Proof. The map $\Phi(t, \boldsymbol{x}) = \varphi(t) + x_1 t^{k-1} + x_2 t^{k-2} + \cdots + x_k$ gives an unfolding of φ. We show this Φ gives the desired versal unfolding. The matrix in (8.21) can be written in the form

$$\begin{pmatrix} A \\ O_{(r-k) \times k} \end{pmatrix},$$

where A is a $k \times k$ anti-diagonal regular matrix whose $(j, k - j + 1)$-element is $(k - j)!$, and $O_{(r-k) \times k}$ is the $(r - k) \times k$ zero matrix. Thus, $\Phi(t, \boldsymbol{x})$ is an r-dimensional versal unfolding of φ. $\qquad \square$

Here, we give an example of a one-dimensional versal unfolding:

Proposition 8.3.5. *Let $\gamma : (\boldsymbol{R}, 0) \to (\boldsymbol{R}^2, \boldsymbol{0})$ be a C^∞-map-germ satisfying the conditions of the criteria for the cusp (Theorem 1.3.2 in Chapter 1),* *namely,* $\gamma'(0) = \mathbf{0}$ *and*

$$\det\big(\gamma''(0), \gamma'''(0)\big) \neq 0. \tag{8.24}$$

Then, there exists a two-dimensional versal unfolding whose discriminant set coincides with the image of γ.

Proof. By Proposition B.1.2 in Appendix B, γ is a front. We take the unit normal vector field $\mathbf{n}(t)$. We set $\Phi(t, \mathbf{x})$ by (8.16) in Example 8.2.10. Then Φ is an unfolding of the function

$$\varphi(t) := -\gamma(t) \cdot \mathbf{n}(t)$$

of one variable. We already saw that the discriminant set of Φ coincides with the image of γ in Example 8.2.10, and we can show Φ is a versal unfolding. In fact, setting $\mathbf{x} = (x_1, x_2)$, we have

$$(\Phi_{x_1}, \Phi_{x_2})|_{(t,\mathbf{x})=(0,\mathbf{0})} = \mathbf{n}(0), \qquad (\Phi'_{x_1}, \Phi'_{x_2})|_{(t,\mathbf{x})=(0,\mathbf{0})} = \mathbf{n}'(0),$$

and then we see

$$\operatorname{rank} \begin{pmatrix} \Phi_{x_1} & \Phi'_{x_1} \\ \Phi_{x_2} & \Phi'_{x_2} \end{pmatrix} \bigg|_{(t,\mathbf{x})=(0,\mathbf{0})} = 2.$$

To show Φ is a versal unfolding, it is enough to show that Φ has regularity of order 3 in the variable t. Since γ' and $\mathbf{n}$ are perpendicular, we have

$$\varphi' = -\gamma \cdot \mathbf{n}', \qquad \varphi'' = -\gamma' \cdot \mathbf{n}' - \gamma \cdot \mathbf{n}''.$$

Then we have $\varphi'(0) = \varphi''(0) = 0$, because $\gamma(0) = \gamma'(0) = \mathbf{0}$. Thus we have to prove $\varphi'''(0) = -\gamma''(0)\cdot\mathbf{n}'(0) \neq 0$. Since $\gamma''\cdot\mathbf{n}+\gamma'\cdot\mathbf{n}'$ vanishes identically, $\gamma''(0) \cdot \mathbf{n}(0) = 0$. On the other hand, since γ is a front, $\mathbf{n}'(0) \neq \mathbf{0}$ holds. Since $\mathbf{n}'(0)$ is perpendicular to $\mathbf{n}(0)$, the vector $\gamma''(0)(\neq \mathbf{0})$ is a non-zero multiple of $\mathbf{n}'(0)$. Thus, $\gamma''(0) \cdot \mathbf{n}'(0) \neq 0$ holds. $\square$

We next prove the following proposition.

Proposition 8.3.6. *Let* Ψ *be an* r-*dimensional versal unfolding of* ψ *satisfying*

$$(\psi(0) =)\psi'(0) = \cdots = \psi^{(k)}(0) = 0, \quad \psi^{(k+1)}(0) \neq 0 \tag{8.25}$$

for $k(\leq r)$, *and let* Φ *be an* r-*dimensional unfolding of* φ. *If* Φ *and* Ψ *are* $P\mathcal{K}$-*equivalent, then* φ *satisfies*

$$(\varphi(0) =)\varphi'(0) = \cdots = \varphi^{(k)}(0) = 0, \quad \varphi^{(k+1)}(0) \neq 0. \tag{8.26}$$

Moreover, Φ is an r-dimensional versal unfolding of φ. Namely, whether an unfolding is versal or not is invariant under the $P\mathcal{K}$-equivalence.

To prove this proposition, we introduce a relation "$\equiv$".

Definition 8.3.7. Let U be an open set of $\boldsymbol{R} \times \boldsymbol{R}^r$, and $\alpha, \beta : U \to \boldsymbol{R}$ two C^∞-functions. Let $r_i : U \to \boldsymbol{R}$ $(i = 1, \ldots, l)$ be C^∞-functions. Then,

$$\alpha \equiv \beta \quad \mathrm{mod}\ (r_1, \ldots, r_l) \tag{8.27}$$

if there exist C^∞-functions $q_1, \ldots, q_l$ on U such that $\alpha - \beta = \sum_{i=1}^{l} r_i q_i$.

Let U, V be open subsets of $\boldsymbol{R} \times \boldsymbol{R}^r$. Consider a map given by

$$\Xi : V \ni (t, \boldsymbol{x}) \mapsto (\vartheta(t, \boldsymbol{x}), \Theta(\boldsymbol{x})) \in U,$$

and assume it is a diffeomorphism. We set

$$s := \vartheta(t, \boldsymbol{x}), \quad y := \Theta(\boldsymbol{x}).$$

Then $(s, \boldsymbol{y}) \in \boldsymbol{R} \times \boldsymbol{R}^r$ can be considered as a coordinate system of U. We recall that the prime $(')$ stands for the derivative with respect to the variable t, and the dot $(\dot{\ })$ stands for the derivative with respect to the variable s. We have the following lemma.

Lemma 8.3.8. *Let $\alpha : V \to \boldsymbol{R}$ and $\beta : U \to \boldsymbol{R}$ be two C^∞-functions. Moreover, let $p_i : V \to \boldsymbol{R}$ $(i = 1, \ldots, l)$ be C^∞-functions. Suppose*

$$\alpha(t, \boldsymbol{x}) \equiv \zeta(t, \boldsymbol{x})\beta(\vartheta(t, \boldsymbol{x}), \Theta(\boldsymbol{x})) \quad \mathrm{mod}\ (p_1(\vartheta, \Theta), \ldots, p_l(\vartheta, \Theta)) \tag{8.28}$$

holds for a function $\zeta : V \to \boldsymbol{R}$, where $\mu(\vartheta, \Theta)$ is the function defined by

$$\mu(\vartheta, \Theta)(t, \boldsymbol{x}) := \mu(\vartheta(t, \boldsymbol{x}), \Theta(\boldsymbol{x}))$$

for a given function $\mu : U \to \boldsymbol{R}$. Then

$$\alpha'(t, \boldsymbol{x}) \equiv \zeta(t, \boldsymbol{x})\vartheta'(t, \boldsymbol{x})\,\dot{\beta}(\vartheta(t, \boldsymbol{x}), \Theta(\boldsymbol{x}))$$
$$\mathrm{mod}\ (p_1(\vartheta, \Theta), \ldots, p_l(\vartheta, \Theta), \dot{p}_1(\vartheta, \Theta), \ldots, \dot{p}_l(\vartheta, \Theta), \beta(\vartheta, \Theta), \Theta)$$

holds.

Proof. By the assumption (8.28), there exist C^∞-functions $q_1, \ldots, q_l : V \to \boldsymbol{R}$ such that

$$\alpha - \zeta\beta(\vartheta, \Theta) = \sum_{i=1}^{l} p_i(\vartheta, \Theta) q_i.$$

Differentiating this equation with respect to t, we have

$$\alpha' - \zeta'\beta(\vartheta, \Theta) - \zeta\dot{\beta}(\vartheta, \Theta)\vartheta' = \sum_{i=1}^{l}(\vartheta'\dot{p}_i(\vartheta, \Theta)q_i + p_i(\vartheta, \Theta)q_i'),$$

and this shows the assertion. □

We extend the lemma using the same notation when α, β are vector valued C^∞-functions as follows.

Definition 8.3.9. Let U be an open set of $\boldsymbol{R} \times \boldsymbol{R}^r$, and $\alpha, \beta : U \to \boldsymbol{R}^r$ two vector valued C^∞-functions. Let $r_i : U \to \boldsymbol{R}$ $(i = 1, \ldots, l)$ be C^∞-functions, and let $X_j : U \to \boldsymbol{R}^r$ $(j = 1, \ldots, h)$ be vector valued C^∞-functions. Then,

$$\alpha \equiv \beta \quad \mod (r_1, \ldots, r_l\,; X_1, \ldots, X_h) \qquad (8.29)$$

if there exist vector valued C^∞-functions $Y_1, \ldots, Y_l : U \to \boldsymbol{R}^r$ and C^∞-functions $q_1, \ldots, q_h : U \to \boldsymbol{R}$ such that

$$\alpha - \beta = \sum_{i=1}^{l} r_i Y_i + \sum_{j=1}^{h} q_j X_j.$$

Also in this case, we have a similar lemma to Lemma 8.3.8.

Lemma 8.3.10. *Let* $\alpha : V \to \boldsymbol{R}^r$ *and* $\beta : U \to \boldsymbol{R}^r$ *be two* $\boldsymbol{R}^r$-*valued* C^∞-*functions. Moreover, let* $p_1, \ldots, p_l$ *and* $X_1, \ldots, X_l$ *be* C^∞-*functions on* U *and* $\boldsymbol{R}^r$-*valued* C^∞-*functions on* U. *Suppose that*

$$\alpha(t, \boldsymbol{x}) \equiv \zeta(t, \boldsymbol{x})\beta(\vartheta(t, \boldsymbol{x}), \Theta(\boldsymbol{x})) \qquad (8.30)$$
$$\mod (p_1(\vartheta, \Theta), \ldots, p_l(\vartheta, \Theta)\,; X_1(\vartheta, \Theta), \ldots, X_h(\vartheta, \Theta))$$

holds for a C^∞-*function* $\zeta : V \to \boldsymbol{R}$, *where* $X(\vartheta, \Theta)$ *is defined by*

$$X(t, \boldsymbol{x}) := X(\vartheta(t, \boldsymbol{x}), \Theta(\boldsymbol{x}))$$

for a given $X : U \to \boldsymbol{R}^r$. *Then*

$$\alpha'(t, \boldsymbol{x}) \equiv \zeta(t, \boldsymbol{x})\vartheta'(t, \boldsymbol{x})\dot{\beta}(\vartheta(t, \boldsymbol{x}), \Theta(\boldsymbol{x}))$$
$$\mod (p_1(\vartheta, \Theta), \ldots, p_l(\vartheta, \Theta), \dot{p}_1(\vartheta, \Theta), \ldots, \dot{p}_l(\vartheta, \Theta)\,;$$
$$\beta(\vartheta, \Theta), X_1(\vartheta, \Theta), \ldots, X_h(\vartheta, \Theta), \dot{X}_1(\vartheta, \Theta), \ldots, \dot{X}_h(\vartheta, \Theta)).$$

Proof. By the assumption (8.30), there exist vector valued C^∞-functions $Y_1, \ldots, Y_l : V \to \boldsymbol{R}^r$ and C^∞-functions $q_1, \ldots, q_h : V \to \boldsymbol{R}$ such that

$$\alpha - \zeta\beta(\vartheta, \Theta) = \sum_{i=1}^{l} p_i(\vartheta, \Theta)Y_i + \sum_{j=1}^{h} q_j X_j(\vartheta, \Theta).$$

Differentiating this equation by t, we have

$$\alpha' - \zeta'\beta(\vartheta,\Theta) - \zeta\dot{\beta}(\vartheta,\Theta)\vartheta' = \sum_{i=1}^{l}(\vartheta'\dot{p}_i(\vartheta,\Theta)Y_i + p_i(\vartheta,\Theta)Y_i')$$

$$+ \sum_{j=1}^{h}(q_j'X_j(\vartheta,\Theta) + q_j\vartheta'\dot{X}_j(\vartheta,\Theta)),$$

and this shows the assertion. $\qquad\square$

Proof of Proposition 8.3.6. Let Ψ be an r-dimensional versal unfolding of ψ, and let $(s, \boldsymbol{y}) = (s, y_1, \ldots, y_r)$ be the variables of Ψ. Since Φ and Ψ are $P\mathcal{K}$-equivalent, there exist a diffeomorphism $\Xi : (\boldsymbol{R} \times \boldsymbol{R}^r, (0, \boldsymbol{0})) \to (\boldsymbol{R} \times \boldsymbol{R}^r, (0, \boldsymbol{0}))$ of the form $\Xi(t, \boldsymbol{x}) = (\vartheta(t, \boldsymbol{x}), \Theta(\boldsymbol{x}))$ and a function ρ satisfying $\rho(0, \boldsymbol{0}) \neq 0$ such that (cf. (8.3))

$$\Phi(t, \boldsymbol{x}) = \rho(t, \boldsymbol{x})\Psi(\vartheta(t, \boldsymbol{x}), \Theta(\boldsymbol{x})). \qquad (8.31)$$

Here, we regard $s = \vartheta(t, \boldsymbol{x})$ and $\boldsymbol{y} = \Theta(\boldsymbol{x})$. The formula (8.31) can be written as $\Phi = \rho\Psi(\vartheta, \Theta)$. Differentiating this with respect to t, we have (cf. (8.6)) $\Phi' = \rho'\Psi(\vartheta, \Theta) + \rho\dot{\Psi}(\vartheta, \Theta)\vartheta'$. So, we have

$$\Phi' \equiv \rho\vartheta'\dot{\Psi}(\vartheta, \Theta) \quad \mathrm{mod}\,\Psi(\vartheta, \Theta).$$

By Lemma 8.3.8, we have

$$\Phi'' \equiv \rho\,(\vartheta')^2\ddot{\Psi}(\vartheta, \Theta) \quad \mathrm{mod}\,(\Psi(\vartheta, \Theta), \dot{\Psi}(\vartheta, \Theta)). \qquad (8.32)$$

Let $\Psi^{[3]}$ be the third order derivative of $\Psi(s, \boldsymbol{y})$ with respect to the variable s, which is different from $\Psi^{(3)}$, the third order derivative with respect to t. Applying Lemma 8.3.8 to (8.32), we have

$$\Phi''' \equiv \rho\,(\vartheta')^3\Psi^{[3]}(\vartheta, \Theta) \quad \mathrm{mod}\,(\Psi(\vartheta, \Theta), \dot{\Psi}(\vartheta, \Theta), \ddot{\Psi}(\vartheta, \Theta)).$$

Repeating similar calculations, we have

$$\Phi^{(j)} \equiv \rho\,(\vartheta')^j\Psi^{[j]}(\vartheta, \Theta) \qquad (8.33)$$
$$\mathrm{mod}\,(\Psi(\vartheta, \Theta), \dot{\Psi}(\vartheta, \Theta), \ldots, \Psi^{[j-1]}(\vartheta, \Theta)) \quad (j = 1, \ldots, k+1),$$

where $\Psi^{[j-1]}$ is the $(j-1)$st order derivative of $\Psi(s, \boldsymbol{y})$ with respect to s. The condition (8.20) is now

$$\Psi(0, \boldsymbol{0}) = \dot{\Psi}(0, \boldsymbol{0}) = \cdots = \Psi^{[k]}(0, \boldsymbol{0}) = 0, \quad \Psi^{[k+1]}(0, \boldsymbol{0}) \neq 0. \qquad (8.34)$$

Thus, by (8.33), we have

$$\Phi(0, \boldsymbol{0}) = \Phi'(0, \boldsymbol{0}) = \cdots = \Phi^{(k)}(0, \boldsymbol{0}) = 0, \quad \Phi^{(k+1)}(0, \boldsymbol{0}) \neq 0,$$

and this is equivalent to (8.26). This proves the first assertion.

On the other hand, differentiating (8.31) with respect to $x_1, \ldots, x_r$, we have (8.8). Using the notation "mod", we have

$$\nabla\Phi \equiv \rho\, J\nabla\Psi(\vartheta, \Theta) \qquad \mathrm{mod}\ (\Psi(\vartheta, \Theta), \dot\Psi(\vartheta, \Theta)),$$

where J is the matrix as in (8.12). Setting

$$\alpha(t, \boldsymbol{x}) := \nabla\Phi(t, \boldsymbol{x}), \quad \zeta(t, \boldsymbol{x}) := \rho(t, \boldsymbol{x}), \quad \beta(s, \boldsymbol{y}) := J(\Theta^{-1}(\boldsymbol{y}))\nabla\Psi(s, \boldsymbol{y})$$

and applying Lemma 8.3.10, we have

$$\nabla\Phi' \equiv \rho\,\vartheta' J\nabla\dot\Psi(\vartheta, \Theta) \quad \mathrm{mod}\ (\Psi(\vartheta, \Theta), \dot\Psi(\vartheta, \Theta), \ddot\Psi(\vartheta, \Theta)\,;\, J\nabla\Psi(\vartheta, \Theta)). \tag{8.35}$$

By Lemma 8.3.10, we have

$$\nabla\Phi'' \equiv \rho\,(\vartheta')^2 J\nabla\ddot\Psi(\vartheta, \Theta) \tag{8.36}$$
$$\mathrm{mod}\ (\Psi(\vartheta, \Theta), \dot\Psi(\vartheta, \Theta), \ddot\Psi(\vartheta, \Theta), \dddot\Psi(\vartheta, \Theta)\,;\, J\nabla\Psi(\vartheta, \Theta), J\nabla\dot\Psi(\vartheta, \Theta)).$$

With similar calculations, for $j = 1, \ldots, k-1$, we have

$$\nabla\Phi^{(j)} \equiv \rho\,(\vartheta')^j J\nabla\Psi^{[j]}(\vartheta, \Theta) \tag{8.37}$$
$$\mathrm{mod}\ (\Psi(\vartheta, \Theta), \ldots, \Psi^{[j+1]}(\vartheta, \Theta)\,;\, J\nabla\Psi(\vartheta, \Theta), \ldots, J\nabla\Psi^{[j-1]}(\vartheta, \Theta)).$$

Then by (8.34), for each $j = 1, \ldots, k-1$, there exist

$$a_{j,0},\ a_{j,1},\ \ldots,\ a_{j,j-1} \in \boldsymbol{R}$$

such that

$$\nabla\Phi^{(j)}(0, 0) = \rho(0, 0)(\vartheta'(0, 0))^j J(0)\nabla\Psi^{[j]}(0, 0)$$
$$+ \sum_{i=0}^{j-1} a_{j,i} J(0)\nabla\Psi^{[i]}(0, 0).$$

Since the matrix rank does not change under the elementary column operations, (8.37) implies

$$\mathrm{rank}\left(\nabla\Phi, \nabla\Phi', \ldots, \nabla\Phi^{(k-1)}\right)\big|_{(t,\boldsymbol{x})=(0,0)}$$
$$= \mathrm{rank}\left(J\nabla\Psi, J\nabla\dot\Psi, \ldots, J\nabla\Psi^{[k-1]}\right)\big|_{(s,\boldsymbol{y})=(0,0)}$$

holds. Furthermore, since $J(0)$ is regular, we have

$$\mathrm{rank}\left(\nabla\Phi, \nabla\Phi', \ldots, \nabla\Phi^{(k-1)}\right)\big|_{(t,\boldsymbol{x})=(0,0)}$$
$$= \mathrm{rank}\left(\nabla\Psi, \nabla\dot\Psi, \ldots, \nabla\Psi^{[k-1]}\right)\big|_{(s,\boldsymbol{y})=(0,0)} = k.$$

This implies that Φ is a versal unfolding. □

The following theorem plays a key role in showing the criterion for swallowtails.

Theorem 8.3.11. *Assume that a function φ satisfies*

$$\varphi(0) = \varphi'(0) = \cdots = \varphi^{(k)}(0) = 0, \quad \varphi^{(k+1)}(0) \neq 0. \quad (8.38)$$

Then for $r \geq k$, any r-dimensional versal unfolding of φ is $P\mathcal{K}$-equivalent to the versal unfolding $\Psi_{k,r}$ given in (8.22).

Proof. Let $\Phi(t, \boldsymbol{x})$ be an r-dimensional versal unfolding of $\varphi(t)$. We show Φ is $P\mathcal{K}$-equivalent to the unfolding $\Psi_{k,r}$ of t^{k+1} given in (8.22). Since Φ is an unfolding of φ, and φ satisfies (8.38) by assumption, Φ has regularity of order $k + 1$ in the variable t. By Theorem 8.2.6 (Malgrange preparation theorem III), we may assume that Φ has the form

$$\Phi(t, \boldsymbol{x}) = t^{k+1} + H_1(\boldsymbol{x})t^{k-1} + \cdots + H_{k-1}(\boldsymbol{x})t + H_k(\boldsymbol{x}). \quad (8.39)$$

By Lemma 8.3.6, this unfolding $\Phi(t, \boldsymbol{x})$ is a versal unfolding of t^{k+1}. Differentiating (8.39) j times for $j = 1, \ldots, k-1$, we have $\Phi^{(j)}(0, \boldsymbol{x}) = j! H_{k-j}(\boldsymbol{x})$. Thus,

$$k = \mathrm{rank}(\nabla\Phi, \nabla\Phi', \ldots, \nabla\Phi^{(k-1)})|_{(t,\boldsymbol{x})=(0,0)}$$
$$= \mathrm{rank}\,(\nabla H_k, \nabla H_{k-1}, \ldots, \nabla H_1)|_{(t,\boldsymbol{x})=(0,0)}.$$

Since $k \leq r$, permuting the order of $\boldsymbol{x} = (x_1, \ldots, x_r)$ if necessary, we may assume the rank of the matrix considering only derivatives with respect to $x_1, \ldots, x_k$ is k. Namely,

$$\mathrm{rank}\begin{pmatrix} (H_1)_{x_1} & (H_2)_{x_1} & \cdots & (H_k)_{x_1} \\ \vdots & \vdots & \ddots & \vdots \\ (H_1)_{x_k} & (H_2)_{x_k} & \cdots & (H_k)_{x_k} \end{pmatrix}\Bigg|_{(t,\boldsymbol{x})=(0,0)} = k.$$

We denote the $k \times k$ matrix in the left-hand side by $\mathcal{H}$. Let us set new variables $(\hat{t}, \hat{\boldsymbol{x}}) = (\hat{t}, \hat{x}_1, \ldots, \hat{x}_r)$ as

$$\begin{cases} \hat{t} := t, \\ \hat{x}_i := H_i(\boldsymbol{x}) \quad (i = 1, \ldots, k), \\ \hat{x}_j := x_j \quad (j = k+1, \ldots, r). \end{cases}$$

Since the Jacobian matrix of this transformation at the origin is

$$\begin{pmatrix} 1 & 0 & 0 \\ 0 & \mathcal{H}^T & * \\ 0 & 0 & E_{r-k} \end{pmatrix},$$

where E_{r-k} stands for the $(r-k) \times (r-k)$ identity matrix, $(t, \boldsymbol{x}) \mapsto (\hat{t}, \hat{\boldsymbol{x}})$ is a coordinate transformation on a neighborhood of the origin in $\boldsymbol{R} \times \boldsymbol{R}^r$. Writing $\Phi(t, \boldsymbol{x})$ using $(\hat{t}, \hat{\boldsymbol{x}})$, we have

$$\Phi(t, \boldsymbol{x}) = \hat{t}^{k+1} + \hat{x}_1 \hat{t}^{k-1} + \cdots + \hat{x}_{k-1} \hat{t} + \hat{x}_k.$$

Thus, Φ is $P\mathcal{K}$-equivalent to $\Psi_{k,r}$. $\qquad\square$

Remark 8.3.12. Amongst many unfoldings of a given function of one variable, there is a special unfolding for which all such unfoldings can be obtained via that unfolding. Versal unfoldings have this property. In fact, let Φ_0 be an r-dimensional versal unfolding of φ, and n a positive integer. We can show that for an n-dimensional unfolding Φ of φ, there exist

- a C^∞-function-germ $\vartheta : (\boldsymbol{R} \times \boldsymbol{R}^n, (0, \boldsymbol{0})) \to (\boldsymbol{R}, 0)$ such that $\vartheta'(0, \boldsymbol{0}) \neq 0$,
- a map-germ $\Theta : (\boldsymbol{R}^n, \boldsymbol{0}) \to (\boldsymbol{R}^r, \boldsymbol{0})$, and
- a C^∞-function-germ $\rho \in C_0^\infty(\boldsymbol{R} \times \boldsymbol{R}^n)$ such that $\rho(0) \neq 0$,

so that $\Phi(t, \boldsymbol{x}) = \rho(t, \boldsymbol{x}) \Phi_0(\vartheta(t, \boldsymbol{x}), \Theta(\boldsymbol{x}))$, by using Theorem 8.2.6 (see Exercise 2 in this section). Namely, all unfoldings can be obtained via a versal unfolding. The word "versal" comes from dropping "uni" from "universal". The reason seems to be that the uniqueness of versal unfoldings does not hold for given φ.

Here, we give a proof of the criterion for cusps (Theorem 1.3.2 in Chapter 1), one which is different from that in Section 3.2 of Chapter 3.

Proof of Theorem 1.3.2. Let $\gamma : (\boldsymbol{R}, 0) \to (\boldsymbol{R}^2, \boldsymbol{0})$ have a singularity at the origin, namely, $\gamma'(0) = \boldsymbol{0}$, and let $\det(\gamma''(0), \gamma'''(0)) \neq 0$. By Proposition B.1.2 in Appendix B, γ is a front. Thus we have a unit normal vector field $\boldsymbol{n}$. We define a C^∞-function Φ by (8.16). Then as is shown in Example 8.2.10, the discriminant set D_Φ coincides with the image of γ. By Proposition 8.3.5, Φ is a versal unfolding, thus by Theorem 8.3.11, Φ is $P\mathcal{K}$-equivalent to the unfolding $\Psi := t^3 + x_1 t + x_2$ of t^3 whose discriminant set is the image of the cusp $\gamma_C(t) := (-3t^2, 2t^3)$. Since Ψ is admissible (cf. Remark 8.3.2), by applying Theorem 8.2.11 for any sufficiently small choice of neighborhood $V(\subset \boldsymbol{R}^2)$ of the origin $\boldsymbol{0} := (0, 0)$, there exists a diffeomorphism $\Theta : \Theta^{-1}(V) \to V$ such that

$$D_\Psi \cap V = \Theta(D_\Phi) \cap V.$$

Without loss of generality, we may assume that V is a disk $B^2(\mathbf{0}, r)$ of radius $r (> 0)$ centered at $\mathbf{0} \in \mathbf{R}^2$, where r is a sufficiently small positive number. Taking sufficiently small $\varepsilon (> 0)$, we have $\Theta(\gamma(-\varepsilon, \varepsilon)) \subset B^2(\mathbf{0}, r)$. Since $D_\Psi = \gamma_C(\mathbf{R})$, we have

$$\gamma_C(\mathbf{R}) \cap B^2(\mathbf{0}, r) \supset \Theta \circ \gamma((-\varepsilon, \varepsilon)) \cap B^2(\mathbf{0}, r).$$

Since $\gamma_C^{-1}(\mathbf{0}) = \{0\}$, Corollary E.1.8 in Appendix E implies that the map γ_C is $\mathbf{R}$-proper at $t = 0$. By the Zakalyukin-type lemma (Theorem E.2.1) in Appendix E, there exists a local diffeomorphism φ of $\mathbf{R}$ such that $\Theta \circ \gamma \circ \varphi = \gamma_C$. Hence γ has a cusp at the origin.   □

Remark 8.3.13. If the image of γ is contained in the image of γ_C, then we can show that γ is right–left equivalent to γ_C without using the Zakalyukin-type lemma: By $\gamma''(0) \neq \mathbf{0}$, the half-arc-length parameter (cf. Proposition B.2.1 in Appendix B) is well defined. By Corollary B.2.8 in Appendix B, γ has the same normalized curvature with respect to the half-arc-length parameter as that of the cusp γ_C. Then, by Theorem B.2.3 in Appendix B, the expression of γ coincides with the integral form of the cusp γ_C by the half-arc-length parameter. Thus γ is right–left equivalent to γ_C. Hence γ has a cusp at the origin.

Exercises 8.3

1   Consider a space curve γ with non-zero curvature, and suppose $\gamma(0) = \mathbf{0}$. Let t be the arc-length parameter and $\mathbf{b}$ the bi-normal vector. Then set a C^∞-function Φ by $\Phi(t, \mathbf{x}) = \mathbf{b}(t) \cdot \mathbf{x}$ for $\mathbf{x} \in \mathbf{R}^3$, where "$\cdot$" stands for the Euclidean inner product of $\mathbf{R}^3$. Show that

   (1) Φ satisfies $\Phi_t(0, \mathbf{0}) = 0$, and does not have any critical point at the origin,
   (2) Φ being admissible as an unfolding is equivalent to the torsion τ not vanishing, and
   (3) on a neighborhood where $\tau(t) \neq 0$, the discriminant set D_Φ coincides with the image of the tangent developable surface of γ. (This fact corresponds to the fact that the tangent developable of γ has cuspidal edges along γ.)

2   Let $\Phi_0 : (\mathbf{R} \times \mathbf{R}^r, 0) \to (\mathbf{R}, 0)$ be a versal unfolding of $\varphi(t) = t^{k+1}$ $(r \geq k)$. Then for any n, $\varphi(t)$ and any unfolding $\Phi : (\mathbf{R} \times \mathbf{R}^n, 0) \to (\mathbf{R}, 0)$, there exist a C^∞-function $\vartheta : (\mathbf{R} \times \mathbf{R}^n, (0, 0)) \to (\mathbf{R}, 0)$, a map $\Theta : (\mathbf{R}^n, 0) \to (\mathbf{R}^r, 0)$, and a C^∞-function $\rho \in C_0^\infty(\mathbf{R} \times \mathbf{R}^n)$ satisfying $\rho(0) \neq 0$ such that $\Phi(t, \mathbf{x}) = \rho(t, \mathbf{x}) \Phi_0(\vartheta(t, \mathbf{x}), \Theta(\mathbf{x}))$. (Hint: Without loss of generality, we may set $\Phi_0 = t^{k+1} + x_1 t^{k-1} + \cdots + x_{k-1} t + x_k$. Then apply Theorem 8.2.6.)

8.4. Proof of the Criterion for Swallowtails

8.4.1. Admissible unfoldings constructed from fronts with non-degenerate singular points

Let $f : (\mathbf{R}^2, \mathbf{0}) \to (\mathbf{R}^3, \mathbf{0})$ be a front, and ν a unit normal vector field of f. We introduce a useful coordinate system for f.

Lemma 8.4.1. *Let $f : (\mathbf{R}^2, \mathbf{0}) \to (\mathbf{R}^3, \mathbf{0})$ be a wave front as a map-germ, and $\mathbf{0}$ a non-degenerate singular point. Then there exists a coordinate system (u, v) near the origin and a linear transformation T on $\mathbf{R}^3$ such that*

(1) $T \circ f(u, v) = (u, f_2(u, v), f_3(u, v))$, *and* $(T \circ f)_u(0, 0) = (1, 0, 0)$,
(2) $\partial_v := \partial/\partial v$ *gives the null vector field along the singular curve.*

Proof. Since $\mathbf{0} := (0, 0)$ is a non-degenerate singular point, the rank of the matrix (f_u, f_v) is one. So, by a suitable linear transformation of the domain of f, we may assume that $f_v(0, 0) = \mathbf{0}$ (cf. Lemma 2.3.5 in Chapter 2). Moreover, by a suitable linear transformation T of $\mathbf{R}^3$, we may assume $T \circ f_u(0, 0) = (1, 0, 0)$. We set

$$\hat{f}(u, v) := T \circ f(u, v) = (\hat{f}_1(u, v), \hat{f}_2(u, v), \hat{f}_3(u, v)).$$

Then $(\hat{f}_1)_u(0, 0) \neq 0$, and so we can define a diffeomorphism-germ $\varphi : (\mathbf{R}^2, \mathbf{0}) \to (\mathbf{R}^2, \mathbf{0})$ by

$$\varphi(u, v) := (\hat{f}_1(u, v), v).$$

Setting $(\xi, \eta) := \varphi(u, v)$, we have

$$\hat{f} \circ \varphi^{-1}(\xi, \eta) = \hat{f}(u, v) = (\hat{f}_1(u, v), \hat{f}_2(u, v), \hat{f}_3(u, v))$$
$$= (\xi, \hat{f}_2 \circ \varphi^{-1}(\xi, \eta), \hat{f}_3 \circ \varphi^{-1}(\xi, \eta)).$$

Thus, (1) holds for a new coordinate system (ξ, η). Now we change the notation (ξ, η) to (u, v), and show (2). Let (u_0, v_0) be an arbitrary point on the singular curve. Since the rank of $d\hat{f}(u_0, v_0)$ is one, there exists $(a, b) \neq (0, 0)$ such that

$$a\hat{f}_u(u_0, v_0) + b\hat{f}_v(u_0, v_0) = \mathbf{0}.$$

By $(\hat{f}_1)_u(u_0, v_0) = 1$ and $(\hat{f}_1)_v(u_0, v_0) = 0$, the first component of the above equation gives $a = 0$. Thus ∂_v gives a null vector at each singular point (u_0, v_0). $\square$

Let $f : (\mathbf{R}^2, \mathbf{0}) \to (\mathbf{R}^3, \mathbf{0})$ be a front, and $(0, 0)$ a non-degenerate singular point, and let the coordinate system (u, v) on the source satisfy (cf. Lemma 8.4.1)

(a) $f(u, v) = (u, f_2(u, v), f_3(u, v))$ and $f_u(0, 0) = (1, 0, 0)$, and
(b) $f_v(u, v) = \mathbf{0}$ if (u, v) is a singular point of f.

Let $\nu(u, v)$ be a smooth unit normal vector field of f defined near the origin $(0, 0)$. We set

$$\Phi(v, \boldsymbol{x}) := (\boldsymbol{x} - f(x_1, v)) \cdot \nu(x_1, v), \quad \varphi(v) := \Phi(v, \mathbf{0}), \qquad (8.40)$$

where $\boldsymbol{x} = (x_1, x_2, x_3)$ and "$\cdot$" stands for the Euclidean inner product of $\boldsymbol{R}^3$. This function Φ satisfies the following:

Proposition 8.4.2. *The function Φ is an unfolding of the function defined by $\varphi(v) := -f(0, v) \cdot \nu(0, v)$, and D_Φ coincides with the image of f near the origin.*

Proof. Since $f_v \cdot \nu$ vanishes identically, $\Phi(v, \boldsymbol{x}) = \Phi_v(v, \boldsymbol{x}) = 0$ is equivalent to

$$(\boldsymbol{x} - f(x_1, v)) \cdot \nu(x_1, v) = (\boldsymbol{x} - f(x_1, v)) \cdot \nu_v(x_1, v) = 0. \qquad (8.41)$$

If $\boldsymbol{x}$ lies in the image of f, then there exists (x_1, v) such that $\boldsymbol{x} = f(x_1, v)$. Since (8.41), we have $\Phi(v, \boldsymbol{x}) = \Phi_v(v, \boldsymbol{x}) = 0$. Thus, $\boldsymbol{x} \in D_\Phi$.

Conversely, suppose that $\boldsymbol{x} \in D_\Phi$, then there exists v such that $\Phi(v, \boldsymbol{x}) = \Phi_v(v, \boldsymbol{x}) = 0$. By (a), the first component of $\boldsymbol{x} - f(x_1, v)$ vanishes, and so we have

$$(\boldsymbol{x} - f(x_1, v)) \cdot \boldsymbol{e}_1 = 0, \qquad \boldsymbol{e}_1 := (1, 0, 0). \qquad (8.42)$$

By (8.41) and (8.42), the vector $\boldsymbol{x} - f(x_1, v)$ is perpendicular to three vectors $\boldsymbol{e}_1, \nu(x_1, v)$, and $\nu_v(x_1, v)$. Thus, if we show that these vectors are linearly independent, then $\boldsymbol{x} - f(x_1, v) = \mathbf{0}$, namely, $\boldsymbol{x}$ is contained in the image of f, proving the assertion.

So, we now show the linear independence as follows: Since f is a front at $\mathbf{0}$, the rank of

$$\begin{pmatrix} f_u(0, 0) & f_v(0, 0) \\ \nu_u(0, 0) & \nu_v(0, 0) \end{pmatrix} = \begin{pmatrix} \boldsymbol{e}_1 & \mathbf{0} \\ \nu_u(0, 0) & \nu_v(0, 0) \end{pmatrix}$$

is two, and in particular,

$$\nu_v(0, 0) \neq \mathbf{0}. \qquad (8.43)$$

On the other hand, since

$$f_u \cdot \nu_v = -f_{uv} \cdot \nu = f_v \cdot \nu_u \qquad (8.44)$$

and $f_v(0, 0) = \mathbf{0}$, the two vectors $f_u(0, 0)$ and $\nu_v(0, 0)$ are perpendicular. Thus, $\{\boldsymbol{e}_1, \nu(x_1, v), \nu_v(x_1, v)\}$ is a set of linearly independent vectors near the origin. $\square$

The following assertion follows immediately.

Corollary 8.4.3. *It holds that*

$$f_u(0,0) = e_1, \quad \nu(0,0) \cdot e_1 = \nu_v(0,0) \cdot e_1 = 0.$$

We also have the following:

Proposition 8.4.4. *The function* Φ *defined in* (8.40) *is admissible as an unfolding.*

Proof. Since $f(0) = 0$, and $\nu(0)$ is a unit normal vector of f at 0, $\Phi'(0,0) = 0$ holds. By (8.4), we have

$$\Phi_{x_1}(v, x) = (e_1 - f_u(x_1, v)) \cdot \nu(x_1, v) + (x - f(x_1, v)) \cdot \nu_u(x_1, v), \quad (8.45)$$

$$\Phi_{x_i}(v, x) = e_i \cdot \nu(x_1, v) \quad (i = 2, 3), \quad (8.46)$$

where $e_2 := (0,1,0)$ and $e_3 := (0,0,1)$. Substituting $(v, x) = (0,0)$ into (8.45), since $f(0,0) = 0$ and $f_u(0,0) = e_1$, we have $\Phi_{x_1}(0,0) = 0$. Differentiating (8.45) by v and writing $' = \partial/\partial v$, we have

$$\Phi'_{x_1} = -f'_u \cdot \nu + (e_1 - f_u) \cdot \nu' - f' \cdot \nu_u + (x - f) \cdot \nu'_u. \quad (8.47)$$

Substituting $(v, x) = (0,0)$ into it, we see that the third and fourth terms in the right hand side are zero, since

$$f(0,0) = f'(0,0) = 0. \quad (8.48)$$

As the first term is $f'_u \cdot \nu = -f' \cdot \nu_u$, and $f'(0) = 0$, we see that this also vanishes at the origin, and we have

$$\Phi'_{x_1}(0,0) = (e_1 - f_u(0,0)) \cdot \nu'(0,0).$$

Since $e_1 - f_u$ vanishes at $(x_1, v) = (0,0)$, we have $\Phi'_{x_1}(0,0) = 0$. On the other hand, by (8.46), we have

$$\Phi_{x_i}(0,0) = e_i \cdot \nu(0,0) = \nu_i(0,0) \quad (i = 2, 3),$$

where $\nu = (\nu_1, \nu_2, \nu_3)$, and next differentiating (8.46) with respect to v and substituting $(0,0)$, we have

$$\Phi'_{x_i}(0,0) = e_i \cdot \nu'(0,0) = \nu'_i(0,0) \quad (i = 2, 3).$$

By Corollary 8.4.3, the first component of $\nu(0,0)$, $\nu'(0,0)$ are zero. Thus

$$\operatorname{rank} \begin{pmatrix} \Phi_{x_1} & \Phi'_{x_1} \\ \Phi_{x_2} & \Phi'_{x_2} \\ \Phi_{x_3} & \Phi'_{x_3} \end{pmatrix} \Bigg|_{(v,x)=(0,0)} = \operatorname{rank} \begin{pmatrix} 0 & 0 \\ \nu_2(0,0) & \nu'_2(0,0) \\ \nu_3(0,0) & \nu'_3(0,0) \end{pmatrix}$$

$$= \operatorname{rank}\left(\nu(0,0), \nu'(0,0)\right). \quad (8.49)$$

By (8.43), we have $\nu'(0,0) \neq \mathbf{0}$. Since ν' is perpendicular to ν, the rank of the right hand side of (8.49) is 2. □

By using the above results, we show the criterion for cuspidal edges.

Corollary 8.4.5. *If a front-germ* $f : (\mathbf{R}^2, \mathbf{0}) \to (\mathbf{R}^3, \mathbf{0})$ *satisfies the condition in the criterion for cuspidal edges ((1) of Theorem 4.2.3 in Chapter 4), then the function* φ *defined by* (8.40) *satisfies*

$$\varphi(0) = \varphi'(0) = \varphi''(0) = 0, \qquad \varphi'''(0) \neq 0, \tag{8.50}$$

and Φ *is a three dimensional versal unfolding of* φ.

Proof. Since

$$\operatorname{rank} \begin{pmatrix} \Phi_{x_1} & \Phi'_{x_1} \\ \Phi_{x_2} & \Phi'_{x_2} \\ \Phi_{x_3} & \Phi'_{x_3} \end{pmatrix} \Bigg|_{(v,\boldsymbol{x})=(0,0)} = 2$$

is already shown, it is sufficient to show (8.50). Since $f_v \cdot \nu = 0$ and $f' := f_v$, differentiating $\varphi(v) = -f(0,v) \cdot \nu(0,v)$, we have

$$-\varphi' = f' \cdot \nu + f \cdot \nu' = f \cdot \nu'. \tag{8.51}$$

By (8.48), we have $\varphi(0) = \varphi'(0) = 0$. Next, differentiating (8.51), we have

$$-\varphi'' = f' \cdot \nu' + f \cdot \nu''. \tag{8.52}$$

By (8.48), we also have $\varphi''(0) = 0$. Differentiating (8.52) again, we have

$$-\varphi''' = f'' \cdot \nu' + 2f' \cdot \nu'' + f \cdot \nu''' = f'' \cdot \nu' + f \cdot \nu'''. \tag{8.53}$$

By (8.48) again,

$$\varphi'''(0,0) = -f''(0,0) \cdot \nu'(0,0) \tag{8.54}$$

holds. Since $f_u(0,0) = \boldsymbol{e}_1$ and we showed the last part of the proof of Lemma 8.4.2, $\{\boldsymbol{e}_1, \nu, \nu'\}$ is a linearly independent set at the origin, we have

$$f''(0,0) = a f_u(0,0) + b\nu(0,0) + c\nu'(0,0) \qquad (a,b,c \in \mathbf{R}).$$

Since $f_1 = u$, we have $(f_1)''(0,0) = 0$. From this together with Corollary 8.4.3, the first components of $f''(0,0)$, $\nu(0,0)$, $\nu'(0,0)$ are zero, and we can conclude $a = 0$. On the other hand, since

$$f''(0,0) \cdot \nu(0,0) = -f'(0,0) \cdot \nu'(0,0) = 0$$

and ν is perpendicular to ν', we have $b = 0$. Thus,

$$f''(0,0) = c\nu'(0,0). \tag{8.55}$$

By (8.54), $c = -\varphi'''(0,0)$ holds, and so it is enough to show $c \neq 0$. The condition (1) in Theorem 4.2.3 does not depend on the choice of the identifier of singularities, and we may regard $\Lambda := \det(f_u, f', \nu)$ as such an identifier. Since ∂_v is a null vector field, the non-degeneracy condition $\Lambda_\eta \neq 0$ of f at $(0,0)$ reduces to $\Lambda'(0,0) \neq 0$. Since $\varepsilon := \det(e_1, \nu, \nu') \in \{\pm 1\}$, we have

$$0 \neq \Lambda'(0,0) = \det(f_u(0,0), f''(0,0), \nu(0,0)) = c\varepsilon, \qquad (8.56)$$

proving $\varphi'''(0) \neq 0$. □

We can show the criteria for the cuspidal edges as follows:

Proofs of (1) of Theorem 2.6.3 and (1) of Theorem 4.2.3. As we showed in Section 4.2 in Chapter 4, "(1) of Theorem 2.6.3" is equivalent to "(1) of Theorem 4.2.3". So, we show (1) of Theorem 4.2.3. By Corollary 8.4.5, if f satisfies the condition of Theorem 4.2.3 in Chapter 4, then Φ is a versal unfolding satisfying (8.50), and by Theorem 8.3.11, it is is $P\mathcal{K}$-equivalent to $\Psi := t^3 + x_1 t + x_2$. By Example 8.3.3, we can write $D_\Psi = g_C(R^2)$, where $g_C(u,v) = (-3u^2, 2u^3, v)$.

Since Ψ is admissible (cf. Remark 8.3.2), by applying Theorem 8.2.11 for any sufficiently small choice of neighborhood $V(\subset R^3)$ of the origin $\mathbf{0} := (0,0,0)$, there exists a diffeomorphism $\Theta : \Theta^{-1}(V) \to V$ such that

$$D_\Psi \cap V = \Theta(D_\Phi) \cap V.$$

Without loss of generality, we may assume that V is the ball $B^3(\mathbf{0}, r)$ of radius $r(> 0)$ centered at $\mathbf{0}$ in R^3, where r is a sufficiently small positive number. We can take a neighborhood W of the source space of f such that $\Theta(f(W)) \subset B^3(\mathbf{0}, r)$. Then we have

$$g_C(R^2) \cap B^3(\mathbf{0}, r) \supset \Theta \circ f(W) \cap B^3(\mathbf{0}, r).$$

By Corollary E.1.12 in Appendix E, g_C is R^2-proper at $(0,0)$. Since $g_C^{-1}(g_C(0,0)) = \{(0,0)\}$, by the Zakalyukin-type lemma (Theorem E.2.1) in Appendix E, there exists a local diffeomorphism φ of R^2 such that $\Theta \circ f \circ \varphi = g_C$. Hence f has a cuspidal edge at $(0,0)$. □

8.4.2. The criterion for swallowtails

To prove the criterion for swallowtails, we first prepare the following assertion:

Proposition 8.4.6. *Let* $f : (\mathbf{R}^2, 0) \to (\mathbf{R}^3, 0)$ *be a wave-front-germ. If* f *satisfies the conditions in the criteria for swallowtails, namely, conditions (2) in Theorem 4.2.3 of Chapter 4, then* φ *in* (8.40) *satisfies*

$$\varphi(0) = \varphi'(0) = \varphi''(0) = \varphi'''(0) = 0, \qquad \varphi^{(4)}(0) \neq 0, \tag{8.57}$$

and Φ *is a three-dimensional versal unfolding of* φ.

Proof. First, we show (8.57). We can show $\varphi(0) = \varphi'(0) = \varphi''(0) = 0$ by the same method as in the proof of Corollary 8.4.5. If f satisfies the conditions of the criteria for swallowtails, we have $\Lambda_\eta(0, 0) = 0$, that is, $\Lambda'(0, 0) = 0$. Thus, by (8.56), $c = 0$ holds and then $\varphi'''(0) = 0$. Moreover, by (8.55), we have

$$f''(0, 0) = \mathbf{0}. \tag{8.58}$$

Differentiating (8.53) and substituting $(x_3, v) = (0, \mathbf{0})$, and by (8.58), we have

$$-\varphi^{(4)}(0) = f'''(0, 0) \cdot \nu'(0, 0). \tag{8.59}$$

Thus, it is enough to show $f''' \cdot \nu' \neq 0$ at the origin in order to show $\varphi^{(4)}(0) \neq 0$. Since $f_u(0, 0) = e_1$ and $\{e_1, \nu, \nu'\}$ is a set of mutually orthogonal vectors at the origin (cf. Corollary 8.4.3), $f'''(0, 0)$ can be written as

$$f'''(0, 0) = a_1 f_u(0, 0) + b_1 \nu(0, 0) + c_1 \nu'(0, 0). \tag{8.60}$$

Differentiating $\lambda = \det(f_u, f', \nu)$ with respect to v twice, since $f'(0, 0) = f''(0, 0) = \mathbf{0}$, we have

$$0 \neq \det(f_u(0, 0), f'''(0, 0), \nu(0, 0)) = c_1 \det(f_u(0, 0), \nu'(0, 0), \nu(0, 0)).$$

Thus, $c_1 \neq 0$. Since

$$f'''(0, 0) \cdot \nu'(0, 0) = c_1 \nu'(0, 0) \cdot \nu'(0, 0),$$

we obtain $\varphi^{(4)}(0) \neq 0$. Next, we show

$$\mathrm{rank} \begin{pmatrix} \Phi_{x_1} & \Phi'_{x_1} & \Phi''_{x_1} \\ \Phi_{x_2} & \Phi'_{x_2} & \Phi''_{x_2} \\ \Phi_{x_3} & \Phi'_{x_3} & \Phi''_{x_3} \end{pmatrix} \Bigg|_{(v, \boldsymbol{x}) = (0, 0)} = 3. \tag{8.61}$$

By (8.49), it is enough to show $\Phi''_{x_1}(0, \mathbf{0}) \neq 0$ under the condition (2) of Theorem 4.2.3 in Chapter 4. Since $f'' = f' = f = \mathbf{0}$ at $(0, 0)$ (cf. (8.48) and (8.58)) and $e_1 - f_u = \mathbf{0}$, by differentiating (8.47), we have

$$\Phi''_{x_1} = -f''_u \cdot \nu - 2f'_u \cdot \nu' = -f'_u \cdot \nu'$$

at the origin. Thus, it is enough to show $f'_u \cdot \nu' \neq 0$ at the origin. Since f_u, ν, ν' are linearly independent near the origin, we set

$$f'_u = A f_u + B\nu + C\nu'. \tag{8.62}$$

Since $\Lambda' = 0$ and $d\Lambda \neq \mathbf{0}$ at $(0,0)$, we have $\Lambda_u(0,0) \neq 0$. By (8.62), we have

$$0 \neq \Lambda_u(0,0) = \det(f_u(0,0), f'_u(0,0), \nu(0,0))$$
$$= C(0,0) \det(f_u(0,0), \nu'(0,0), \nu(0,0)),$$

which implies $C(0,0) \neq 0$. Thus, by (8.62), we have $f'_u \cdot \nu' \neq 0$ at the origin. Hence (8.61) is shown. □

Using the above result, we show the criterion for swallowtails.

Proofs of (2) **of Theorem 2.6.3 and** (2) **of Theorem 4.2.3.** As we showed in Section 4.2 in Chapter 4, "(2) of Theorem 2.6.3(1)" is equivalent to "(2) of Theorem 4.2.3", so we show (2) of Theorem 4.2.3 in Chapter 4. We assume that $f : (\mathbf{R}^2, \mathbf{0}) \to (\mathbf{R}^3, \mathbf{0})$ satisfies the condition (2) of Theorem 4.2.3 in Chapter 4. By Proposition 8.4.6, Φ is a versal unfolding satisfying (8.57), and by Theorem 8.3.11, it is $P\mathcal{K}$-equivalent to $\Psi := t^4 + x_1 t^2 + x_2 t + x_3$. By Example 8.3.3, we can write $D_\Psi = g_S(\mathbf{R}^2)$, where $g_S(u,v) = (u, -4v^3 - 2uv, 3v^4 + uv^2)$. Since Ψ is admissible (cf. Remark 8.3.2), by applying Theorem 8.2.11 for any sufficiently small choice of neighborhood $V(\subset \mathbf{R}^3)$ of the origin $\mathbf{0} := (0,0,0)$, there exists a diffeomorphism $\Theta : \Theta^{-1}(V) \to V$ such that

$$D_\Psi \cap V = \Theta(D_\Phi) \cap V.$$

Without loss of generality, we may assume that V is the ball $B^3(\mathbf{0}, r)$ of radius $r(> 0)$ centered at $\mathbf{0}$ in $\mathbf{R}^3$, where r is a sufficiently small positive number. We can take a neighborhood W of the origin in the source space of f, $\Theta(f(W)) \subset B^3(\mathbf{0}, r)$. So we have

$$g_S(\mathbf{R}^2) \cap B^3(\mathbf{0}, r) \supset \Theta \circ f(W) \cap B^3(\mathbf{0}, r).$$

By Corollary E.1.8, g_S is $\mathbf{R}^2$-proper at $(0,0)$. Since $g_S^{-1}(g_S(0,0)) = \{(0,0)\}$, by the Zakalyukin-type lemma (Theorem E.2.1) in Appendix E, there exists a local diffeomorphism φ of $\mathbf{R}^2$ such that $\Theta \circ f \circ \varphi = g_S$. Hence $(0,0)$ is a swallowtail singular point of f. □

8.5.   An Introduction to Criteria for Other Types of Singularities Appearing on Wave Fronts

Fronts have various singularities in addition to those we discussed so far.

Example 8.5.1. We set map-germs $f_1, f_2, f_3, f_{4,\pm}$

$$f_1(u,v) := (u, v(u^2+v^2), v^2(2u^2+3v^2)),$$
$$f_2(u,v) := (u, v(u^2-v^2), v^2(2u^2-3v^2)),$$
$$f_3(u,v) := (u, 5v^4+2uv, uv^2+4v^5),$$
$$f_{4,+}(u,v) := (uv, u^2+3v^2, v(u^2+v^2)),$$
$$f_{4,-}(u,v) := (uv, u^2-3v^2, v(u^2-v^2)),$$

respectively. All of them are fronts, and have a singular point at the origin. We call them *cuspidal lips, cuspidal beaks, cuspidal butterfly,* $D_{4,\pm}$ *singularity* respectively (Figs. 8.1 and 8.2).

The cuspidal butterfly is a non-degenerate singularity. On the other hand, the cuspidal lips and the cuspidal beaks are not non-degenerate, but the rank of the matrix (f_u, f_v) at the singular point is one, so one can apply the proof of Lemma 8.4.1, we see that there exists a coordinate system (u, v) centered at p such that $\tilde{\eta} := \partial_v$ is an extended null vector field. We state criteria for these singularities by using the notion of the null vector field.

Fig. 8.1.   cuspidal lips, cuspidal beaks, cuspidal butterfly.

Fig. 8.2. $D_{4,+}$ singularity, $D_{4,-}$ singularity.

Fact 8.5.2 ([40, Theorem A.1]; [39, Theorem 8.2]). Let $f : U \to \mathbb{R}^3$ be a front, and let p be a singular point such that $\mathrm{rank}(f_u(p), f_v(p)) = 1$. Let Λ be an identifier of singularities, and $\tilde{\eta}$ an extended null vector field. Then

- f at p is a cuspidal lips if and only if $d\Lambda(p) = 0$ and $\det H_\Lambda(p) > 0$.
- f at p is a cuspidal beaks if and only if $d\Lambda(p) = 0$, $\det H_\Lambda(p) < 0$ and $\Lambda_{\eta\eta}(p) \neq 0$.
- f at p is a cuspidal butterfly if and only if p is a non-degenerate singular point, $\Lambda_\eta(p) = \Lambda_{\eta\eta}(p) = 0$ and $\Lambda_{\eta\eta\eta}(p) \neq 0$.

Here,

$$\Lambda_\eta := d\Lambda(\tilde{\eta}), \qquad \Lambda_{\eta\eta} := d\Lambda_\eta(\tilde{\eta}), \qquad \Lambda_{\eta\eta\eta} := d\Lambda_{\eta\eta}(\tilde{\eta}),$$

and

$$H_\Lambda(p) := \begin{pmatrix} \dfrac{\partial^2 \Lambda}{\partial u^2}(p) & \dfrac{\partial^2 \Lambda}{\partial v \partial u}(p) \\ \dfrac{\partial^2 \Lambda}{\partial u \partial v}(p) & \dfrac{\partial^2 \Lambda}{\partial v^2}(p) \end{pmatrix}$$

stands for the Hessian of the function Λ at p.

For the $D_{4,\pm}$ singularity, the following criteria is known.

Fact 8.5.3 ([67, Theorem 1.1]). Let $f : U \to \mathbb{R}^3$ be a front, and let p be a singular point such that $f_u(p) = f_v(p) = \mathbf{0}$. Let Λ be an identifier of singularities. Then f at p is a $D_{4,+}$ singularity ($D_{4,-}$ singularity) if and only if $\det H_\Lambda(p) < 0$ ($\det H_\Lambda(p) > 0$).

One can prove these criteria like as for cuspidal edges and swallowtails, by constructing an unfolding such that the discriminant set coincides with the image of the front [39, 40, 67].

Exercises 8.5

1 Show the following unfoldings are admissible. Moreover, find their discriminant sets, and show the parameterizations of them are the same as Φ_1 (cuspidal lips), Φ_2 (cuspidal beaks) and Φ_3 (cuspidal butterfly) given in Example 8.5.1, respectively. Here we set $(x, y, z) \in \mathbb{R}^3$.

(1) $\Phi_1(t, x, y, z) := t^4/4 + t^2 x^2/2 - ty + z/4$,
(2) $\Phi_2(t, x, y, z) := t^4/4 - t^2 x^2/2 - ty + z/4$,
(3) $\Phi_3(t, x, y, z) := t^5 + t^2 x - ty + z$.

Chapter 9

Coherent Tangent Bundles

In this chapter, we formulate the Gauss–Bonnet type formula given in Chapter 6 without assuming the existence of ambient spaces. For this purpose, we introduce the concept of "coherent tangent bundle", assuming that readers are familiar with vector bundles and their (fiber-wise) inner products.

9.1. Definition of Coherent Tangent Bundles

We first give the definition of coherent tangent bundles. Let $\mathcal{M}^m$ be an m-manifold. Consider a vector bundle

$$\pi : \mathcal{E} \longrightarrow \mathcal{M}^m$$

over $\mathcal{M}^m$, and set

$$\Gamma(\mathcal{E}) := \{ s \in C^\infty(\mathcal{M}^m, \mathcal{E}) \, ; \, \pi \circ s = \mathrm{id}_{\mathcal{M}^m} \},$$

where $C^\infty(\mathcal{M}^m, \mathcal{E})$ denotes the C^∞ maps from $\mathcal{M}^m$ to the manifold $\mathcal{E}$ and $\mathrm{id}_{\mathcal{M}^m}$ is the identity map of $\mathcal{M}^m$. (Fundamental properties of vector bundles are found in [86, 58].) Each element of $\Gamma(\mathcal{E})$ is called a *section* of $\mathcal{E}$. In particular, $\Gamma(T\mathcal{M}^m)$ is the set of C^∞-vector fields defined on $\mathcal{M}^m$. Let $r(< \infty)$ be the rank of the vector bundle $\mathcal{E}$. We let denote $\mathcal{E}|_U$ the restriction of the vector bundle $\mathcal{E}$ to an open subset $U(\subset \mathcal{M}^m)$. Then, for each point $p \in \mathcal{M}^m$, there exist a neighborhood U of p and a diffeomorphism $\tau : \mathcal{E}|_U \to U \times \mathbf{R}^r$ such that the restriction $\tau_q : \mathcal{E}_q \to \mathbf{R}^r$ of τ to $q \in U$ gives a linear isomorphism. Such a τ is called a *local trivialization* of $\mathcal{E}$.

We let $\mathcal{E}^*$ be the dual vector bundle $\mathcal{E}$, that is, each fiber $\mathcal{E}^*_p$ is the dual vector space of $\mathcal{E}_p$ for each $p \in \mathcal{M}^m$ (see [58, Chapter 3], where $\mathcal{E}^*$ is denoted by $\mathrm{Hom}(\mathcal{E}, \mathbf{R})$). Consider the tensor product $\mathcal{E}^* \otimes \mathcal{E}^*$. A section

$\langle \, , \, \rangle \in \Gamma(\mathcal{E}^* \otimes \mathcal{E}^*)$ is called an *inner product* or *metric* on $\mathcal{E}$ if $\langle \, , \, \rangle_p$ is a symmetric positive definite bi-linear form on $\mathcal{E}_p$ for each $p \in \mathcal{M}^m$. A Riemannian metric on the manifold $\mathcal{M}^m$ is an example of a metric on the tangent bundle $T\mathcal{M}^m$.

We denote by $C^\infty(\mathcal{M}^m)$ the set of C^∞-functions on $\mathcal{M}^m$. A *covariant derivative* or a (linear) *connection* on the vector bundle $\mathcal{E}$ is a map

$$D : \Gamma(T\mathcal{M}^m) \times \Gamma(\mathcal{E}) \ni (X, \xi) \mapsto D_X\xi \in \Gamma(\mathcal{E})$$

satisfying the following properties:

(1) For each $f, h \in C^\infty(\mathcal{M}^m)$ and each $X, Y \in \Gamma(T\mathcal{M}^m)$,

$$D_{fX+gY}\xi = fD_X\xi + gD_Y\xi \qquad (\xi \in \Gamma(\mathcal{E}))$$

holds. In particular, if $(\boldsymbol{v} :=)X_p = Y_p$, then we have

$$(D_X\xi)_p = (D_Y\xi)_p \qquad (\xi \in \Gamma(\mathcal{E})). \tag{9.1}$$

We denote this by $D_{\boldsymbol{v}}\xi$.

(2) For each $\xi, \eta \in \Gamma(\mathcal{E})$, we have

$$D_X(\xi + \eta) = D_X\xi + D_X\eta \qquad (X \in \Gamma(T\mathcal{M}^m)).$$

(3) For each $f \in C^\infty(\mathcal{M}^m)$ and $\xi \in \Gamma(\mathcal{E})$, we have

$$D_X(f\xi) = (Xf)\xi + fD_X\xi \qquad (X \in \Gamma(T\mathcal{M}^m)).$$

We fix an inner product $\langle \, , \, \rangle$ on the vector bundle $\mathcal{E}$. A covariant derivative D on $\mathcal{E}$ is said to be *compatible with respect to the inner product* if it satisfies

$$X\langle \xi, \eta \rangle = \langle D_X\xi, \eta \rangle + \langle \xi, D_X\eta \rangle \quad (X \in \Gamma(T\mathcal{M}^m), \ \xi, \eta \in \Gamma(\mathcal{E})). \tag{9.2}$$

For example, let g be a Riemannian metric on the manifold $\mathcal{M}^m$, then the Levi-Civita connection ∇ is a typical example of a covariant derivative on the tangent bundle $T\mathcal{M}^m$ having this compatibility (cf. Example 9.1.3).

Let $\pi \colon \mathcal{E} \to \mathcal{M}^m$ (resp. $\pi' \colon \mathcal{E}' \to \mathcal{N}^n$) be a vector bundle on a manifold $\mathcal{M}^m$ (resp. $\mathcal{N}^n$). Then a C^∞-map $\varphi \colon \mathcal{E} \to \mathcal{E}'$ is called a *homomorphism of vector bundles* if there exists a C^∞-map $f : \mathcal{M}^m \to \mathcal{N}^n$ such that for each $p \in \mathcal{M}^m$, φ induces a linear map $\varphi|_{\pi^{-1}(p)} \colon \mathcal{E}_p \to \mathcal{E}'_{f(p)}$. (In particular, φ maps each fiber $\mathcal{E}_p$ to the fiber $\mathcal{E}'_{f(p)}$.) We now consider the case that $\mathcal{M}^m = \mathcal{N}^n$ and $f : \mathcal{M}^m \to \mathcal{M}^m$ is the identity map. In this case, if $\varphi \colon \mathcal{E} \to \mathcal{E}'$ be a homomorphism between two vector bundles, then a section $X \in \Gamma(\mathcal{E})$ induces a section

$$\varphi(X) := \varphi \circ X \in \Gamma(\mathcal{E}').$$

Definition 9.1.1. Suppose that $\pi : \mathcal{E} \to \mathcal{M}^m$ is a vector bundle of rank $m := \dim \mathcal{M}^m$. Let D be a covariant derivative of $\mathcal{E}$ which is compatible with a given inner product $\langle \ , \ \rangle$ on $\mathcal{E}$, and $\varphi \colon T\mathcal{M}^m \to \mathcal{E}$ a homomorphism of vector bundles. Then a 5-tuple $(\mathcal{M}^m, \mathcal{E}, \langle \ , \ \rangle, D, \varphi)$ is called a *coherent tangent bundle* if it satisfies the identity
$$D_X \varphi(Y) - D_Y \varphi(X) - \varphi([X, Y]) = 0 \qquad (X, Y \in \Gamma(T\mathcal{M}^m)), \qquad (9.3)$$
where 0 means the zero section of $\mathcal{E}$ and $[X, Y]$ denotes the bracket of two vector fields X and Y.

For the sake of simplicity, we often write just $\mathcal{E}$ to express a given coherent tangent bundle instead to the full set of data $(\mathcal{M}^m, \mathcal{E}, \langle \ , \ \rangle, D, \varphi)$.

Definition 9.1.2. A point $p \in \mathcal{M}^m$ is called a *singular point* or *semidefinite point* of φ if $\varphi_p : T_p\mathcal{M}^m \to \mathcal{E}_p$ is not a linear isomorphism.

Example 9.1.3. Let $(\mathcal{M}^m, g)$ be a Riemannian manifold, and ∇ the Levi-Civita connection. For a local coordinate neighborhood $(U; x_1, \ldots, x_m)$, we can write
$$\nabla_{\partial/\partial x_i} \frac{\partial}{\partial x_j} = \sum_{k=1}^{m} \Gamma_{ij}^k \frac{\partial}{\partial x_k} \qquad (i, j = 1, \ldots, m).$$
Each C^∞-function Γ_{jk}^i $(i, j, k = 1, \ldots, m)$ on U is called the *Christoffel symbol*. It is well known that the identities
$$\Gamma_{ij}^k = \Gamma_{ji}^k \qquad (i, j, k = 1, \ldots, m) \qquad (9.4)$$
hold. By regarding the map φ as the identity map of $\mathcal{M}^m$, ∇ satisfies (9.3). It can be easily checked that the identity (9.2) is equivalent to the following condition
$$\frac{\partial g_{ij}}{\partial x_k} = \sum_{h=1}^{m} \Gamma_{kj}^h g_{ih} + \sum_{h=1}^{m} \Gamma_{ik}^h g_{hj} \qquad (i, j, k = 1, \ldots, m), \qquad (9.5)$$
where $g_{ij} := g(\partial/\partial x_i, \partial/\partial x_j)$ $(i, j = 1, \ldots, m)$. Thus $(\mathcal{M}^m, T\mathcal{M}^m, g, \nabla, \mathrm{id})$ is a coherent tangent bundle, where $\mathrm{id} : T\mathcal{M}^m \to T\mathcal{M}^m$ is the identity map. Taking another coordinate system $(V; y_1, \ldots, y_m)$, we set
$$\nabla_{\partial/\partial y_i} \frac{\partial}{\partial y_j} = \sum_{k=1}^{m} \Lambda_{ij}^k \frac{\partial}{\partial y_k} \qquad (i, j = 1, \ldots, m).$$
It is then well-known that the identity
$$\Lambda_{ij}^k = \sum_{a=1}^{m} \frac{\partial^2 x_a}{\partial y_i \partial y_j} \frac{\partial y_k}{\partial x_a} + \sum_{a,b,c=1}^{m} \frac{\partial x_a}{\partial y_i} \frac{\partial x_b}{\partial y_j} \frac{\partial y_k}{\partial x_c} \Gamma_{ab}^c \qquad (9.6)$$
holds on $U \cap V$, which gives the coordinate-change-formula for the Christoffel symbols (cf. [41, Proposition 1.7.2]).

Let $f : \mathcal{M}^m \to \mathcal{N}^n$ be a smooth map from an m-dimensional manifold $\mathcal{M}^m$ to an n-dimensional manifold $\mathcal{N}^n$. We now define the "pull-back" of a given vector bundle on $\mathcal{N}^n$ as a vector bundle on $\mathcal{M}^m$. Let $\pi : \mathcal{E} \to \mathcal{N}^n$ be a vector bundle of rank r. We set

$$f^*\mathcal{E} := \{(p, \xi) \in \mathcal{M}^m \times \mathcal{E}\,;\, f(p) = \pi(\xi)\},$$

and define its projection by

$$\tilde{\pi} : f^*\mathcal{E} \ni (p, \xi) \mapsto p \in \mathcal{M}^m.$$

Each fiber $\tilde{\pi}^{-1}(p)$ ($p \in \mathcal{M}^m$) has the structure of an r-dimensional vector space. We fix a bundle isomorphism $\tau : \mathcal{E}|_U \to U \times \mathbf{R}^r$ as a local trivialization, where U is an open subset of $\mathcal{N}^n$. For each $p \in \mathcal{M}^m$ satisfying $f(p) \in U$, the continuity of f implies that there exists a neighborhood V of p such that $f(V) \subset U$. Then τ induces a map given by

$$\tilde{\tau} : f^*\mathcal{E}|_V \ni \big(q, \tau^{-1}(f(q), \boldsymbol{v})\big) \longmapsto (q, \boldsymbol{v}) \in V \times \mathbf{R}^r,$$

which gives a local trivialization of $f^*\mathcal{E}$ as a vector bundle. Thus, $f^*\mathcal{E}$ has the structure of a vector bundle of rank r over $\mathcal{M}^m$, and is called the *pullback* (vector bundle) of $\mathcal{E}$ by f (cf. [86, Definition 2.11] or [58, Chapter 3]). In this situation, the map defined by

$$\iota_f : f^*\mathcal{E} \ni (p, \xi) \longmapsto \xi_{f(p)} \in \mathcal{E}$$

gives a homomorphism between the vector bundles $f^*\mathcal{E}$ and $\mathcal{E}$, whose restriction to each fiber is a linear isomorphism. We call ι_f the *canonical vector bundle homomorphism*.

If we take an inner product $\langle\ ,\ \rangle$ on $\mathcal{E}$, then

$$\langle (p, \xi_1), (p, \xi_2) \rangle := \langle \xi_1, \xi_2 \rangle_{f(p)} \qquad (p \in \mathcal{M}^m)$$

gives an inner product on $f^*\mathcal{E}$, which is called the *induced metric* on $f^*\mathcal{E}$ with respect to f.

For each $\xi \in \Gamma(\mathcal{E})$, we set

$$\tilde{\xi}(p) := (p, \xi_{f(p)}) \qquad (p \in \mathcal{M}^m), \tag{9.7}$$

then a vector bundle homomorphism

$$\Gamma(\mathcal{E}) \ni \xi \longmapsto \tilde{\xi} \in \Gamma(f^*\mathcal{E}) \tag{9.8}$$

is canonically induced.

Theorem 9.1.4. *Let $(\mathcal{N}^n, g)$ be a Riemannian manifold, and let ∇ be the Levi-Civita connection of $\mathcal{N}^n$. For each smooth map $f : \mathcal{M}^m \to \mathcal{N}^n$ from a manifold $\mathcal{M}^m$, there exists a unique covariant derivative $\tilde{\nabla}$ on the induced bundle $f^*T\mathcal{N}^n$ satisfying the following three properties:*

(1) *For each $\xi \in \Gamma(T\mathcal{N}^n)$ and $\boldsymbol{v} \in T\mathcal{M}^m$, the identity $\tilde{\nabla}_{\boldsymbol{v}}\tilde{\xi} = \nabla_{df(\boldsymbol{v})}(\xi \circ f)$ holds (cf. (9.1)).*

(2) *$\tilde{\nabla}$ is compatible with respect to the inner product induced by f (cf. (9.2)).*

(3) *The homomorphism*

$$\tilde{\varphi}_f : T\mathcal{M}^m \ni \boldsymbol{v}_p \mapsto (p, df(\boldsymbol{v}_p)) \in f^*T\mathcal{N}^n \qquad (p \in \mathcal{M}^m, \ \boldsymbol{v}_p \in T_p\mathcal{M}^m)$$

of vector bundles induced by f satisfies

$$\tilde{\nabla}_X\tilde{\varphi}_f(Y) - \tilde{\nabla}_Y\tilde{\varphi}_f(X) - \tilde{\varphi}_f([X,Y]) = 0 \qquad (X, Y \in \Gamma(T\mathcal{M}^m)).$$

This covariant derivative $\tilde{\nabla}$ is called the *connection induced by f*.

Proof. We take a local coordinate system $(U; x_1, \ldots, x_n)$ of $\mathcal{N}^n$ and a local coordinate system $(V; u_1, \ldots, u_m)$ of $\mathcal{M}^m$ such that $f(V) \subset U$. We set

$$X_j := \widetilde{\frac{\partial}{\partial x_j}} \left(= \frac{\partial}{\partial x_j} \circ f \right) \in \Gamma(f^*T\mathcal{N}^n|_V) \qquad (j = 1, \ldots, n).$$

Then $X_1, \ldots, X_n$ gives a basis of $f^*\mathcal{E}|_p$ at each point p of V. We fix a section $\tilde{\xi} \in \Gamma(f^*\mathcal{E}|_U)$. Then we can write

$$\tilde{\xi} = \xi_1 X_1 + \cdots + \xi_n X_n \qquad \left(\xi_1, \ldots, \xi_n \in C^\infty(V) \right). \tag{9.9}$$

If the desired $\tilde{\nabla}$ exists,

$$\tilde{\nabla}_{\boldsymbol{v}}\tilde{\xi} = \sum_{i=1}^n \left(\boldsymbol{v}(\xi_i)X_i + \xi_i\tilde{\nabla}_{\boldsymbol{v}}X_i \right) \qquad (\boldsymbol{v} \in TV)$$

must hold. Since $\boldsymbol{v}$ is expressed as a linear combination of $\partial/\partial u_a$ ($a = 1, \ldots, m$), it is sufficient to define $\tilde{\nabla}_{(\partial/\partial u_a)}X_j$. Here,

$$df\left(\frac{\partial}{\partial u_a} \right) = \sum_{k=1}^n \frac{\partial f_k}{\partial u_a} \frac{\partial}{\partial x_k} \tag{9.10}$$

holds, where we set $f = (x_1, \ldots, x_n)$, that is, we denote $x_i \circ f$ by x_i. We set $\nabla_{\partial/\partial x_i}\partial/\partial x_j = \sum_{k=1}^n \Gamma_{ij}^k \partial/\partial x_k$. If $\tilde{\nabla}_{\partial/\partial u_a}X_j$ exists, (1) implies that

$$\tilde{\nabla}_{\partial/\partial u_a}X_j = \left(\nabla_{df(\partial/\partial u_a)}\frac{\partial}{\partial x_j} \right) \circ f$$

$$= \sum_{i=1}^n \frac{\partial x_i}{\partial u_a} \left(\nabla_{\partial/\partial x_i}\frac{\partial}{\partial x_j} \right) \circ f = \sum_{k=1}^n \left(\sum_{i=1}^n \frac{\partial x_i}{\partial u_a}\Gamma_{ij}^k \circ f \right) X_k. \tag{9.11}$$

So we set

$$\tilde{\nabla}_{\partial/\partial u_a} X_j := \sum_{k=1}^{n} \left(\sum_{i=1}^{n} \frac{\partial x_i}{\partial u_a} \Gamma_{ij}^k \circ f \right) X_k. \tag{9.12}$$

To see that this $\tilde{\nabla}$ is well defined, it is sufficient to show that the right-hand side does not depend on the choice of local coordinate system $(x_1, \ldots, x_n)$ of $\mathcal{N}^n$. Take another choice of local coordinates $(y_1, \ldots, y_n)$ on U. We set

$$Y_j := \frac{\widetilde{\partial}}{\partial y_j} = \frac{\partial}{\partial y_j} \circ f \in \Gamma(f^* \mathcal{E}|_V) \qquad (j = 1, \ldots, n).$$

Then, like as in (9.12), the connection $\check{\nabla}$ for $(y_1, \ldots, y_n)$ corresponding to $\tilde{\nabla}$ should be

$$\check{\nabla}_{\partial/\partial u_a} Y_j := \sum_{k=1}^{n} \left(\sum_{i=1}^{n} \frac{\partial y_i}{\partial u_a} \Lambda_{ij}^k \circ f \right) Y_k,$$

where $\{\Lambda_{ij}^k\}_{i,j,k=1}^n$ is the family of Christoffel symbols with respect to the local coordinate system $(y_1, \ldots, y_n)$, and we write $f = (y_1, \ldots, y_n)$. It is sufficient to show that $\check{\nabla} = \tilde{\nabla}$ holds on U. Making the abbreviations

$$\Gamma_{ij}^k := \Gamma_{ij}^k \circ f, \qquad \Lambda_{ij}^k := \Lambda_{ij}^k \circ f,$$

we have

$$\check{\nabla}_{\partial/\partial u_a} X_j = \sum_{\beta=1}^{n} \check{\nabla}_{\partial/\partial u_a} \left(\frac{\partial y_\beta}{\partial x_j} Y_\beta \right) = \sum_{\beta=1}^{n} \left(\frac{\partial}{\partial u_a} \left(\frac{\partial y_\beta}{\partial x_j} \right) Y_\beta + \frac{\partial y_\beta}{\partial x_j} \check{\nabla}_{\frac{\partial}{\partial u_a}} Y_\beta \right)$$

$$= \sum_{i,\beta=1}^{n} \frac{\partial x_i}{\partial u_a} \left(\frac{\partial^2 y_\beta}{\partial x_i \partial x_j} Y_\beta + \sum_{\alpha,\gamma=1}^{n} \frac{\partial y_\beta}{\partial x_j} \frac{\partial y_\alpha}{\partial x_i} \Lambda_{\alpha\beta}^\gamma Y_\gamma \right)$$

$$= \sum_{i,\beta,l=1}^{n} \frac{\partial x_i}{\partial u_a} \left(\frac{\partial^2 y_\beta}{\partial x_i \partial x_j} \frac{\partial x_l}{\partial y_\beta} X_l + \sum_{\alpha,\gamma=1}^{n} \frac{\partial y_\beta}{\partial x_j} \frac{\partial y_\alpha}{\partial x_i} \Lambda_{\alpha\beta}^\gamma \frac{\partial x_l}{\partial y_\gamma} X_l \right)$$

$$= \sum_{i,\beta,l=1}^{n} \frac{\partial x_i}{\partial u_a} \left(\frac{\partial^2 y_\beta}{\partial x_i \partial x_j} \frac{\partial x_l}{\partial y_\beta} + \sum_{\alpha,\gamma=1}^{n} \frac{\partial y_\beta}{\partial x_j} \frac{\partial y_\alpha}{\partial x_i} \frac{\partial x_l}{\partial y_\gamma} \Lambda_{\alpha\beta}^\gamma \right) X_l$$

$$= \sum_{i,l=1}^{n} \frac{\partial x_i}{\partial u_a} \Gamma_{ij}^l X_l = \tilde{\nabla}_{\partial/\partial u_a} X_j,$$

where we used the formula (9.6) in the last line of the above computation. This implies $\check{\nabla} = \tilde{\nabla}$, and we have succeeded in proving the well-definedness of the induced covariant derivative on $f^* T\mathcal{N}^n$, proving (1).

Next we shall prove (2). In fact, we have

$$\langle \tilde{\nabla}_{\partial/\partial u_a} X_i, X_j \rangle + \langle X_i, \tilde{\nabla}_{\partial/\partial u_a} X_j \rangle$$

$$= \sum_{k=1}^{n} \frac{\partial x_k}{\partial u_a} \left(\sum_{h=1}^{n} \langle \Gamma_{ki}^h X_h, X_j \rangle + \sum_{h=1}^{n} \langle X_i, \Gamma_{kj}^h X_h \rangle \right) \circ f$$

$$= \sum_{k=1}^{n} \frac{\partial x_k}{\partial u_a} \left(\sum_{h=1}^{n} g \left(\Gamma_{ki}^h \frac{\partial}{\partial x_h}, \frac{\partial}{\partial x_j} \right) + \sum_{h=1}^{n} g \left(\frac{\partial}{\partial x_i}, \Gamma_{kj}^h \frac{\partial}{\partial x_h} \right) \right) \circ f$$

$$= \sum_{k=1}^{n} \frac{\partial x_k}{\partial u_a} \sum_{h=1}^{n} \left(\Gamma_{ki}^h g_{hj} + \Gamma_{kj}^h g_{ih} \right) \circ f$$

$$= \sum_{k=1}^{n} \frac{\partial x_k}{\partial u_a} \left(\frac{\partial g_{ij}}{\partial x_k} \circ f \right) = \sum_{k=1}^{n} \frac{\partial g_{ij} \circ f}{\partial u_a} = \frac{\partial}{\partial u_a} \langle X_i, X_j \rangle,$$

where we applied (9.5) in the last line of this computation. This proves the compatibility of $\tilde{\nabla}$ with the induced metric on $f^* T \mathcal{N}^n$.

We finally check the condition (3). We set

$$T(X, Y) := \tilde{\nabla}_X \tilde{\varphi}_f(Y) - \tilde{\nabla}_Y \tilde{\varphi}_f(X) - \tilde{\varphi}_f([X, Y]).$$

Then it can be easily checked that

$$T(X, Y) = -T(Y, X),$$

$$T(hX, Y) = hT(X, Y) \qquad (X, Y \in \Gamma(T\mathcal{M}^m), \ h \in C^\infty(\mathcal{M}^m)).$$

So, to prove $T = 0$, it is sufficient to show that $T(\partial/\partial u_a, \partial/\partial u_b) = 0$ $(a, b = 1, \ldots, n)$. In fact, we have

$$T \left(\frac{\partial}{\partial u_a}, \frac{\partial}{\partial u_b} \right) = \tilde{\nabla}_{\frac{\partial}{\partial u_a}} \tilde{\varphi}_f \left(\frac{\partial}{\partial u_b} \right) - \tilde{\nabla}_{\frac{\partial}{\partial u_b}} \tilde{\varphi}_f \left(\frac{\partial}{\partial u_a} \right).$$

Noting that

$$\tilde{\varphi}_f \left(\frac{\partial}{\partial u_b} \right) = \sum_{k=1}^{n} \frac{\partial x_k}{\partial u_b} X_k, \tag{9.13}$$

and applying (9.4), we have that

$$\tilde{\nabla}_{\partial/\partial u_a} \tilde{\varphi}_f \left(\frac{\partial}{\partial u_b} \right) = \sum_{k=1}^{n} \frac{\partial^2 x_k}{\partial u_a \partial u_b} X_k + \sum_{i,j,k=1}^{n} \frac{\partial x_i}{\partial u_a} \frac{\partial x_j}{\partial u_b} \Gamma_{ij}^k X_k$$

$$= \sum_{k=1}^{n} \frac{\partial^2 x_k}{\partial u_b \partial u_a} X_k + \sum_{i,j,k=1}^{n} \frac{\partial x_j}{\partial u_b} \frac{\partial x_i}{\partial u_a} \Gamma_{ji}^k X_k = \tilde{\nabla}_{\partial/\partial u_b} \tilde{\varphi}_f \left(\frac{\partial}{\partial u_a} \right).$$

$\square$

When $n = m$, the following corollary holds:

Corollary 9.1.5. *Let $\mathcal{M}^m$, $\mathcal{N}^m$ be two manifolds of dimension m, and let $f \colon \mathcal{M}^m \to \mathcal{N}^m$ be a C^∞-map. For each Riemannian metric g on $\mathcal{N}^m$, a coherent tangent bundle $(\mathcal{M}^m, f^*T\mathcal{N}^m, \langle\ ,\ \rangle, \tilde{\nabla}, \varphi_f)$ is canonically induced.*

The inner product $\langle\ ,\ \rangle$ is the pull-back of g by f.

Definition 9.1.6. Two coherent tangent bundles

$$(\mathcal{M}^m, \mathcal{E}, \langle\ ,\ \rangle_1, D, \varphi), \qquad (\mathcal{M}^m, \tilde{\mathcal{E}}, \langle\ ,\ \rangle_2, \tilde{D}, \tilde{\varphi})$$

are said to be *isomorphic* if there exists a smooth map $\iota \colon \mathcal{E} \to \tilde{\mathcal{E}}$ such that

- for each $p \in \mathcal{M}^m$, $\iota(\mathcal{E}_p) = \tilde{\mathcal{E}}_p$ holds, and ι gives a linear isomorphism between $\mathcal{E}_p$ and $\tilde{\mathcal{E}}_p$,
- $\tilde{\varphi} = \iota \circ \varphi$,
- ι preserves the inner product, that is, for each $p \in \mathcal{M}^m$ and each ξ, $\eta \in \mathcal{E}_p$, the identity

$$\langle \xi, \eta \rangle_1 = \langle \iota(\xi), \iota(\eta) \rangle_2$$

holds, and
- for each $v \in T_p\mathcal{M}^m$ and $\xi \in \Gamma(\mathcal{E})$, the identity $\iota(D_v \xi) = \tilde{D}_v \iota(\xi)$ holds.

The map ι is called an *isomorphism of coherent tangent bundles*.

Obviously, the following assertion holds, which will be applied in Section 9.4.

Lemma 9.1.7. *Let $(\mathcal{M}^m, \mathcal{E}, \langle\ ,\ \rangle_1, D, \varphi)$ be a coherent tangent bundle and $\mathcal{W}^m$ an open subset of $\mathcal{M}^m$. If the restriction $\varphi|_{\mathcal{W}^m}$ of φ to $\mathcal{W}^m$ has no singular points (cf. Definition 9.1.2), then $\varphi|_{\mathcal{W}^m}$ gives an isomorphism of coherent tangent bundles between $T\mathcal{W}^m$ and the restriction $\mathcal{E}|_{\mathcal{W}^m}$ of the coherent tangent bundle $\mathcal{E}$.*

9.2.　Wave Fronts and Coherent Tangent Bundles

Let $f \colon \mathcal{M}^m \to \mathcal{N}^{m+1}$ be a frontal map (cf. Definition 10.3.4 in Chapter 10) from an m-dimensional manifold $\mathcal{M}^m$ to an $(m+1)$-dimensional Riemannian manifold $(\mathcal{N}^{m+1}, g)$, and let $\nu(\in \Gamma(f^*T\mathcal{N}^{m+1}))$ be a *unit normal vector field* of f defined on $\mathcal{M}^m$, that is, it satisfies

$$g(df(v), \nu) = 0 \qquad (v \in T\mathcal{M}^m).$$

In this section, we show that f induces a coherent tangent bundle $(\mathcal{M}^m, \mathcal{E}_f, \langle\ ,\ \rangle, D^f, \varphi_f)$. Moreover, if $(\mathcal{N}^{m+1}, g)$ is a Riemannian manifold of constant curvature (cf. [41]), then f induces also another coherent tangent bundle $(\mathcal{M}^m, \mathcal{E}_f, \langle\ ,\ \rangle, D^f, \psi_f)$ as follows: By Theorem 9.1.4, a coherent tangent bundle $(\mathcal{M}^m, f^*T\mathcal{N}^{m+1}, \langle\ ,\ \rangle, \tilde{\nabla}, \varphi_f)$ is canonically induced. We then define a vector bundle $\mathcal{E}_f$ over $\mathcal{M}^m$ by

$$\mathcal{E}_f := \{\xi \in f^*T\mathcal{N}^{m+1}\,;\, \langle\xi, \tilde{\nu}\rangle = 0\},$$

whose rank is m, where $\tilde{\nu}$ is the section of f^*TN associated with ν (cf. (9.7)). Moreover, the restriction of the canonical inner product of $f^*T\mathcal{N}^{m+1}$ canonically induces an inner product $\langle\ ,\ \rangle$ on the vector bundle $\mathcal{E}_f$. Let $\tilde{\nabla}$ be the Levi-Civita connection. We set

$$D_X^f\xi := \tilde{\nabla}_X\xi - \langle\tilde{\nabla}_X\xi, \tilde{\nu}\rangle\tilde{\nu} \qquad (X \in \Gamma(T\mathcal{M}^m),\ \xi \in \Gamma(\mathcal{E}_f)), \qquad (9.14)$$

and then D^f gives a covariant derivative on $\mathcal{E}_f$, which coincides with the canonical covariant derivative in submanifold theory when f is an immersion. The unit normal vector field ν has $\pm$-ambiguity. However, in the right-hand side of (9.14), ν appears twice. So $D_X^f\xi$ does not depend on the choice of ν. Let

$$\tilde{\varphi}_f : T\mathcal{M}^m \ni \boldsymbol{v} \longmapsto (\pi(\boldsymbol{v}), df(\boldsymbol{v})) \in f^*T\mathcal{N}^{m+1}$$

be the map induced by f as explained in the previous section, where $\pi :$ $T\mathcal{M}^m \to \mathcal{M}^m$ is the canonical projection. Since $df(\boldsymbol{v})$ is perpendicular to ν, $\tilde{\varphi}_f$ canonically induces the following homomorphism of vector bundles

$$\varphi_f : T\mathcal{M}^m \ni \boldsymbol{v} \longmapsto \tilde{\varphi}_f(\boldsymbol{v}) \in \mathcal{E}_f. \qquad (9.15)$$

The following assertion holds.

Theorem 9.2.1. *For a given (co-orientable) frontal f, $(\mathcal{M}^m, \mathcal{E}_f, \langle\ ,\ \rangle, D^f, \varphi_f)$ is a coherent tangent bundle.*

Proof. For each $X \in \Gamma(T\mathcal{M}^m)$, $h \in C^\infty(\mathcal{M}^m)$ and $\xi \in \Gamma(\mathcal{E}_f)$, the equality $\langle\xi, \tilde{\nu}\rangle = 0$ yields

$$D_{hX}^f\xi = \tilde{\nabla}_{hX}\xi - \langle\tilde{\nabla}_{hX}\xi, \tilde{\nu}\rangle\tilde{\nu} = h\tilde{\nabla}_X\xi - \langle h\tilde{\nabla}_X\xi, \tilde{\nu}\rangle\tilde{\nu}$$
$$= h\left(\tilde{\nabla}_X\xi - \langle\tilde{\nabla}_X\xi, \tilde{\nu}\rangle\tilde{\nu}\right) = hD_X^f\xi,$$
$$D_X^f h\xi = \tilde{\nabla}_X h\xi - \langle\tilde{\nabla}_X h\xi, \tilde{\nu}\rangle\tilde{\nu}$$
$$= dh(X)\xi + h\tilde{\nabla}_X\xi - \langle dh(X)\xi + h\tilde{\nabla}_X\xi, \tilde{\nu}\rangle\tilde{\nu}$$
$$= dh(X)\xi + h\tilde{\nabla}_X\xi - h\langle\tilde{\nabla}_X\xi, \tilde{\nu}\rangle\tilde{\nu} = dh(X)\xi + hD_X^f\xi.$$

Thus, D^f gives a covariant derivative on $\mathcal{E}_f$. Since $\tilde{\nabla}$ is compatible with the inner product induced by f (cf. (2) of Theorem 9.1.4), the difference $\tilde{\nabla}_X \xi - D^f_X \xi$ is a scalar multiple of ν, and so for each $\xi, \eta \in \Gamma(\mathcal{E}_f)$, we have

$$X\langle \xi, \eta \rangle = \langle \tilde{\nabla}_X \xi, \eta \rangle + \langle \xi, \tilde{\nabla}_X \eta \rangle = \langle D^f_X \xi, \eta \rangle + \langle \xi, D^f_X \eta \rangle.$$

So, D^f is compatible with the inner product $\langle\ ,\ \rangle$. On the other hand, the definition of D^f (cf. (9.14)) and (3) of Theorem 9.1.4 yield that for each $X, Y \in \Gamma(T\mathcal{M}^m)$, we have

$$D^f_X \varphi_f(Y) - D^f_Y \varphi_f(X) = \tilde{\nabla}_X df(Y) - \tilde{\nabla}_Y df(X) = \varphi_f([X, Y]),$$

that is, (9.3) holds. $\qquad\square$

For a frontal map $f : \mathcal{M}^m \to \mathcal{N}^{m+1}$, we set

$$\psi_f : T\mathcal{M}^m \ni v \longmapsto -\tilde{\nabla}_v \nu \in \mathcal{E}_f,$$

where ν is the unit normal vector field. Then ψ_f is a homomorphism of vector bundles. If $(\mathcal{N}^{m+1}, g)$ has constant sectional curvature, then ψ_f also gives the structure of a coherent tangent bundle, as follows:

Theorem 9.2.2. *If $(\mathcal{N}^{m+1}, g)$ is of constant sectional curvature c, then $(\mathcal{M}^m, \mathcal{E}_f, \langle\ ,\ \rangle, D^f, \psi_f)$ is a coherent tangent bundle.*

Proof. Let R be the curvature tensor of D^f. Then by definition, we have

$$R(X, Y)\nu = \tilde{\nabla}_X \tilde{\nabla}_Y \nu - \tilde{\nabla}_Y \tilde{\nabla}_X \nu - \tilde{\nabla}_{[X,Y]}\nu$$
$$= -\tilde{\nabla}_X \psi_f(Y) + \tilde{\nabla}_X \psi_f(Y) + \psi_f([X, Y]).$$

So, it is sufficient to show that $R(X, Y)\nu = \mathbf{0}$. In fact, since $\mathcal{N}^{m+1}$ is of constant sectional curvature c (cf. [41]), we have

$$R(X, Y)\nu = c\Big(g(Y, \nu)X - g(X, \nu)Y\Big) = \mathbf{0}. \qquad\square$$

Remark 9.2.3. As a consequence, there are two distinct bundle homomorphisms φ_f and ψ_f giving the structure of a coherent tangent bundle for $\mathcal{E}_f$ when $(\mathcal{N}^{m+1}, g)$ is a space of constant sectional curvature. As we will see in Section 9.4, each of φ_f and ψ_f induces two Gauss–Bonnet type formulas (Theorems 9.4.2 and 9.4.3), and so we obtain four Gauss–Bonnet type formulas in total.

9.3.   Area Elements and Singular Curvature for Coherent Tangent Bundles

From here onward, we assume that $\mathcal{M}^2$ is a two-dimensional manifold. We let $(\mathcal{M}^2, \mathcal{E}, \langle\ ,\ \rangle, D, \varphi)$ be a coherent tangent bundle. Then

$$ds^2(v_1, v_2) := \langle \varphi(v_1), \varphi(v_2) \rangle \qquad \big(v_1, v_2 \in T_p\mathcal{M}^2,\ p \in \mathcal{M}^2\big) \qquad (9.16)$$

gives a positive semi-definite metric on $\mathcal{M}^2$. We call this covariant tensor the *first fundamental form* of the bundle homomorphism φ. The metric ds^2 is not positive definite at any point p for which $\varphi : T_p\mathcal{M}^2 \to \mathcal{E}_p$ is not a linear isomorphism. We have called such a point p a *singular point* of the map φ (cf. Definition 9.1.2).

The acceleration vectors along regular curves on $\mathcal{M}^2$. We fix a point $p \in \mathcal{M}^2$ and take a neighborhood U of p. If we choose U to be sufficiently small, then we can take a frame field of $\mathcal{E}$ (i.e., fields forming a basis on U) $\{s_1, s_2\}$. We then set

$$D_{\boldsymbol{v}} \boldsymbol{s}_i = \sum_{j=1}^{2} \omega_i^j(\boldsymbol{v}) \boldsymbol{s}_j \qquad (\boldsymbol{v} \in TU, \ i = 1, 2),$$

where ω_i^j $(i, j = 1, 2)$ are 1-forms on U. Then each $\xi \in \Gamma(\mathcal{E}|_U)$ can be written as $\xi = \xi^1 \boldsymbol{s}_1 + \xi^2 \boldsymbol{s}_2$, and

$$D_{\boldsymbol{v}} \xi = \sum_{j=1}^{2} \left(\boldsymbol{v}(\xi^j) + \sum_{i=1}^{2} \xi^i \omega_i^j(\boldsymbol{v}) \right) \boldsymbol{s}_j \qquad (9.17)$$

holds.

For a regular curve $\gamma : [a, b] \to \mathcal{M}^2$ defined on a closed interval $[a, b](\subset \boldsymbol{R})$, we set

$$\hat{\gamma}' := \varphi \circ \gamma' (\neq \boldsymbol{0}). \qquad (9.18)$$

Using the frame field $\{s_1, s_2\}$, we can write

$$\hat{\gamma}'(t) = \sum_{i=1}^{2} c_i(t) \boldsymbol{s}_i\big(\gamma(t)\big) \qquad (t \in [a, b]),$$

where $c_i : [a, b] \to \boldsymbol{R}$ $(i = 1, 2)$ are two C^∞-functions. We then set

$$\hat{\gamma}'' := D_{\gamma'} \hat{\gamma}'(t),$$

that is,

$$\hat{\gamma}''(t) = \sum_{j=1}^{2} \left(c_j'(t) + \sum_{i=1}^{2} c_i(t) \omega_i^j\big(\gamma'(t)\big) \right) \boldsymbol{s}_j\big(\gamma(t)\big).$$

The definition of $\hat{\gamma}''(t)$ does not depend on the choice of a frame $\{s_1, s_2\}$.

Area elements. Let $(\mathcal{M}^2, \mathcal{E}, \langle\ ,\ \rangle, D, \varphi)$ be a coherent tangent bundle over $\mathcal{M}^2$. If the vector bundle $\mathcal{E}$ of rank 2 is orientable (i.e. there exists a skew-symmetric section $\mu \in \Gamma(\mathcal{E}^* \otimes \mathcal{E}^*)$ without zeros), then we say that $(\mathcal{M}^2, \mathcal{E}, \langle\ ,\ \rangle, D, \varphi)$ is *co-orientable*. The orientability of $\mathcal{M}^2$ is equivalent to the orientability of the tangent bundle $T\mathcal{M}^2$. So, co-orientability is a different notion from, and independent of, the orientability of $\mathcal{M}^2$. Here, we assume that $\mathcal{M}^2$ is orientable and we fix an orientation of $\mathcal{M}^2$. A local coordinate system which is compatible with the orientation of $\mathcal{M}^2$ is said to be *positive*.

We now suppose that $(\mathcal{M}^2, \mathcal{E}, \langle\ ,\ \rangle, D, \varphi)$ is co-orientable, and fix an orientation of $\mathcal{E}$ (i.e., a co-orientation). We can then take an open covering of $\{U_\alpha\}_{\alpha \in A}$ of $\mathcal{M}^2$ and an orthonormal frame field (e_1^α, e_2^α) on U_α compatible with the orientation of $\mathcal{E}$, with its dual frame fields $\omega_1^\alpha, \omega_2^\alpha$ in $\mathcal{E}^*|_{U_\alpha}$ satisfying

$$\omega_i^\alpha(e_j^\alpha) = \delta_{ij} \qquad (i, j = 1, 2), \tag{9.19}$$

where δ_{ij} is the Kronecker delta. We then set

$$\det{}_\mathcal{E} := \omega_1^\alpha \wedge \omega_2^\alpha. \tag{9.20}$$

For any pair of sections $\xi_i \in \Gamma(\mathcal{E}|_{U_\alpha})$ $(i = 1, 2)$, we can write

$$\xi_i = a_{1i} e_1^\alpha + a_{2i} e_2^\alpha.$$

Then, by definition, it holds that

$$\det{}_\mathcal{E}(\xi_1, \xi_2) = \det \begin{pmatrix} a_{11} & a_{12} \\ a_{21} & a_{22} \end{pmatrix}. \tag{9.21}$$

In particular, $\det_\mathcal{E} \in \Gamma(\mathcal{E}^* \wedge \mathcal{E}^*)$ can be considered as a determinant function. By this formula, one can easily observe that $\det_\mathcal{E}(\xi_1, \xi_2)$ does not depend on the choice of orthonormal frame field (e_1^α, e_2^α) on U_α, which is compatible to the orientation of $\mathcal{E}$. In particular, one can show that $\det_\mathcal{E}$ is independent of the choice of the indices α. In fact, choosing indices α, β so that $U_\alpha \cap U_\beta \neq \emptyset$, we can write

$$\begin{cases} (e_1^\beta, e_2^\beta) = (e_1^\alpha, e_2^\alpha) \begin{pmatrix} \cos\theta & -\sin\theta \\ \sin\theta & \cos\theta \end{pmatrix}, \\ \begin{pmatrix} \omega_1^\beta \\ \omega_2^\beta \end{pmatrix} = \begin{pmatrix} \cos\theta & \sin\theta \\ -\sin\theta & \cos\theta \end{pmatrix} \begin{pmatrix} \omega_1^\alpha \\ \omega_2^\alpha \end{pmatrix}, \end{cases} \tag{9.22}$$

where θ is a smooth function on $U_\alpha \cap U_\beta$ (see [84, (12.5)]). By (9.21) and (9.22), one can check that $\omega_1^\beta \wedge \omega_2^\beta$ coincides with $\omega_1^\alpha \wedge \omega_2^\alpha$. Thus, by (9.20), $\det_\mathcal{E}$ can be considered as a global section

$$\det{}_\mathcal{E} \in \Gamma(\mathcal{E}^* \wedge \mathcal{E}^*),$$

which is called the *area element* of the oriented vector bundle $\mathcal{E}$. In particular, for an orthonormal frame (e_1, e_2) of $\mathcal{E}$ on U_α, we have

$$\det_{\mathcal{E}}(e_1, e_2) = \pm 1. \tag{9.23}$$

Here, $\det_{\mathcal{E}}(e_1, e_2) = 1$ implies that (e_1, e_2) is compatible with the orientation of $\mathcal{E}$ (i.e., the co-orientation). Such an (e_1, e_2) is called a *positive frame field* on $\mathcal{E}|_{U_\alpha}$. We then set

$$d\hat{A}(v_1, v_2) := \det_{\mathcal{E}}\big(\varphi(v_1), \varphi(v_2)\big) \qquad (v_1, v_2 \in T_p\mathcal{M}^2, \ p \in \mathcal{M}^2),$$

which is the pull-back of $\det_{\mathcal{E}}$ by φ, giving a smooth 2-form on $\mathcal{M}^2$. We call this $d\hat{A}$ the *signed area-element* of φ.

Since the base manifold $\mathcal{M}^2$ is oriented, we may assume that each U_α has positively oriented local coordinates (u^α, v^α). By setting

$$\lambda^\alpha := \det_{\mathcal{E}}\left(\varphi\left(\frac{\partial}{\partial u^\alpha}\right), \varphi\left(\frac{\partial}{\partial v^\alpha}\right)\right),$$

it holds that

$$d\hat{A} = \lambda^\alpha \, du^\alpha \wedge dv^\alpha. \tag{9.24}$$

We call the function λ^α on U_α the *signed area density function* on U_α. We now set

$$dA := |\lambda^\alpha| \, du^\alpha \wedge dv^\alpha.$$

Since U_α is positively oriented, the fact that the expression as in (9.24) does not depend on α implies that dA defines a continuous 1-form on $\mathcal{M}^2$, which is called the *(unsigned) area element*. The following assertion holds:

Proposition 9.3.1. *Let $f : \mathcal{M}^2 \to \mathbf{R}^3$ be a co-orientable frontal. Then the signed area element $d\hat{A}$ and the unsigned area element dA of the coherent tangent bundle $\mathcal{E}_f$ coincide with the signed area element and the unsigned area element induced by the first fundamental form of f, respectively.*

Proof. The coherent tangent bundle $\mathcal{E}_f$ induced by f can be canonically considered as a subbundle of $f^*T\mathbf{R}^3$. Let ν be the unit normal vector field of f. It can be easily checked that

$$\det_{\mathcal{E}_f}(\xi, \eta) := \det(\xi, \eta, \nu) \qquad (\xi, \eta \in \Gamma(\mathcal{E}_f))$$

gives the area element of $\mathcal{E}_f$. Then we have

$$\det_{\mathcal{E}_f}\left(\varphi\left(\frac{\partial}{\partial u^\alpha}\right), \varphi\left(\frac{\partial}{\partial v^\alpha}\right)\right) = \det(f_{u^\alpha}, f_{v^\alpha}, \nu),$$

for each α, which proves the assertion. $\qquad\square$

We return to the general setting: Since

$$\langle D_{\boldsymbol{v}} e_1^\alpha, e_1^\alpha \rangle = \frac{1}{2} \boldsymbol{v} \langle e_1^\alpha, e_1^\alpha \rangle = \boldsymbol{v}(1) = 0,$$

we can write

$$D_{\boldsymbol{v}} e_1^\alpha = -\mu^\alpha(\boldsymbol{v}) e_2^\alpha, \qquad (9.25)$$

where μ^α is a smooth 1-form on U_α. This μ^α is called a *connection form*[1] with respect to the orthonormal frame $\{e_1^\alpha, e_2^\alpha\}$. By the compatibility of D with the inner product, we have

$$\langle D_{\boldsymbol{v}} e_1^\alpha, e_2^\alpha \rangle = -\langle e_1^\alpha, D_{\boldsymbol{v}} e_2^\alpha \rangle,$$

and so we can write

$$D_{\boldsymbol{v}} e_2^\alpha = \mu^\alpha(\boldsymbol{v}) e_1^\alpha. \qquad (9.26)$$

The following assertion holds:

Lemma 9.3.2. *The exterior derivative $d\mu^\alpha$ of the connection form μ^α does not depend on the indices α, and a smooth 2-form defined on $\mathcal{M}^2$ is induced.*

Proof. Let θ be the function given in (9.22). Then

$$\mu^\beta = -d\theta + \mu^\alpha \qquad (9.27)$$

holds on $U_\alpha \cap U_\beta$ (see [84, Lemma 13.2]), and taking the exterior derivative of (9.27), we get the conclusion $d\mu^\alpha = d\mu^\beta$.   □

By Lemma 9.3.2, if we set

$$\hat{\Omega} := d\mu^\alpha, \qquad (9.28)$$

then this does not depend on the index α, and defines a smooth 2-form on $\mathcal{M}^2$. We call $\hat{\Omega}$ the *Euler form* of $\mathcal{E}$. Then we have the following proposition:

Proposition 9.3.3. *Let O be an open subset of $\mathcal{M}^2$ such that $\varphi|_O : T\mathcal{M}^2|_O \to \mathcal{E}|_O$ is injective. Then the identity*

$$\hat{\Omega} = K \, d\hat{A}$$

holds on O, where K is the Gaussian curvature (cf. (2.17) in Chapter 2) of the first fundamental form (cf. (9.16)) of φ.

---

[1] In the theory of connections, for each $(U_\alpha \, ; u_\alpha, v_\alpha)$ the 2-form $\begin{pmatrix} 0 & \mu^\alpha \\ -\mu^\alpha & 0 \end{pmatrix}$ with values in the anti-symmetric matrices is called a *connection form*, Here, instead of saying we have a matrix-valued 2-form, we call μ^α a connection form.

Proof. Since the restriction $\varphi|_O$ of the bundle homomorphism φ is injective, it is an isomorphism of vector bundles. So, we can then identify $T\mathcal{M}^2|_O$ with $\mathcal{E}|_O$, and the condition (9.3) can be interpreted as D being a torsion free connection on $T\mathcal{M}^2|_O$. Since D is compatible with the inner product, D must coincide with the Levi-Civita connection (cf. [84, (13.15)]) with respect to the first fundamental form $ds^2 := \varphi^*\langle\,,\,\rangle$ on O. □

We let $(\mathcal{M}^2, \mathcal{E}, \langle\,,\,\rangle, D, \varphi)$ be a coherent tangent bundle such that $T\mathcal{M}^2$ is oriented and $\mathcal{E}$ is also oriented (i.e., $\mathcal{E}$ is co-oriented as a coherent tangent bundle). Let p be a point on $\mathcal{M}^2$ and $(U; u, v)$ a local coordinate neighborhood of p. We can take a local signed area density function $\lambda : U \to \mathbf{R}$, that is, λ satisfies

$$d\hat{A} = \lambda\, du \wedge dv.$$

Then p is a singular point of φ (cf. Definition 9.1.2) if and only if $\lambda(p) = 0$ holds. Like as in the case of the non-degenerate singular points given in Section 2.6 of Chapter 2, we say that p is *non-degenerate* if

$$\big(\lambda_u(p), \lambda_v(p)\big) \neq \mathbf{0}.$$

If p is a non-degenerate singular point, then the implicit function theorem implies that the singular set can be parametrized as a regular curve $\gamma : (-\varepsilon, \varepsilon) \to U$ $(\gamma(0) = p)$, where ε is a positive number. We call this curve γ the *singular curve* or the *characteristic curve* passing through p. The direction of $\gamma'(t)$ (i.e., the one-dimensional vector subspace of $T_{\gamma(t)}\mathcal{M}^2$ spanned by $\gamma'(t)$) is called the *singular direction* at the singular point $\gamma(t)$. Since $\gamma(t)$ is a singular point, there exists a non-zero vector $v \in T_{\gamma(t)}\mathcal{M}^2$ such that $\varphi(v) = \mathbf{0}$. Such a vector v is called the *null vector* at $\gamma(t)$, and the direction of v is called the *null direction* at $\gamma(t)$. Since the null direction is uniquely determined for each t, there exists a vector field $\eta(t)$ along γ such that each $\eta(t) \in T_{\gamma(t)}\mathcal{M}^2$ gives a null vector at $\gamma(t)$.

From now on, we assume that the singular points of φ consist only of non-degenerate singular points. Like for frontal maps, a non-degenerate singular point p is called a *singular point of the first kind* if the null direction is linearly independent of the singular direction $\gamma'(0)$. Otherwise, p is called a *singular point of the second kind*. (This definition is parallel to the same notion for frontal maps given in Section 5.2 of Chapter 5.) Moreover, a singular point p of the second kind is said to be *admissible* if there exists a neighborhood U of p such that the singular points on $U \setminus \{p\}$ consist only of singular points of the first kind. The following assertion is useful for discussing non-degenerate singular points.

Lemma 9.3.4. *Let $p \in \mathcal{M}^2$ be a non-degenerate singular point of φ, then there exists a local coordinate system (u, v) centered at p such that*

(1) *the u-axis gives the singular curve,*
(2) *if p is of the first kind, then $\partial/\partial v$ gives the null vector field along the u-axis.*

Proof. The proof is the same as for the case of non-degenerate singular points of a front. See the proof of Lemma 5.2.10 in Section 5.2 of Chapter 5. $\qquad\square$

We set

$$\hat{\gamma}'(t) = \varphi(\gamma'(t)), \qquad \hat{\gamma}'' := D_{\gamma'(t)}\hat{\gamma}'(t),$$

and we let $p := \gamma(0)$ be a non-degenerate singular point of the first kind. If $\varepsilon(> 0)$ is sufficiently small, then $\gamma'(t)$ ($|t| < \varepsilon$) does not coincide with the null direction for each t, and $\hat{\gamma}'(t) \neq \mathbf{0}$ for each t. Since p is non-degenerate, we may assume that $d\lambda \neq 0$ on U. Since $\lambda = 0$ along γ, the condition that p is of the first kind implies that

$$\lambda_\eta(t) := d\lambda(\eta(t))$$

does not vanish for each t, where $\eta(t)$ is a null vector field along γ, so we can define the sign

$$\operatorname{sgn}(\lambda_\eta)$$

of the function λ_η. By using the $\pm$-ambiguity of the null direction, we may assume that $(\gamma'(t), \eta(t))$ gives a positive frame along γ, that is, it is compatible with the orientation of $\mathcal{M}^2$. We then set (cf. (5.18))

$$\kappa_s := \operatorname{sgn}(\lambda_\eta)\frac{\det_{\mathcal{E}}(\hat{\gamma}', \hat{\gamma}'')}{|\hat{\gamma}'|^3} \qquad \left(|\hat{\gamma}'| := \sqrt{\langle \hat{\gamma}', \hat{\gamma}'\rangle}\right),$$

which is called the *singular curvature function* along the singular curve γ (consisting only of singularities of the first kind). Here, we call the value $\kappa_s(0)$ the *singular curvature* at p (of φ). Moreover, if γ lies in a coordinate neighborhood $(U; u, v)$ which is compatible with the orientation of $\mathcal{M}^2$, then (cf. (5.19))

$$\operatorname{sgn}(\lambda_\eta) = \begin{cases} 1 & (\text{if } \lambda > 0 \text{ on the left-hand side of } \gamma), \\ -1 & (\text{if } \lambda > 0 \text{ on the right-hand side of } \gamma). \end{cases} \tag{9.29}$$

If we do not assume $(\gamma'(t), \eta(t))$ is a positive frame along γ, then

$$\kappa_s(t) = \varepsilon_\gamma(t) \frac{\det_{\mathcal{E}}(\hat{\gamma}'(t), \hat{\gamma}''(t))}{|\hat{\gamma}'(t)|^3}, \quad \varepsilon_\gamma(t) := \mathrm{sgn}(\lambda_\eta \det(\gamma'(t), \eta(t)))$$

holds (cf. (5.17)).

Proposition 9.3.5. *The singular curvature $\kappa_s(0)$ at a singular point p of the first kind does not depend on the choice of the parameter t for the singular curve, the orientation of $\mathcal{M}^2$, the orientation of $\mathcal{E}$ (i.e., the co-orientation), nor the orientation of the singular curve γ.*

Proof. It can be easily checked that the value $\kappa_s(0)$ does not depend on the choice of the parametrization of the singular curve γ. We take a local coordinate system (u, v) centered at p as in Lemma 9.3.4. Then the u-axis gives the singular curve. If we replace (u, v) by $(u, -v)$, then we can observe that the sign change

$$(\lambda(\gamma(t)), \eta(t)) \quad \longmapsto \quad (-\lambda(\gamma(t)), -\eta(t))$$

does not affect the sign in the definition of the singular curvature, that is, the definition of singular curvature does not depend on the orientation of $\mathcal{M}^2$.

On the other hand, if we reverse the orientation of $\mathcal{E}$, then instead of $\det_{\mathcal{E}}$, $-\det_{\mathcal{E}}$ gives the area element of $\mathcal{E}$, and then $-\lambda$ gives the signed area density function, rather than λ, so the value $\kappa_s(0)$ does not change. Finally, if we replace $\gamma(t)$ by $\gamma(-t)$, then the following sign change occurs:

$$(\hat{\gamma}', \eta) \quad \longmapsto \quad (-\hat{\gamma}', -\eta),$$

and this does not affect the sign in the definition of the singular curvature $\kappa_s(t)$. $\qquad\square$

Remark 9.3.6. Let $(\mathcal{M}^2, \mathcal{E}, \langle \ , \ \rangle, D, \varphi)$ be a coherent tangent bundle. Suppose that p is a singular point of the first kind. Using the same argument as in the proof of Lemma 5.5.1, we can take a local coordinate system $(U; u, v)$ such that

- the u-axis parametrizes the singular set,
- $\partial_v := \partial/\partial v$ points in the null direction, and
- setting $\varphi(\partial_v) = v\xi$ ($\xi \in \Gamma(\mathcal{E}|_U)$), then $\{\varphi(\partial_u), \xi\}$ gives an orthogonal frame field along the u-axis.

We can write $ds^2 = Edu^2 + 2Fdudv + Gdv^2$ on U. Since ∂_v gives the null direction of φ, we have $F(u, 0) = 0$. By the same argument as in

Theorem 5.5.4, the singular curvature function κ_s is given by

$$\kappa_s(u) = \frac{-E_{vv}(u,0)}{\sqrt{2}E(u,0)\sqrt{G_{vv}(u,0)}}. \tag{9.30}$$

This formula is a special case of (9.37).

We next consider the case that p is an admissible singular point of the second kind, and $\gamma : I \to \mathcal{M}^2$ is a singular curve satisfying $\gamma(0) = p$, where I is an interval containing the origin 0 of $\mathbf{R}$. By admissibility, $\gamma(t)$ $(t \neq 0)$ is a singular point of the first kind. Then the following assertion holds:

Proposition 9.3.7. *We set* $ds := |\hat{\gamma}'(t)|dt$, *and then* $\kappa_s \, ds$ *gives a smooth 1-form on I. In particular, it is smooth at $t = 0$.*

Proof. The proof is obtained by imitating the proof of Corollary 5.5.6, as follows: By Lemma 9.3.4, we can take a local coordinate system (u,v) centered at p such that $\gamma(t) = (u(t), v(t))$ and $\eta(t) = (\partial_v)_{\gamma(t)}$ gives the null direction at $\gamma(t)$, where $\partial_u := \partial/\partial u$ and $\partial_v := \partial/\partial v$. Since $p = \gamma(0)$ is an admissible singular point of the second kind, $\gamma'(0)$ points in the same direction as $\eta(0) = (\partial_v)_p$, so $u'(0) = 0$, $v'(0) \neq 0$ holds. Since ∂_v gives the null direction at $\gamma(t)$, we have that

$$\hat{\gamma}' = u'\varphi(\partial_u) + v'\varphi(\partial_v) = u'\varphi(\partial_u). \tag{9.31}$$

Differentiating this, we obtain

$$\hat{\gamma}'' = u''\varphi(\partial_u) + (u')^2 D_{\partial_u}\varphi(\partial_u) + u'v'D_{\partial_u}\varphi(\partial_v).$$

Here we may assume that $\lambda > 0$ on the left-hand side of γ, without loss of generality. By (9.29) and by the fact $\eta(t) = \partial_v$, we have

$$(\varepsilon_\gamma(t) =) \operatorname{sgn}(\det(\gamma'(t), \eta(t))) \operatorname{sgn}(\lambda_v(u(t), v(t))) = 1.$$

Since p is non-degenerate, the facts $\varphi(\partial_u) \neq \mathbf{0}$ and $ds/dt = |\hat{\gamma}'|$ yield (cf. (5.31))

$$\kappa_s(t)ds = \frac{u' \det_{\mathcal{E}}(\varphi(\partial_u), D_{\partial_u}\varphi(\partial_u))}{|\varphi(\partial_u)|^2} + \frac{v' \det_{\mathcal{E}}(\varphi(\partial_u), D_{\partial_u}\varphi(\partial_v))}{|\varphi(\partial_u)|^2}, \tag{9.32}$$

proving the assertion. $\square$

9.4.   Gauss–Bonnet Theorem for Coherent Tangent Bundles

We let $(\mathcal{M}^2, \mathcal{E}, \langle \ , \ \rangle, D, \varphi)$ be a coherent tangent bundle over a two-dimensional oriented manifold $\mathcal{M}^2$, and assume that $\mathcal{E}$ is oriented, that is, the coherent tangent bundle is co-orientable. Moreover, we assume that $\mathcal{M}^2$

is compact and φ admits only singular points of the first kind or admissible singular points of the second kind. Then the signed area element $d\hat{A}$ is induced by φ, which gives a smooth 2-form on $\mathcal{M}^2$. The Gaussian curvature K of the first fundamental form $ds^2 := \varphi^*\langle \, , \, \rangle$ is defined on the regular set on $\mathcal{M}^2$. Then, by Proposition 9.3.3, the Euler form of φ satisfies (cf. (9.28))

$$\hat{\Omega} = K \, d\hat{A}$$

on $\mathcal{M}^2 \setminus \Sigma(\varphi)$, where $\Sigma(\varphi)$ is the singular set of φ. By Proposition 9.3.3, the integration

$$\int_{\mathcal{M}^2} \hat{\Omega}$$

is determined. From now on, we assume that φ admits only non-degenerate singular points. Then the singular set $\Sigma(\varphi)$ of φ on $\mathcal{M}^2$ consists of a disjoint union of regular curves. In particular, the set of regular points O is an open dense subset of $\mathcal{M}^2$. We set

$$\mathcal{M}^+ := \{p \in \mathcal{M}^2 \, ; \, dA_p = d\hat{A}_p\}, \quad \mathcal{M}^- := \{p \in \mathcal{M}^2 \, ; \, dA_p = -d\hat{A}_p\},$$

which are closed subsets, and $\Sigma(\varphi) := \mathcal{M}^+ \cap \mathcal{M}^-$ holds, and set

$$\int_{\mathcal{M}^2} \Omega := \int_{\mathcal{M}^+} \hat{\Omega} - \int_{\mathcal{M}^-} \hat{\Omega}. \tag{9.33}$$

The following is a key fact for proving Gauss–Bonnet formulas for coherent tangent bundles, which is a generalization of Propositions 4.6.3 and 5.4.3 in Chapters 4 and 5, respectively for frontal maps:

Lemma 9.4.1. *Let p be a singular point of the first kind with respect to the map φ, and let $(U; u, v)$ be a positively oriented local coordinate system centered at p such that ∂_v gives a null direction along the u-axis as a singular curve (cf. Lemma 9.3.4). Suppose that $\{v > 0\}$ lies on $\mathcal{M}^+$, and set*

$$\gamma_c(t) := (t, c) \qquad (|t| < \varepsilon),$$

where $|c|$ is sufficiently small enough that the image of γ_c lies in U. We set

$$\gamma_c'' := \nabla_{\gamma_c'} \gamma_c',$$

where ∇ is the Levi-Civita connection with respect to ds^2. Define the geodesic curvature along γ_c ($c \neq 0$) by

$$\kappa_g(t, c) = \frac{dA\big(\gamma_c'(t), \gamma_c''(t)\big)}{|\gamma_c'(t)|^3},$$

then the singular curvature function along γ is expressed by

$$\kappa_s(t) = \lim_{c \to 0^+} \kappa_g(t, c) = -\lim_{c \to 0^-} \kappa_g(t, c),$$

where dA is the unsigned area element of $\mathcal{M}^2$ with respect to ds^2.

Proof. We let O be the regular set of φ. As mentioned before, $\mathcal{E}|_O$ can be identified with $T\mathcal{M}^2|_O$. Since dA coincides with the pull-back of $\det_{\mathcal{E}}$ (respectively, $-\det_{\mathcal{E}}$) by φ on $\mathcal{M}^+$ (respectively, $\mathcal{M}^-$), for each $c \neq 0$, we have

$$\frac{dA\big(\gamma_c'(t), \gamma_c''(t)\big)}{|\gamma_c'(t)|^3} = \operatorname{sgn}(c) \frac{\det_{\mathcal{E}}\big(\hat{\gamma}_c'(t), \hat{\gamma}_c''(t)\big)}{|\hat{\gamma}_c'(t)|^3}, \qquad (9.34)$$

where

$$\hat{\gamma}_c' := \varphi(\gamma_c'), \quad \hat{\gamma}_c'' := D_{\gamma_c'} \varphi(\gamma_c').$$

Letting $c \to 0$ in (9.34), we get the conclusion. $\qquad \square$

In Chapter 6, we defined the sign $\operatorname{sgn}(p)$ of an admissible singular point p of the second kind (cf. (6.15)) with respect to the first fundamental form of a frontal map. Similarly, we can define $\operatorname{sgn}(p)$ of an admissible singular point p of the second kind with respect to the first fundamental form of φ. We now assume that $\mathcal{M}^2$ is compact and oriented. In Chapter 6, we gave a triangulation of $\mathcal{M}^2$ associated with the first fundamental form of a frontal. In the same way, we can give the same triangulation of $\mathcal{M}^2$ associated with the first fundamental form of φ. Since φ has no singular points on $\mathcal{M}_*^+ := \mathcal{M}^+ \backslash \partial\mathcal{M}^+$ (respectively, $\mathcal{M}_*^- := \mathcal{M}^- \backslash \partial\mathcal{M}^-$), by Lemma 9.1.7, $T\mathcal{M}_*^+$ (respectively, $T\mathcal{M}_*^-$) can be identified with $\mathcal{E}|_{\mathcal{M}_*^+}$ (respectively, $\mathcal{E}|_{\mathcal{M}_*^-}$). Thus, we have that

$$\int_{\mathcal{M}^+} \hat{\Omega} + \int_{\Sigma(f)} \kappa_s \, ds = 2\pi\chi(\mathcal{M}^+) + \pi \sum_{i=1}^n \operatorname{sgn}(\mathrm{P}_i),$$

$$-\int_{\mathcal{M}^-} \hat{\Omega} + \int_{\Sigma(f)} \kappa_s \, ds = 2\pi\chi(\mathcal{M}^-) - \pi \sum_{i=1}^n \operatorname{sgn}(\mathrm{P}_i).$$

Here, $\Sigma(f)$ is the set of singular curves oriented so that $\mathcal{M}^+$ is to the left-hand side, with respect to a coordinate system compatible with the orientation of $\mathcal{M}^2$, and $\{\mathrm{P}_1, \dots, \mathrm{P}_n\}$ is the set of singular points of the second kind in $\Sigma(f)$. Imitating the proof of Theorem 6.7.5, we obtain the following two assertions.

Theorem 9.4.2 (Intrinsic Gauss-Bonnet type formula I [69]). *Let $\mathcal{M}^2$ be a compact oriented 2-manifold, and let $(\mathcal{M}^2, \mathcal{E}, \langle \ , \ \rangle, D, \varphi)$ be a co-oriented coherent tangent bundle on $\mathcal{M}^2$. Suppose that φ admits only singular points of the first kind or admissible singular points of the second kind. Then the identity*

$$\chi_\mathcal{E} = \frac{1}{2\pi} \int_{\mathcal{M}^2} \hat{\Omega} = \chi(\mathcal{M}^+) - \chi(\mathcal{M}^-) + \#\mathcal{P}^+ - \#\mathcal{P}^-$$

holds, where the integer $\chi_\mathcal{E}$ is the Euler characteristic of the oriented vector bundle $\mathcal{E}$, and $\#\mathcal{P}^\pm$ is the number of positive (respectively, negative) singular points of the second kind.

Theorem 9.4.3 (Intrinsic Gauss-Bonnet type formula II [69]). *Let $\mathcal{M}^2$ be a compact oriented 2-manifold, and let $(\mathcal{M}^2, \mathcal{E}, \langle , \rangle, D, \varphi)$ be a coherent tangent bundle over $\mathcal{M}^2$. Suppose that φ admits only singular points of the first kind or admissible singular points of the second kind. Then the identity*

$$\int_{\mathcal{M}^2} \Omega + 2 \int_{\Sigma(\varphi)} \kappa_s \, ds = 2\pi \chi(\mathcal{M}^2) \tag{9.35}$$

holds, where $\Sigma(\varphi)$ is the set of the singular points of φ, and κ_s is the singular curvature along each singular curve parametrizing $\Sigma(\varphi)$.

Remark 9.4.4. If we think of dA (ds) as a measure on $\mathcal{M}^2$ (respectively, on the singular curves) and

$$\int_{\mathcal{M}^2} \Omega = \int_{\mathcal{M}^2} K \, dA$$

as an integral of the Gaussian curvature K on $\mathcal{M}^2$, the formula (9.35) holds. In fact, if $\mathcal{M}^2$ is not orientable, the formula is obtained by taking the double covering of $\mathcal{M}^2$.

As a consequence, we have obtained two Gauss–Bonnet formulas. We let

$$f : \mathcal{M}^2 \to \mathcal{N}^3$$

be a frontal map from an oriented compact 2-manifold into a Riemannian 3-manifold of constant (sectional) curvature. Since $\mathcal{N}^3$ has constant curvature, there are two vector bundle homomorphisms

$$\varphi_f, \psi_f : T\mathcal{M}^2 \to \mathcal{E}_f$$

giving the structures of a coherent tangent bundle on $\mathcal{E}_f$ (cf. Remark 9.2.3 in Section 9.2). So each φ_f, ψ_f induces two Gauss–Bonnet type formulas, and then we obtain a total of four distinct Gauss–Bonnet type formulas. These four formulas correspond to the formulas given in Chapter 6. We have already introduced these four formulas when $(\mathcal{N}^3, g)$ is the Euclidean 3-space. More precisely, (6.27), (6.28) (respectively, (6.30), (6.31)) are induced by φ_f (respectively, ψ_f). Similarly, when $\mathcal{N}^3$ is the 3-sphere or the hyperbolic 3-space, we again get four distinct formulas, for which there are several applications, see [73] for details.

9.5. Kossowski Metrics

In this section, we introduce Kossowski metrics, which are closely related to the concept of coherent tangent bundles.

Let $\mathcal{M}^2$ be a 2-manifold and ds^2 a positive semi-definite metric on $\mathcal{M}^2$. A point $p \in \mathcal{M}^2$ is called a *singular point* or a *semi-definite point* of ds^2 if the metric is not positive definite at p. A point which is not a singular point is called a *regular point* of ds^2. A singular point p of ds^2 is said to be of *rank* 1 if the vector space

$$V_p := \{ \boldsymbol{v} \in T_p\mathcal{M}^2 \,;\, (ds^2)_p(\boldsymbol{v}, T_p\mathcal{M}^2) = \boldsymbol{0} \}$$

is of dimension 1. A non-zero vector belonging to V_p is called a *null vector*.

Definition 9.5.1. Let ds^2 be a positive semi-definite metric on $\mathcal{M}^2$ and $p \in \mathcal{M}^2$ a singular point of ds^2. A local coordinate neighborhood $(U; u, v)$ of p is said to be *admissible* if ∂_v gives a null vector at p.

The following assertion holds:

Lemma 9.5.2. *Let $f : \mathcal{M}^2 \to \boldsymbol{R}^3$ be a smooth map, and let ds^2 be the first fundamental form of f. Let $(U; u, v)$ be a local coordinate neighborhood of a singular point $p \in \mathcal{M}^2$ such that $f_v(p) = 0$. If we set*

$$ds^2 = E\,du^2 + 2F\,du\,dv + G\,dv^2,$$

then the identities

$$E_v(p) - 2F_u(p) = F(p) = G(p) = G_u(p) = G_v(p) = 0$$

hold. In particular, $(U; u, v)$ is an admissible coordinate.

Proof. Since $f_v(p) = \boldsymbol{0}$, $F(p) = G(p) = G_u(p) = G_v(p) = 0$ obviously holds. So it is sufficient to show the first identity $E_v(p) - 2F_u(p) = 0$. In fact,

$$F_u(p) = (f_u \cdot f_v)_u(p) = f_{uu}(p) \cdot f_v(p) + f_u(p) \cdot f_{uv}(p)$$
$$= f_u(p) \cdot f_{uv}(p) = \frac{1}{2}(f_u \cdot f_u)_v(p). \qquad \square$$

Regarding the above lemma, we give the following definition:

Definition 9.5.3. A positive semi-definite metric ds^2 is called a *Kossowski metric* if for each singular point p, there exists an admissible local coordinate neighborhood $(U; u, v)$ at p such that

$$ds^2 = E \, du^2 + 2F \, du \, dv + G \, dv^2$$

satisfies the following properties:

(1) $E(p) \neq 0$, and $E_v(p) = 2F_u(p)$ holds,
(2) there exists a C^∞-function λ on U such that $EG - F^2 = \lambda^2$, and
(3) $\big(\lambda_u(p), \lambda_v(p)\big) \neq \mathbf{0}$.

The Kossowski metric was introduced by Kossowski [48], and although the above definition is different from the one given in [48], the two concepts are equivalent (see the first section of [33]). We prove the following:

Lemma 9.5.4. *The above definition of the Kossowski metric does not depend on the choice of admissible local coordinates at a singular point p.*

Proof. Let (u, v) be admissible coordinates at p, then $F(p) = G(p) = 0$ holds at this singular point p. Since p is of rank 1, it holds that $E(p) \neq 0$. Differentiating $EG - F^2 = \lambda^2$, we have

$$dE \, G + E \, dG - 2F \, dF = 2\lambda \, d\lambda.$$

Since $\lambda(p) = 0$ and $F(p) = G(p) = 0$ holds, we have $E \, dG = 0$ at p and

$$G_u(p) = G_v(p) = 0. \qquad (9.36)$$

Let (x, y) be another choice of admissible coordinates, then we have

$$(\partial_x :=)\frac{\partial}{\partial x} = \frac{\partial u}{\partial x}\frac{\partial}{\partial u} + \frac{\partial v}{\partial x}\frac{\partial}{\partial v}, \quad (\partial_y :=)\frac{\partial}{\partial y} = \frac{\partial u}{\partial y}\frac{\partial}{\partial u} + \frac{\partial v}{\partial y}\frac{\partial}{\partial v}.$$

Since (x, y) is admissible, we have $u_y(p) = 0$. Also, the Jacobian $u_x v_y - u_y v_x$ does not vanish, and we have $u_x(p)v_y(p) \neq 0$. Since (9.36) implies $dG = 0$, we have $G_x(p) = G_y(p) = 0$. Here, we set $\tilde{E} := ds^2(\partial_x, \partial_x)$, then we have

$$\tilde{E} = u_x^2 E + 2u_x v_x F + v_x^2 G.$$

Differentiating this by y,

$$\tilde{E}_y(p) = 2u_{xy}(p)u_x(p)E(p) + u_x(p)^2 E_y(p) + 2u_x(p)v_x(p)F_y(p)$$

holds. We next set $\tilde{F} := ds^2(\partial_x, \partial_y)$, and then $\tilde{F} = u_x u_y E + (u_x v_y + u_y v_x)F + v_x v_y G$ holds. Similarly, we get

$$\tilde{F}_x(p) = u_x(p)u_{xy}(p)E(p) + u_x(p)v_y(p)F_x(p).$$

As a consequence, we have

$$
\begin{aligned}
\tilde{E}_y(p) - 2\tilde{F}_x(p) &= u_x(p)^2 E_y(p) + 2u_x(p)v_y(p)F_y(p) - 2u_x(p)v_y(p)F_x(p) \\
&= u_x(p)^2 E_v(p)v_y(p) + 2u_x(p)v_x(p)F_v(p)v_y(p) \\
&\quad - 2u_x(p)v_y(p)(F_u(p)u_x(p) + F_v(p)v_x(p)) \\
&= u_x(p)^2 v_y(p)(E_v(p) - 2F_u(p)).
\end{aligned}
$$

Since $u_x(p)v_y(p) \neq 0$, we can prove $\tilde{E}_y(p) - 2\tilde{F}_x(p) = 0$. Since

$$
\begin{pmatrix} \tilde{E} & \tilde{F} \\ \tilde{F} & \tilde{G} \end{pmatrix} = J^T \begin{pmatrix} E & F \\ F & G \end{pmatrix} J, \qquad J := \begin{pmatrix} u_x & u_y \\ v_x & v_y \end{pmatrix}
$$

and J is a regular matrix, the local coordinate (x, y), as well as (u, v), also satisfies (2). $\qquad\qquad\square$

Remark 9.5.5. The smooth function λ satisfying $EG - F^2 = \lambda^2$ is uniquely determined up to $\pm$-ambiguity, see [33, Proof of Proposition 3].

We are interested in Kossowski metrics, because the first fundamental forms of frontal maps $f : \mathcal{M}^2 \to \mathbf{R}^3$ are typical examples, as follows:

Proposition 9.5.6. *Let* $(\mathcal{M}^2, \mathcal{E}, \langle\ ,\ \rangle, D, \varphi)$ *be a coherent tangent bundle over a 2-manifold* $\mathcal{M}^2$. *If all singular points of* φ *are non-degenerate, then the first fundamental form induced by* ds^2 *is a Kossowski metric.*

Proof. Let (u, v) be a local coordinate system centered at a singular point p satisfying $\varphi(\partial_v) = \mathbf{0}$. We set $ds^2 = Edu^2 + 2Fdudv + Gdv^2$. Then, we have

$$
\begin{aligned}
E_v &= ds^2(\partial_u, \partial_u)_v = \langle \varphi(\partial_u), \varphi(\partial_u) \rangle_v = 2\langle D_{\partial_v}\varphi(\partial_u), \varphi(\partial_u) \rangle \\
&= 2\langle D_{\partial_u}\varphi(\partial_v), \varphi(\partial_u) \rangle = 2\langle \varphi(\partial_v), \varphi(\partial_u) \rangle_u - 2\langle \varphi(\partial_v), D_{\partial_u}\varphi(\partial_u) \rangle.
\end{aligned}
$$

Since $\varphi(\partial_v) = \mathbf{0}$ at p, we have

$$E_v(p) = 2\langle \varphi(\partial_v), \varphi(\partial_u) \rangle_u = ds^2(\partial_u, \partial_v)_u = 2F_u(p),$$

which implies that (1) of Definition 9.5.1 holds.

On the other hand, the signed area form can be written as $d\hat{A} = \lambda du \wedge dv$ ($\lambda := d\hat{A}(\partial_u, \partial_v)$), where λ is a smooth function defined on U. Let $U^*(\subset U)$ be the set of regular points. Then

$$e_1 := \frac{1}{\sqrt{E}} \frac{\partial}{\partial u}, \qquad e_2 := \frac{-1}{\sqrt{EG - F^2}} \left(\frac{F}{\sqrt{E}} \frac{\partial}{\partial u} - \sqrt{E} \frac{\partial}{\partial v} \right)$$

give an orthonormal frame field on U^*. Since $\{\varphi(e_1), \varphi(e_2)\}$ is an orthonormal frame field of $\mathcal{E}|_{U^*}$ (cf. (9.20)), we have

$$\pm 1 = \det_{\mathcal{E}}(\varphi(e_1), \varphi(e_2)) = d\hat{A}(e_1, e_2) = \frac{d\hat{A}(\partial_u, \partial_v)}{\sqrt{EG - F^2}} = \frac{\lambda}{\sqrt{EG - F^2}}.$$

Then $\lambda^2 = EG - F^2$ holds on U^*, and this identity holds on U, since U^* is dense in U. □

In the setting of Proposition 9.5.6, the singular curvature function along the singular curve consisting of singular points of the first kind is induced. If (u, v) is a C^r-differentiable local coordinate system so that the u-axis is the singular curve and ∂_v points in the null-direction along the u-axis, then, modifying the argument in [69], one can show the singular curvature function is given by

$$\kappa_s = \frac{-F_v E_u + 2E F_{uv} - E E_{vv}}{2\lambda_v E^{3/2}}, \qquad (9.37)$$

where $ds^2 = E du^2 + 2F du dv + G dv^2$. The following important fact was proved by Kossowski [48].

Fact 9.5.7 ([48]). Let $\mathcal{M}^2$ be a real analytic manifold and ds^2 a real analytic Kossowski metric on $\mathcal{M}^2$, and let p be a singular point of ds^2 satisfying $\hat{\Omega}(p) \neq 0$. Then there exists a neighborhood U of p and a wave front $f : U \to \mathbf{R}^3$ such that the first fundamental form of f coincides with ds^2 on U.

This fact implies that Kossowski metrics characterize the first fundamental forms of wave fronts. An explanation, a generalization and a refinement of this fact is given in [33].

Let $\{(U_\alpha; u^\alpha, v^\alpha)\}_{\alpha \in A}$ (A is the index set) be an open covering of $\mathcal{M}^2$ consisting of local coordinate neighborhoods. A Kossowski metric ds^2 is called *co-orientable*, if for each $\alpha \in A$, there exists λ^α satisfying the condition (2) of Definition 9.5.3 such that $d\hat{A} = \lambda^\alpha du^\alpha \wedge dv^\alpha$ gives a smooth 2-form on $\mathcal{M}^2$. The following fact can be proved:

Fact 9.5.8 ([28]). Let ds^2 be a Kossowski metric on an orientable 2-manifold $\mathcal{M}^2$. Then there exists a coherent tangent bundle on $\mathcal{M}^2$ such that the first fundamental form coincides with ds^2. Moreover, $\mathcal{E}$ is orientable if and only if ds^2 is co-orientable.

Remark 9.5.9. The fundamental theorem for surface theory gives a necessary and sufficient condition for a given pair of first and second fundamental forms to be realized as the actual fundamental forms of a regular surface. In Appendix G, such a theorem for frontals is given in terms of coherent tangent bundles. By the above fact, one can formulate the fundamental theorem for frontals in terms of Kossowski metrics, see [33, Proposition 8] for details.

Suppose that $\mathcal{M}^2$ is compact. Then a co-orientable Kossowski metric on $\mathcal{M}^2$ induces two Gauss–Bonnet formulas as in Theorems 9.4.2 and 9.4.3 of Chapter 9.

The above coherent tangent bundle induced by a Kossowski metric is uniquely determined up to an isomorphism of coherent tangent bundles. In fact, in [28], for a given Kossowski metric, a family $\{g_{\alpha,\beta}\}$ of orthogonal matrix-valued transition functions is constructed, which induces a unique coherent tangent bundle up to isomorphisms (cf [27, Theorem 3.2.1]). In [33], it was shown that the distance function induced by a given Kossowski metric on $\mathcal{M}^2$ is compatible with respect to the topology of the manifold $\mathcal{M}^2$. The definition of Kossowski metrics on n-manifolds and the generalization of Fact 9.5.8 in Chapter 9 to them are given in [74]. The fundamental theorem for surface theory in terms of coherent tangent bundles is given in [72]. The Gauss–Bonnet theorem on coherent tangent bundles is discussed in [69, 73, 8]. It should be remarked that the first fundamental form defined on a neighborhood of cross caps is not a Kossowski metric. Such metrics admit only isolated semi-definite points and belong to the class of Whitney metrics (see [28]).

Chapter 10

Contact Structure and Wave Fronts

In this chapter, we give a general theory for wave fronts as hypersurfaces in an n-dimensional manifold $\mathcal{N}^n$. So far we have treated the case $\mathcal{N}^n = \boldsymbol{R}^n$ ($n = 2, 3$), and readers can understand the cases of $\boldsymbol{R}^2$ and $\boldsymbol{R}^3$ more deeply by reading this chapter in the context of the contents up through Chapter 9. After reading this chapter, for further study of singularities of fronts, we refer the reader to [1] and [37].

10.1. Non-degenerate Skew-symmetric Forms

A map $\alpha : V \times V \to \boldsymbol{R}$ for a finite-dimensional real vector space V is a *skew-symmetric form* if α is bilinear and

$$\alpha(\boldsymbol{v}, \boldsymbol{w}) = -\alpha(\boldsymbol{w}, \boldsymbol{v}) \quad (\boldsymbol{v}, \boldsymbol{w} \in V)$$

holds. A skew-symmetric form α is said to be *non-degenerate* if there is no non-zero vector $\boldsymbol{v} \in V$ satisfying

$$\alpha(\boldsymbol{v}, \boldsymbol{w}) = 0 \quad \text{for all } \boldsymbol{w} \in V.$$

When such a vector $\boldsymbol{v} \neq \boldsymbol{0}$ exists, it is called a *null vector*.

For a non-degenerate skew-symmetric form α and a subspace W of V, we set

$$W' := \{\boldsymbol{v} \in V \,;\, \alpha(\boldsymbol{v}, \boldsymbol{w}) = 0 \text{ for any } \boldsymbol{w} \in W\}.$$

Proposition 10.1.1. *Let α be a non-degenerate skew-symmetric form on V, and let W be a subspace of V. Then*

(1) $\dim W + \dim W' = \dim V$,
(2) $(W')' = W$.

Here, $\dim W$ stands for the dimension of the vector space W.

Proof. Let V^* be the dual vector space of V, namely, the vector space formed by all linear functions on V. Then, by the non-degeneracy of α, the map $\hat{\alpha} : V \ni \boldsymbol{v} \mapsto \alpha(\boldsymbol{v}, *) \in V^*$ is a linear isomorphism. On the other hand, $\varphi \in V^*$ defines a linear map $\varphi : V \to \boldsymbol{R}$. We denote the restriction of φ to W by $\varphi|_W$. Then the linear map $\pi : V^* \ni \varphi \mapsto \varphi|_W \in W^*$ is a surjection. (In fact, by using the direct sum decomposition $V = W \oplus W_1$, for $\psi \in W^*$ we can choose $\varphi \in V^*$ so that $\varphi(W_1) = \{0\}$ and $\varphi|_W = \psi$. So we have $\pi(\varphi) = \psi$.) The kernel of the surjective linear map

$$\pi \circ \hat{\alpha} : V \longrightarrow W^*$$

coincides with W'. By the rank-nullity theorem for linear maps, we have (1). By $\alpha(W, W') = 0$, we have $W \subset (W')'$. By (1),

$$\dim(W')' = \dim V - \dim W' = \dim W$$

holds. Hence $W = (W')'$, namely (2) is proved. $\qquad \square$

Proposition 10.1.2. *If there exists a non-degenerate skew-symmetric form α on V, then the dimension of V is even. Moreover, if a subspace W satisfies*

$$\alpha(\boldsymbol{u}, \boldsymbol{v}) = 0 \quad (\boldsymbol{u}, \boldsymbol{v} \in W),$$

then the dimension of W is less than or equal to $(\dim V)/2$.

Proof. By taking a basis of V, we can identify V with $\boldsymbol{R}^l$, where $l := \dim V$. Let $\{\boldsymbol{u}_1, \ldots, \boldsymbol{u}_l\}$ be the usual basis of $\boldsymbol{R}^l$ as vertical vectors, that is,

$$\boldsymbol{u}_1 = (1, 0, \ldots, 0)^T, \ldots, \boldsymbol{u}_l = (0, 0, \ldots, 1)^T.$$

By setting $a_{jk} := \alpha(\boldsymbol{u}_j, \boldsymbol{u}_k)$, and

$$\boldsymbol{v} = (v_1, \ldots, v_l)^T = \sum_{j=1}^{l} v_j \boldsymbol{u}_j, \quad \boldsymbol{w} = (w_1, \ldots, w_l)^T = \sum_{j=1}^{l} w_j \boldsymbol{u}_j,$$

α satisfies the identity

$$\alpha(\boldsymbol{v}, \boldsymbol{w}) = \boldsymbol{v}^T A \boldsymbol{w},$$

where the right-hand side is a product of matrices, and $A = (a_{ij})_{i,j=1}^{l}$ is the $l \times l$-matrix whose (i, j) component is a_{ij}. Since α is a non-degenerate skew-symmetric form, A is a skew-symmetric regular matrix. Then the identity

$$\det A = \det A^T = \det(-A) = (-1)^l \det A$$

implies that the number l must be even.

If $\alpha(W, W) = 0$, then $W \subset W'$ holds. By Proposition 10.1.1, it holds that

$$\dim V = \dim W + \dim W' \geq 2 \dim W.$$

This proves the assertion. □

10.2. Contact Manifolds

Let $l(= 2m + 1)$ be a positive odd integer and $\mathcal{V}^l$ be an l-dimensional manifold. A *hyperplane field* is a correspondence

$$K : \mathcal{V}^l \ni p \mapsto K_p \subset T_p \mathcal{V}^l$$

from a point $p \in \mathcal{V}^l$ to a $2m$-dimensional subspace K_p of the tangent space $T_p \mathcal{V}^l$.

A vector field X defined on an open set of $\mathcal{V}^l$ is said to *belong to K* if for any $p \in \mathcal{V}^l$, X_p belongs to K_p. The hyperplane field K is *smooth* if for each $p \in \mathcal{V}^l$, there exist $2m$ linearly independent C^∞-vector fields $X_1, \ldots, X_{2m}$ defined on some neighborhood U of p such that these vector fields span K. This family of vector fields $\{X_1, \ldots, X_{2m}\}$ is called a (local) *frame field* (i.e., a basis field) of K.

Example 10.2.1. Let θ be a 1-form without zeros on a $(2m + 1)$-dimensional manifold $\mathcal{V}^{2m+1}$. For each $p \in \mathcal{V}^{2m+1}$, we let K_p be the kernel of the linear map $\theta_p : T_p \mathcal{V}^{2m+1} \to \mathbf{R}$. Then K gives a hyperplane field as follows: We take a Riemannian metric g on $\mathcal{V}^{2m+1}$ and a local coordinate neighborhood $(U; x_1, \ldots, x_l)$ of an arbitrary fixed point $p \in \mathcal{V}^l$ ($l := 2m+1$). If U is sufficiently small, there exists a vector field X without zeros on U so that for any vector field Y on U, $\theta(Y) = g(X, Y)$ holds. By choosing the order of $x_1, \ldots, x_l$, if necessary, we may assume that

$$X, \; \frac{\partial}{\partial x_2}, \ldots, \frac{\partial}{\partial x_l}$$

are linearly independent at each point on U. Applying Gram–Schmidt orthogonalization to these vector fields, we obtain an orthonormal frame field $\{e_1, e_2, \ldots, e_l\}$ ($e_1 := X/|X|$) on U such that $\{e_2, \ldots, e_l\}$ spans K_p at each point $p \in U$. Thus K is a hyperplane field.

Definition 10.2.2. Let K be a smooth hyperplane field of $\mathcal{V}^{2m+1}$. A *differential form associated to K* defined on an open subset $U(\subset \mathcal{V}^{2m+1})$ is a 1-form θ defined on U such that the kernel of the linear function $\theta_q : T_q \mathcal{V}^{2m+1} \to \mathbf{R}$ ($q \in U$) is K_q.

By definition, if such a 1-form θ associated to K exists on U, it is determined up to a non-zero functional multiple. The following assertion holds:

Proposition 10.2.3. *Let K be a smooth hyperplane field on $\mathcal{V}^{2m+1}$. Then for each $p \in \mathcal{V}^{2m+1}$, there exists a 1-form associated to K defined on some neighborhood of p.*

Proof. For a local frame $\{X_1, \ldots, X_{2m}\}$ of K defined on an open subset U of $\mathcal{V}^{2m+1}$, we can find a vector field Y on U such that $\{X_1, \ldots, X_{2m}, Y\}$ gives a basis of the tangent space $T_q \mathcal{V}^{2m+1}$ for each $q \in U$. We set $X_{2m+1} := Y$. Let ξ_j $(j = 1, \ldots, 2m+1)$ be a 1-form on U such that

$$\xi_j(X_k) = \delta_{j,k} \quad (j, k = 1, \ldots, 2m+1),$$

where $\delta_{j,k}$ is the Kronecker delta. Then

$$\theta := \xi_{2m+1}$$

gives a differential form associated with K on U. □

Based on the above discussions, we define the contact structure as follows.

Definition 10.2.4. A smooth hyperplane field K is called a *contact structure* or *contact hyperplane field* if, for each $p \in \mathcal{V}^{2m+1}$, there exists a 1-form θ associated to K defined on an open neighborhood U of p satisfying

$$\theta \wedge \left(\bigwedge^m d\theta \right) \neq 0 \tag{10.1}$$

on U, where the $2m$-form $\bigwedge^m d\theta$ is the mth exterior power of the exterior derivative $d\theta$ of θ. In this situation, the pair $(\mathcal{V}^{2m+1}, K)$ is called a *contact manifold*, and θ is called a (local) *contact form*.

The property (10.1) does not change by non-zero functional multiplication of θ. Some textbooks define contact structure by the existence of a 1-form satisfying (10.1) defined on $\mathcal{V}^{2m+1}$. However, such a definition of contact structure can be considered as a special case of ours, as follows:

Proposition 10.2.5. *If a 1-form θ on a $(2m+1)$-manifold $\mathcal{V}^{2m+1}$ satisfies (10.1) on $\mathcal{V}^{2m+1}$, then for each $p \in \mathcal{V}^{2m+1}$, by taking the kernel K_p of the linear map $\theta_p : T_p \mathcal{V}^{2m+1} \to \mathbf{R}$, the pair $(\mathcal{V}^{2m+1}, K)$ is a contact manifold.*

Proof. As shown in Example 10.2.1, K gives a hyperplane field. So we have obtained the assertion. □

We prove the following lemma.

Lemma 10.2.6. *Let* $(\mathcal{V}^{2m+1}, K)$ *be a* $(2m+1)$-*dimensional contact manifold, and let* θ *be a contact form of* K. *Then for each* $p \in \mathcal{V}^{2m+1}$, $(d\theta)_p$ *defines a non-degenerate skew-symmetric form on* K_p.

Proof. We fix $p \in \mathcal{V}^{2m+1}$ arbitrarily. If $d\theta_p$ is a degenerate skew-symmetric form on the subspace K_p of $T_p\mathcal{V}^{2m+1}$, then there exists $\boldsymbol{w} \in K_p \setminus \{\boldsymbol{0}\}$ such that

$$d\theta_p(\boldsymbol{v}, \boldsymbol{w}) = 0 \quad (\boldsymbol{v} \in K_p).$$

We take vectors $\boldsymbol{u}, \boldsymbol{v}_1, \ldots, \boldsymbol{v}_{2m-1} \in T_p\mathcal{V}^{2m+1}$ such that

- $\boldsymbol{u} \notin K_p$,
- $\boldsymbol{v}_1, \ldots, \boldsymbol{v}_{2m-1} \in K_p$, and
- $\{\boldsymbol{u}, \boldsymbol{v}_1, \ldots, \boldsymbol{v}_{2m-1}, \boldsymbol{w}\}$ is a basis of $T_p\mathcal{V}^{2m+1}$.

Since

$$\theta_p(\boldsymbol{w}) = \theta_p(\boldsymbol{v}_j) = 0 \quad (j = 1, \ldots, 2m-1),$$

we have

$$\theta_p \wedge d\theta_p \wedge \cdots \wedge d\theta_p(\boldsymbol{u}, \boldsymbol{v}_1, \ldots, \boldsymbol{v}_{2m-1}, \boldsymbol{w})$$
$$= \theta_p(\boldsymbol{u}) \cdot (d\theta_p \wedge \cdots \wedge d\theta_p)(\boldsymbol{v}_1, \ldots, \boldsymbol{v}_{2m-1}, \boldsymbol{w}) = 0.$$

This means that the $(2m+1)$-form $\theta_p \wedge d\theta_p \wedge \cdots \wedge d\theta_p$ is zero on $T_p\mathcal{V}^{2m+1}$, contradicting the fact that $\mathcal{V}^{2m+1}$ is a contact manifold. □

Example 10.2.7. Let $(x_1, \ldots, x_m, y_1, \ldots, y_m, z)$ be a coordinate system of $\boldsymbol{R}^{2m+1}$, and set $\theta := dz + y_1 \, dx_1 + \cdots + y_m \, dx_m$. Then the mth exterior power of $d\theta$ is

$$d\theta \wedge \cdots \wedge d\theta = m! \, dy_1 \wedge dx_1 \wedge \cdots \wedge dy_m \wedge dx_m.$$

So, we have

$$\theta \wedge d\theta \wedge \cdots \wedge d\theta = m! \, dz \wedge dy_1 \wedge dx_1 \wedge \cdots \wedge dy_m \wedge dx_m,$$

which does not vanish for each point of $\boldsymbol{R}^{2m+1}$. In particular, θ is a contact form on $\boldsymbol{R}^{2m+1}$.

The following theorem states that contact structures are all locally identified with the one given in the above example.

Theorem 10.2.8 (The Darboux theorem for contact structures).
Let $(\mathcal{V}^{2m+1}, K)$ be a $(2m+1)$-dimensional contact manifold, and let θ be a local contact form of K defined on a neighborhood U of p in $\mathcal{V}^{2m+1}$.
Then there exist a neighborhood $U' \subset U$ and a local coordinate system $(x_1, \ldots, x_m, y_1, \ldots, y_m, z)$ on U' such that

$$\theta = dz + y_1\, dx_1 + \cdots + y_m\, dx_m.$$

See [1] for a proof of this theorem.

Induced contact structures on projective cotangent bundles.
Let $\mathcal{N}^{m+1}$ be an $(m+1)$-manifold. The projective cotangent bundle $P(T^*\mathcal{N}^{m+1})$ is defined as follows: The fiber of $P(T^*\mathcal{N}^{m+1})$ at each $p \in \mathcal{N}^{m+1}$ is the projective space $P(T_p^*\mathcal{N}^{m+1})$ associated with the vector space $T_p^*\mathcal{N}^{m+1}$. Then $P(T^*\mathcal{N}^{m+1}) := \bigcup_{p\in\mathcal{N}^{m+1}} P(T_p^*\mathcal{N}^{m+1})$ has the structure of a $(2m+1)$-manifold so that the canonical projection

$$T^*\mathcal{N}^{m+1} \setminus \{0\} \ni \eta \longrightarrow [\eta] \in P(T^*\mathcal{N}^{m+1}) \tag{10.2}$$

is a submersion. For each $[\eta_p] \in P(T_p^*\mathcal{N}^{m+1})$, we set

$$K_{[\eta_p]} := (d\hat{\pi})^{-1}(\operatorname{Ker}\eta_p), \tag{10.3}$$

where

$$\hat{\pi} : P(T^*\mathcal{N}^{m+1}) \to \mathcal{N}^{m+1} \tag{10.4}$$

is the canonical projection induced by the projection $\pi : T^*\mathcal{N}^{m+1} \to \mathcal{N}^{m+1}$. Then K gives a hyperplane field on $P(T^*\mathcal{N}^{m+1})$. The following assertion holds:

Proposition 10.2.9. *The pair $\big(P(T^*\mathcal{N}^{m+1}), K\big)$ is a contact manifold.*

Proof. We fix an arbitrary $[\eta_p] \in P(T_p^*\mathcal{N}^{m+1})$. We can take a local coordinate system $(U; x_0, \ldots, x_m)$ of $\mathcal{N}^{m+1}$ centered at p such that $\eta_p = (dx_0)_p$. Writing $(x, y) := (x_0, \ldots, x_m, y_1, \ldots, y_m)$ for brevity, we define a map φ by

$$\varphi : U \times \mathbf{R}^m \ni (x, y) \longmapsto [\eta(x, y)] \in P(T^*U),$$

where

$$\eta(x, y) := (dx_0)_x + y_1(dx_1)_x + \cdots + y_m(dx_m)_x. \tag{10.5}$$

By definition, φ is a diffeomorphism, and so we may regard (x, y) as a local coordinate system of $P(T^*\mathcal{N}^{m+1})$ centered at $[\eta_p]$. We consider a 1-form

$$\theta := dx_0 + y_1\, dx_1 + \cdots + y_m\, dx_m \tag{10.6}$$

on $\varphi(U \times \mathbf{R}^m)$ by using the above coordinate system. Then, by Example 10.2.7,

$$\theta \wedge \left(\bigwedge^m d\theta \right) \neq 0 \tag{10.7}$$

holds. Since the kernel of the linear map $\eta(x, y) : T_{\pi(\eta(x,y))} \mathcal{N}^{m+1} \to \mathbf{R}$ coincides with

$$\left\{ a_0 \frac{\partial}{\partial x_0} + \cdots + a_m \frac{\partial}{\partial x_m} \in T_{\pi(\eta(x,y))} \mathcal{N}^{m+1} \, ; \, a_0 + y_1 a_1 + \cdots + y_m a_m = 0 \right\},$$

we have that

$$K_{[\eta(x,y)]} = (d\hat{\pi})^{-1} \left(\operatorname{Ker} \eta(x, y) \right)$$

$$= \left\{ a_0 \frac{\partial}{\partial x_0} + \cdots + a_m \frac{\partial}{\partial x_m} + b_1 \frac{\partial}{\partial y_1} + \cdots + b_m \frac{\partial}{\partial y_m} \, ; \right.$$

$$\left. \begin{array}{l} a_0, \ldots, a_m, b_1, \ldots, b_m \in \mathbf{R}, \\ a_0 + y_1 a_1 + \cdots + y_m a_m = 0 \end{array} \right\}.$$

By (10.6), $K_{[\eta(x,y)]}$ coincides with the kernel of θ at $[\eta(x, y)]$. $\qquad\square$

Example 10.2.10 (The unit cotangent bundle of a Riemannian manifold).

We fix a Riemannian metric g of $\mathcal{N}^{m+1}$. Then the unit tangent bundle $S(T\mathcal{N}^{m+1})$ is defined by

$$S(T\mathcal{N}^{m+1}) := \{ v \in T\mathcal{N}^{m+1} \, ; \, g(v, v) = 1 \}.$$

We then set

$$S(T^*\mathcal{N}^{m+1}) := \{ g(v, *) \in T^*\mathcal{N}^{m+1} \, ; \, v \in S(T\mathcal{N}^{m+1}) \},$$

where $g(v, *)$ $(v \in T_p\mathcal{N}^{m+1})$ is the linear function defined by $T_p\mathcal{N}^{m+1} \ni w \mapsto g(v, w) \in \mathbf{R}$. We call $S(T^*\mathcal{N}^{m+1})$ the *unit cotangent bundle*. Then the map

$$\flat : S(T\mathcal{N}^{m+1}) \ni v \mapsto g(v, *) \in S(T^*\mathcal{N}^{m+1}) \tag{10.8}$$

is a bijection, and so we can identify $S(T\mathcal{N}^{m+1})$ with $S(T^*\mathcal{N}^{m+1})$. Then

$$\Phi : S(T^*\mathcal{N}^{m+1}) \ni \eta \longmapsto [\eta] \in P(T^*\mathcal{N}^{m+1}) \tag{10.9}$$

is an immersion, and is a two-to-one map. Let K be the canonical contact structure on $P(T^*\mathcal{N}^{m+1})$. We set

$$\tilde{K} := d\Phi^{-1}(K).$$

If θ is a 1-form associated to K defined on an open subset W of $P(T^*\mathcal{N}^{m+1})$, then the pull-back $\Phi^*\theta$ gives a 1-form defined on $\Phi^{-1}(W)$ associated to $\tilde{K}$. Since θ satisfies (10.7), $\Phi^*\theta$ gives a contact form. Thus, the hyperplane field $\tilde{K}$ gives a contact structure on $S(T^*\mathcal{N}^{m+1})$.

Different from the case of $P(T^*\mathcal{N}^{m+1})$, there is a contact form associated to $\tilde{K}$ globally[1] defined on $S(T^*\mathcal{N}^{m+1})$. We will now show this:

Define a 1-form Θ on a manifold $T^*\mathcal{N}^{m+1}$ by

$$\Theta_{\eta_p} := \pi^*\eta_p \tag{10.10}$$

for each cotangent vector $\eta_p \in T_p^*\mathcal{N}^{m+1}$ at $p \in \mathcal{N}^{m+1}$, where $\pi : T^*\mathcal{N}^{m+1} \to \mathcal{N}^{m+1}$ is the canonical projection.

Let $(U; x_0, \ldots, x_m)$ be a local coordinate system of the manifold $\mathcal{N}^{m+1}$. Since any $\eta_p \in T_p^*\mathcal{N}^{m+1}$ can be written as

$$\eta_p = \sum_{j=0}^{m} \xi_j(p)(dx_j)_p \quad (\xi_j(p) \in \mathbf{R}, \ j = 0, \ldots, m), \tag{10.11}$$

$(T^*U; x_0, \ldots, x_m, \xi_0, \ldots, \xi_m)$ gives a local coordinate system of T^*U. Using this, we can write

$$\Theta = \sum_{i=0}^{m} \xi_i dx_i. \tag{10.12}$$

In particular, Θ is a smooth 1-form on the manifold $T^*\mathcal{N}^{m+1}$. Let $\tilde{\theta}$ be the restriction of Θ to $S(T^*\mathcal{N}^{m+1})$. Let $\hat{\pi} : P(T^*\mathcal{N}^{m+1}) \to \mathcal{N}^{m+1}$ be the canonical projection as in (10.4). For each cotangent vector $\eta_p \in T_p^*\mathcal{N}^{m+1}$ at $p \in \mathcal{N}^{m+1}$, $\tilde{\theta}_{\eta_p} = \Phi^*\hat{\pi}^*\eta_p$ holds. Applying (10.3), we have,

$$\mathrm{Ker}\,\tilde{\theta}_{\eta_p} = \mathrm{Ker}(\Phi^*\hat{\pi}^*\eta_p) = d\Phi^{-1}(d\hat{\pi}^{-1}(\mathrm{Ker}\,\eta_p)) = d\Phi^{-1}(K_{[\eta_p]}) = \tilde{K}_{\eta_p}.$$

So $\tilde{\theta}$ is a differential form associated to $\tilde{K}$. Since we have already seen that $\left(S(T^*\mathcal{N}^{m+1}), \tilde{K} \right)$ is a contact manifold, $\tilde{\theta}$ satisfies

$$\tilde{\theta} \wedge \left(\bigwedge^{m} d\tilde{\theta} \right) \neq 0.$$

In particular, $\tilde{\theta}$ is a contact form defined globally on $S(T^*\mathcal{N}^{m+1})$, which is called the *canonical contact form* on $S(T^*\mathcal{N}^{m+1})$.

---

[1] $P(T^*\mathcal{N}^{m+1})$ is a non-orientable manifold if the dimension of $\mathcal{N}^{m+1}$ is even. In this case, there is no contact form defined on $P(T^*\mathcal{N}^{m+1})$.

Definition 10.2.11. For two contact manifolds $(\mathcal{V}^{2m+1}, K)$ and $((\mathcal{V}')^{2m+1}, K')$ of the same dimension, an immersion

$$f : \mathcal{V}^{2m+1} \to (\mathcal{V}')^{2m+1}$$

satisfying $df(K) = K'$ is called a *local contact diffeomorphism*. Moreover, if f is a diffeomorphism, then f is called a *contact diffeomorphism*.

As we saw in Example 10.2.10, the following proposition holds.

Proposition 10.2.12. *The canonical projection* $\Phi : S(T^*\mathcal{N}^{m+1}) \to P(T^*\mathcal{N}^{m+1})$ *is a local contact diffeomorphism. Moreover, if $\mathcal{N}^{m+1}$ is connected, then Φ is a double covering map.*

Furthermore, we show the following theorem.

Theorem 10.2.13. *Let $\psi : \mathcal{N}^{m+1} \to \mathcal{N}^{m+1}$ be a diffeomorphism. Then the map Ψ defined by*

$$\Psi : P(T^*\mathcal{N}^{m+1}) \ni [\eta] \mapsto \left[(\psi^{-1})^*\eta\right] \in P(T^*\mathcal{N}^{m+1})$$

is a contact diffeomorphism satisfying

$$\hat{\pi} \circ \Psi = \psi \circ \hat{\pi}, \tag{10.13}$$

where $\hat{\pi} : P(T^*\mathcal{N}^{m+1}) \to \mathcal{N}^{m+1}$ is the canonical projection.

Proof. Since $(\psi^{-1})^*\eta \in T_{\psi(p)}\mathcal{N}^{m+1}$ if $\eta \in T_p\mathcal{N}^{m+1}$, it is clear that Ψ is a diffeomorphism satisfying (10.13). Let K be the canonical contact hyperplane field on $P(T^*\mathcal{N}^{m+1})$. We fix $\eta \in T_p^*\mathcal{N}^{m+1}$ arbitrarily. Since $K_{[\eta]} = d\hat{\pi}^{-1}(\text{Ker}\,\eta)$,

$$d\hat{\pi} \circ d\Psi(K_{[\eta]}) = d(\hat{\pi} \circ \Psi)(d\hat{\pi}^{-1}(\text{Ker}\,\eta))$$

$$= d(\psi \circ \hat{\pi})(d\hat{\pi}^{-1}(\text{Ker}\,\eta))$$

$$= d\psi \circ d\hat{\pi} \circ d\hat{\pi}^{-1}(\text{Ker}\,\eta) = d\psi(\text{Ker}\,\eta)$$

holds. By the equivalency

$$v \in d\psi(\text{Ker}\,\eta) \iff d\psi^{-1}(v) \in \text{Ker}\,\eta$$

$$\iff \eta(d\psi^{-1}(v)) = 0$$

$$\iff v \in \text{Ker}\left((d\psi^{-1})^*\eta\right),$$

we have $d\hat{\pi} \circ d\Psi(K_{[\eta]}) = \text{Ker}\left((d\psi^{-1})^*\eta\right)$ and

$$d\Psi(K_{[\eta]}) \subset (d\hat{\pi})^{-1}\left(\text{Ker}\left((d\psi^{-1})^*\eta\right)\right) = K_{\Psi([\eta])}.$$

This implies that Ψ is a contact diffeomorphism. $\qquad\qquad\square$

We give a non-trivial example of a contact manifold.

Example 10.2.14 (Odd-dimensional spheres). Let C^{m+1} be an $(m+1)$-dimensional complex space, and $(z_0, \ldots, z_m)$ the canonical complex coordinate system. We divide this into the real and imaginary parts as

$$z_j = x_j + \sqrt{-1}y_j \quad (j = 0, 1, \ldots, m).$$

We define a 1-form θ on $C^m = R^{2m+2}$ by

$$\theta := \sum_{j=0}^{m}(x_j \, dy_j - y_j \, dx_j) = \frac{\sqrt{-1}}{2} \sum_{j=0}^{m} (z_j \, d\overline{z_j} - \overline{z_j} dz_j).$$

Restricting θ to the unit sphere $S^{2m+1} \subset R^{2m+2}$, it then defines a contact structure on S^{2m+1}. (To see this, it is enough to check the property $\theta \wedge (\bigwedge^m d\theta) \neq 0$ only at the point $p = (1, 0, \ldots, 0)$, since θ is invariant under the canonical unitary group action on C^{m+1}.)

10.3.   Legendrian Submanifolds and Wave Fronts

Let $(\mathcal{V}^{2m+1}, K)$ be a $(2m + 1)$-dimensional contact manifold. We define Legendre submanifolds of $\mathcal{V}^{2m+1}$ as follows:

Definition 10.3.1. Let $\mathcal{M}^l$ be a manifold. A C^∞-map $\varphi : \mathcal{M}^l \to \mathcal{V}^{m+1}$ is called *isotropic* if

$$d\varphi(T_p\mathcal{M}^l) \subset K_{\varphi(p)} \tag{10.14}$$

holds for each $p \in \mathcal{M}^l$.

If θ is a (local) contact form associated with the contact hyperplane field K, then (10.14) holds if and only if the pull-back $\varphi^*\theta$ of θ by φ vanishes identically. By Proposition 10.1.2, the following assertion holds.

Proposition 10.3.2. *If $\varphi \colon \mathcal{M}^l \to \mathcal{V}^{m+1}$ is an isotropic immersion into a $(2m + 1)$-dimensional manifold, then $l = \dim \mathcal{M} \leq m$ holds.*

Definition 10.3.3. An isotropic immersion $\varphi \colon \mathcal{M}^l \to \mathcal{V}^{2m+1}$ is called a *Legendrian immersion* if $l = \dim \mathcal{M} = m$ holds. In this case, $\varphi(\mathcal{M}^m)$ is called a *Legendrian submanifold*.

The concept of Legendrian submanifolds is an analogue of that of Lagrangian submanifolds in symplectic manifolds (cf. [15]).

Proof of Proposition 10.3.2. Let θ be a contact 1-form associated with a contact manifold $(\mathcal{V}^{2m+1}, K)$. By the definition of an isotropic map φ, we have

$$0 = d(\varphi^*\theta) = \varphi^* d\theta.$$

In particular,

$$d\theta(\boldsymbol{v}, \boldsymbol{v}) = 0 \quad (\boldsymbol{v} \in d\varphi(T_p\mathcal{M}^l))$$

holds. By Lemma 10.2.6, the restriction of $d\theta$ to $K_{\varphi(p)}$ gives a skew-symmetric non-degenerate bilinear form, and the dimension of the vector space $T_p\mathcal{M}^l$ is less than or equal to half of the dimension of $K_\varphi(p)$ (cf. Proposition 10.1.2).

We next give the definition of wave fronts of an $(m+1)$-manifold $\mathcal{N}^{m+1}$ as follows:

Definition 10.3.4. A C^∞-differentiable map $f : \mathcal{M}^m \to \mathcal{N}^{m+1}$ from an m-manifold $\mathcal{M}^m$ into an $(m + 1)$-manifold $\mathcal{N}^{m+1}$ is called a *frontal* (respectively, *wave front*) if there exists an isotropic map (respectively, a Legendrian immersion) $L : \mathcal{M}^m \to P(T^*\mathcal{N}^{m+1})$ satisfying $f = \hat{\pi} \circ L$, where $\hat{\pi} : P(T^*\mathcal{N}^{m+1}) \to \mathcal{N}^{m+1}$ is the canonical projection. This map L is called the *isotropic lift* (respectively, *Legendrian lift*) of the isotropic map (respectively, the Legendrian immersion) f.

By the definition of wave fronts, the following assertion is obvious:

Proposition 10.3.5. *An immersion of an m-manifold into an $(m + 1)$-manifold is a wave front.*

This definition of wave fronts depends only on the differentiable structure of $\mathcal{N}^{m+1}$. However, the uniqueness of the Legendrian lift may not hold in general: For example, consider the constant map $f : \boldsymbol{R} \ni t \mapsto o := (0, 0) \in \boldsymbol{R}^2$. We let $h(t)$ be a smooth function on $\boldsymbol{R}$ and define a map by

$$L : \boldsymbol{R} \ni t \longmapsto [\cos h(t)(dx)_o + \sin h(t)(dy)_o] \in P(T^*\boldsymbol{R}^2).$$

Then L is a Legendrian lift of f whenever dh/dt has no zeros.

Our definition of wave front (cf. Definition 10.3.4) and Theorem 10.2.13 yield the following:

Proposition 10.3.6. *Let $\varphi : \mathcal{N}^{m+1} \to \mathcal{N}^{m+1}$ be a diffeomorphism on an $(m+1)$-dimensional manifold $\mathcal{N}^{m+1}$, and let $f : \mathcal{M}^m \to \mathcal{N}^{m+1}$ be a frontal (respectively, a wave front). Then $\varphi \circ f$ is also a frontal (respectively, a wave front).*

Remark 10.3.7. Theorems 1.5.6 and 2.4.1 in Chapters 1 and 2, respectively, are special cases of this proposition.

We fix a Riemannian metric g of $\mathcal{N}^{m+1}$. Then, as in (10.8), we can identify $S(T\mathcal{N}^{m+1})$ with $S(T^*\mathcal{N}^{m+1})$ by the map $\flat$. Since $S(T^*\mathcal{N}^{m+1})$ has a canonical contact structure (cf. Example 10.2.10), $S(T\mathcal{N}^{m+1})$ has a contact structure induced by $\flat$.

Definition 10.3.8. A C^∞-map $f : \mathcal{M}^m \to \mathcal{N}^{m+1}$ from an m-manifold $\mathcal{M}^m$ into an $(m+1)$-manifold $\mathcal{N}^{m+1}$ is called a *co-orientable frontal* (respectively, a *co-orientable wave front*) if there exists an isotropic map (respectively, a Legendrian immersion) $\tilde{L} : \mathcal{M}^m \to S(T^*\mathcal{N}^{m+1})$ such that $f = \pi \circ \tilde{L}$, where $\pi : S(T^*\mathcal{N}^{m+1}) \to \mathcal{N}^{m+1}$ is the canonical projection.

This map $\tilde{L}$ is called the *lift* of the map f to $S(T^*\mathcal{N}^{m+1})$. Let $\Phi : S(T^*\mathcal{N}^{m+1}) \to P(T^*\mathcal{N}^{m+1})$ be the projection defined in (10.9). Then $\Phi \circ \tilde{L} : \mathcal{M}^m \to P(T^*\mathcal{N}^{m+1})$ gives a lift of f to $P(T^*\mathcal{N}^{m+1})$ in the sense of Definition 10.3.4. If a frontal (respectively, wave front) is not a co-orientable frontal (respectively, co-orientable wave front), it is said to be *non-co-orientable*.

Remark 10.3.9. Suppose $f : \mathcal{M}^m \to \mathcal{N}^{m+1}$ is a frontal which is not co-orientable. Then, using the general theory of covering spaces, there exists a double covering $\hat{\mathcal{M}}^m$ of $\mathcal{M}^m$ such that f can be lifted to a co-orientable frontal $\hat{f} : \hat{\mathcal{M}}^m \to \mathcal{N}^{m+1}$. In particular, if $\mathcal{M}^m$ is simply connected, then any frontal $f : \mathcal{M}^m \to \mathcal{N}^{m+1}$ is co-orientable.

We can prove the following:

Proposition 10.3.10. *Let $\mathcal{M}^m$ be an m-manifold and $f : \mathcal{M}^m \to \mathcal{N}^{m+1}$ a C^∞-map. Then f is a co-orientable frontal if and only if there exists a C^∞-map $\tilde{\nu} : \mathcal{M}^m \to S(T\mathcal{N}^{m+1})$ such that $\tilde{\nu}(p)$ is perpendicular to $df(T_p\mathcal{M}^m)$ for each $p \in \mathcal{M}^m$. Moreover, f is a wave front if and only if $\tilde{\nu}$ is an immersion.*

Proof. Consider a co-orientable frontal (respectively, a wave front) C^∞-map $f : \mathcal{M}^m \to \mathcal{N}^{m+1}$. Then there exists an isotropic map (respectively, a Legendrian immersion) $\tilde{L} : \mathcal{M}^m \to S(T^*\mathcal{N}^{m+1})$ such that $\pi \circ \tilde{L} = f$, where $\pi : T^*\mathcal{N}^{m+1} \to \mathcal{N}^{m+1}$ is the canonical projection. We set

$$\tilde{\nu} := \sharp \circ \tilde{L} : \mathcal{M}^m \to S(T\mathcal{N}^{m+1}),$$

where $\sharp$ is the inverse operation of $\flat$ as in (10.8). Then we have $\tilde{\pi} \circ \tilde{\nu} = f$, where $\tilde{\pi} : S(T\mathcal{N}^{m+1}) \to \mathcal{N}^{m+1}$ is the canonical projection. Let $\tilde{\theta}$ be the canonical contact form on $S(T^*\mathcal{N}^{m+1})$. Since $\tilde{L}$ is isotropic (respectively, Legendrian), for a given $p \in \mathcal{M}^m$ and $\boldsymbol{v} \in T_p\mathcal{M}^m$, we set $\eta := \tilde{L}(p) \in T^*\mathcal{N}^{m+1}$, and then we have

$$0 = \tilde{\theta}_\eta(d\tilde{L}(\boldsymbol{v})) = \Theta_\eta(d\tilde{L}(\boldsymbol{v})) = \pi^*\eta(d\tilde{L}(\boldsymbol{v}))$$
$$= \tilde{L}^*\pi^*\eta(\boldsymbol{v}) = (\pi \circ \tilde{L})^*\eta(\boldsymbol{v}) = f^*\eta(\boldsymbol{v})$$
$$= \eta(df(\boldsymbol{v})) = \tilde{L}(p)(df(\boldsymbol{v})) = g(\tilde{\nu}(p), df(\boldsymbol{v})).$$

So, $\tilde{\nu}(p)$ is perpendicular to $df(\boldsymbol{v})$.

Conversely, let $\tilde{\nu} : \mathcal{M}^m \to S(T\mathcal{N}^{m+1})$ be a map which is orthogonal to $df(\boldsymbol{v})$ for each $\boldsymbol{v} \in T\mathcal{M}^m$. Setting $\tilde{L} := \flat \circ \tilde{\nu}$, and reading the above computations from the opposite direction, we have

$$0 = g(\tilde{\nu}(p), df(\boldsymbol{v})) = \tilde{\theta}_\eta(d\tilde{L}(\boldsymbol{v})).$$

So, $d\tilde{L}(\boldsymbol{v}) \in \tilde{K}_{\tilde{L}(p)}$, that is, $\tilde{L}$ is Legendrian, proving the assertion. $\square$

When $\mathcal{N}^{m+1} = \boldsymbol{R}^{m+1}$ and $f : \mathcal{M}^m \to \mathcal{N}^{m+1}$ is a frontal, then $S(T\mathcal{N}^{m+1})$ can be identified with $\boldsymbol{R}^{m+1} \times S^m$ and we can write $\tilde{\nu} = (f, \nu)$, where ν is a map into the unit sphere S^m. Here, ν can be identified with the Gauss map of f. So we get the following:

Corollary 10.3.11. *Let $\mathcal{M}^m$ be an m-manifold and $f : \mathcal{M}^m \to \boldsymbol{R}^{m+1}$ a C^∞-map. Then f is a co-orientable frontal if and only if there exists a C^∞-map*

$$\nu : \mathcal{M}^m \to S^m = \{x \in \boldsymbol{R}^{m+1} \,;\, |x| = 1\}$$

such that $\nu(p)$ is perpendicular to $df(T_p\mathcal{M}^m)$ for each $p \in \mathcal{M}^m$. Moreover, f is a wave front if and only if

$$L = (f, \nu) : \mathcal{M}^m \to \boldsymbol{R}^{m+1} \times S^m = S(T\boldsymbol{R}^{m+1})$$

is an immersion. In particular, the frontals and wave fronts into $\boldsymbol{R}^n$ ($n = 2, 3$) discussed in this book are all frontals and wave fronts in the sense of this chapter.

10.4. Wave Fronts and Legendre Fibrations

In this section, we define "Legendre fibrations", and show that projections of Legendrian submanifolds to the base spaces of the Legendre fibrations can be considered as wave fronts.

Definition 10.4.1. Let $(\mathcal{V}^{2m+1}, K)$ be a $(2m + 1)$-dimensional contact manifold and $\mathcal{N}^{m+1}$ an $(m+1)$-manifold. A C^∞-map $\Pi : \mathcal{V}^{2m+1} \longrightarrow \mathcal{N}^{m+1}$

is called *locally trivial* if there exists an m-manifold F^m satisfying the following condition:

- For each $p \in \mathcal{N}^{m+1}$, there exists a diffeomorphism $\Psi \colon \Pi^{-1}(U) \to U \times F^m$ such that $\Pi_U \circ \Psi = \Pi$, where U is a certain neighborhood of p and $\Pi_U \colon U \times F^m \to U$ is the canonical projection.

In this setting, the inverse image $\Pi^{-1}(p)$ is called the *fiber*. If each fiber is a Legendrian submanifold (cf. Definition 10.3.3) of $\mathcal{V}^{2m+1}$, the map $\Pi \colon \mathcal{V}^{2m+1} \to \mathcal{N}^{m+1}$ is called a *Legendrian fibration*.

In Example 10.2.7, we gave a canonical contact structure on $\boldsymbol{R}^{2m+1}$, and it is easily verified that the canonical projection

$$\boldsymbol{R}^{2m+1} \ni (x_1, \ldots, x_m, y_1, \ldots, y_m, z) \mapsto (x_1, \ldots, x_m, z) \in \boldsymbol{R}^{m+1} \quad (10.15)$$

gives a Legendrian fibration. In general, the following assertion holds:

Proposition 10.4.2. *The canonical projection* $\hat{\pi} \colon P(T^*\mathcal{N}^{m+1}) \to \mathcal{N}^{m+1}$ *given in* (10.4) *is a Legendrian fibration.*

Proof. By (10.3), tangent vectors of each fiber are contained in K, so we get the conclusion. $\qquad\square$

By the definition of the contact structure of $S(T^*\mathcal{N}^{m+1})$, we have the following:

Corollary 10.4.3. *The canonical projection* $\pi \colon S(T^*\mathcal{N}^{m+1}) \to \mathcal{N}^{m+1}$ *is a Legendrian fibration.*

As a consequence of the Darboux Theorem (cf. Theorem 10.2.8), we can prove the following:

Proposition 10.4.4. *Any contact manifold* $(\mathcal{V}^{2m+1}, K)$ *locally gives the structure of a Legendrian fibration. More precisely, for each* $p \in \mathcal{V}^{2m+1}$, *there exist a neighborhood U of p, a domain V of* $\boldsymbol{R}^{m+1}$, *and a C^∞-map* $\Pi \colon U \to V$ *giving a Legendrian fibration.*

Proof. We let $\dim \mathcal{V}^{2m+1} = 2m + 1$, and let θ be a 1-form defined on a sufficiently small neighborhood U of p which gives a (local) contact form of $(\mathcal{V}^{2m+1}, K)$. By Theorem 10.2.8, there exists a local coordinate system $(x_1, \ldots, x_m, y_1, \ldots, y_m, z)$ on U such that

$$\theta = dz + y_1 \, dx_1 + \cdots + y_m \, dx_m$$

holds. Then $\Pi \colon (x_1, \ldots, x_m, y_1, \ldots, y_m, z) \longmapsto (x_1, \ldots, x_m, z)$ gives a Legendrian fibration. $\qquad\square$

Remark 10.4.5. A local Legendrian fibration as in Proposition 10.4.4 may not be unique in general. In fact, $\theta := dz + y\,dx$ gives a contact form of $(\boldsymbol{R}^3; x, y, z)$, and taking a new coordinate system (u, v, w) given by

$$u = x - y, \quad v = y, \quad w = z + \frac{y^2}{2},$$

we get a new expression $\theta = dw + v\,du$. However, the map $(x, y, z) \mapsto (u, v, w)$ does not preserve the fibers of the original Legendrian fibration $(x, y, z) \mapsto (x, z)$.

Let $\mathcal{V}^{2m+1}$ be a contact manifold of dimension $2m + 1$. To give a global structure theorem for Legendrian fibrations, we now fix a Legendrian fibration $\Pi : \mathcal{V}^{2m+1} \to \mathcal{N}^{m+1}$. We fix a point $c \in \mathcal{V}^{2m+1}$, and set $p = \Pi(c)$. Since $\Pi^{-1}(p)$ is a Legendrian submanifold, we have

$$T_c\Pi^{-1}(p) \subset K_c.$$

The kernel of the differential $d\Pi$ of Π is m-dimensional, and $d\Pi(K_c)$ is a vector subspace of $T_p\mathcal{N}^{m+1}$ of codimension one (i.e., of dimension $m - 1$). Thus there exists a 1-form η on $\mathcal{N}^{m+1}$ such that $\eta_p \in T_p^*\mathcal{N}^{m+1}$ and

$$\operatorname{Ker}\eta_p = d\Pi(K_c).$$

We then obtain a C^∞-map

$$\Phi : \mathcal{V}^{2m+1} \ni c \mapsto [\eta_p] \in P(T^*\mathcal{N}^{m+1}). \tag{10.16}$$

The following assertion holds:

Theorem 10.4.6. *The map Φ is a local contact diffeomorphism such that*

$$\Pi = \hat{\pi} \circ \Phi, \tag{10.17}$$

where $\hat{\pi} : P(T^*\mathcal{N}^{m+1}) \to \mathcal{N}^{m+1}$ is the canonical projection.

Proof. We let $\dim \mathcal{V}^{2m+1} = 2m + 1$. Since (10.17) is clear, it is sufficient to show that Φ is a local contact diffeomorphism. We fix $c \in \mathcal{V}^{2m+1}$ arbitrarily. By the definition of Legendrian fibrations, there exists a simply connected local coordinate system $(\mathcal{U}; x_0, x_1, \ldots, x_m, y_1, \ldots, y_m)$ centered at c such that the projection Π is expressed by

$$(x_0, x_1, \ldots, x_m, y_1, \ldots, y_m) \mapsto (x_0, x_1, \ldots, x_m).$$

In particular, $(x_0, x_1, \ldots, x_m)$ gives a local coordinate system defined on a neighborhood of $p = \Pi(c) \in \mathcal{N}^{m+1}$. Then a local contact form θ of $\mathcal{V}^{2m+1}$ on $\mathcal{U}$ can be written as

$$\theta = \sum_{i=0}^{m} p_i\,dx_i + \sum_{j=1}^{m} q_j\,dy_j,$$

where p_i, q_j are C^∞-functions defined on $\mathcal{U}$. Suppose that $q_j \neq 0$ for some j. Then $\partial/\partial y_j$ does not belong to the kernel of θ, which contradicts the fact that Π is a Legendrian fibration. Thus we have $q_1 = \cdots = q_m = 0$. Since $\theta \neq 0$ at c (i.e., $\theta_c \neq 0$), by a suitable linear transformation of $(x_0, \ldots, x_m)$, we may assume that $p_0(c) \neq 0$ and $p_1(c) = \cdots = p_m(c) = 0$. Moreover, by rewriting $(1/p_0)\theta$ as θ, we can write

$$\theta_{\tilde{c}} = (dx_0)_{\tilde{c}} + \sum_{i=1}^{m} p_i(\tilde{c})(dx_i)_{\tilde{c}} \qquad (\tilde{c} \in \mathcal{U}). \tag{10.18}$$

Since

$$\theta \wedge \bigwedge^{m} d\theta = m! \, dx_0 \wedge (dp_1 \wedge dx_1) \wedge \cdots \wedge (dp_m \wedge dx_m)$$

does not vanish at each point,

$$dx_0, dx_1, \ldots, dx_m, \, dp_1, \ldots, dp_m$$

are linearly independent. So, taking $\mathcal{U}$ to be sufficiently small,

$$(x_0, x_1, \ldots, x_m, p_1, \ldots, p_m)$$

gives a new local coordinate system on $\mathcal{U}$. Using this, the projection Π can be rewritten as

$$\mathcal{U} \ni \tilde{c} := (x_0, x_1, \ldots, x_m, p_1, \ldots, p_m) \mapsto \Pi(\tilde{c}) := (x_0, x_1, \ldots, x_m) \in \mathcal{N}^{m+1}.$$

By (10.16) and (10.18), we have

$$\Phi(\tilde{c}) := \left[(dx_0)_{\Pi(\tilde{c})} + \sum_{i=1}^{m} p_i(\tilde{c})(dx_i)_{\Pi(\tilde{c})} \right] \in P(T^*_{\Pi(\tilde{c})} \mathcal{N}^{m+1}).$$

Since $(x_0, x_1, \ldots, x_m, p_1, \ldots, p_m)$ can be also considered as a local coordinate system of $P(T^*\mathcal{N}^{m+1})$ centered at $\Phi(c)$ satisfying (10.5), Φ gives a local contact diffeomorphism at c. $\qquad \square$

As a consequence, the following assertion holds, which asserts that projections of Legendrian submanifolds in $\mathcal{V}^{2m+1}$ via a Legendrian fibration can be considered as wave fronts in $\mathcal{N}^{m+1}$.

Corollary 10.4.7. *Let* $\Pi : \mathcal{V}^{2m+1} \to \mathcal{N}^{m+1}$ *be a Legendrian fibration, and let* $L : \mathcal{M}^m \to \mathcal{V}^{2m+1}$ *be a Legendrian immersion. Then* $f := \Pi \circ L$ *can be locally identified with a wave front in* $\mathcal{N}^{m+1}$.

Proof. In the proof of Theorem 10.4.6, we obtained a local contact diffeomorphism $\Phi : \mathcal{V}^{2m+1} \to P(T^*\mathcal{N}^{m+1})$. Then $\tilde{L} := \Phi \circ L$ gives a Legendrian immersion into $P(T^*\mathcal{N}^{m+1})$. Moreover, we have

$$f = \Pi \circ L = \hat{\pi} \circ \Phi \circ L = \hat{\pi} \circ \tilde{L},$$

which implies f can be expressed as a projection of the map $\tilde{L}$. Thus f is a wave front in the sense of Definition 10.3.4. $\square$

Since an n-manifold is locally identified with $\mathbf{R}^n$, we also get the following:

Corollary 10.4.8. *Projections of Legendrian submanifolds in a contact manifold of dimension $2m + 1$ can be locally identified with wave fronts in* $\mathbf{R}^{m+1}$.

Finally, we give the following important example of a Legendrian fibration:

Example 10.4.9 (Hopf fibration). Let $\mathrm{SO}(3)$ be the set of 3×3 special orthogonal matrices. If we set $A = (\boldsymbol{a}_1, \boldsymbol{a}_2, \boldsymbol{a}_3) \in \mathrm{SO}(3)$, then $\{\boldsymbol{a}_1, \boldsymbol{a}_2, \boldsymbol{a}_3\}$ is an orthonormal basis of $\mathbf{R}^3$. We set

$$\pi : \mathrm{SO}(3) \ni A := (\boldsymbol{a}_1, \boldsymbol{a}_2, \boldsymbol{a}_3) \longmapsto \boldsymbol{a}_1 \in S^2.$$

Since $\boldsymbol{a}_2 \in \mathbf{R}^3$ is perpendicular to $\boldsymbol{a}_1$, we may regard $\boldsymbol{a}_2 \in T_{\pi(A)}S^2$. We can write

$$\boldsymbol{a}_3 = \boldsymbol{a}_1 \times \boldsymbol{a}_2,$$

where $\times$ is the vector product, and so the matrix A is determined by $\boldsymbol{a}_1$ and $\boldsymbol{a}_2$. In particular, we can identify $\mathrm{SO}(3)$ with the unit sphere bundle $S(TS^2)$. So there is a canonical contact structure on $\mathrm{SO}(3)$. On the other hand, we set

$$\mathrm{SU}(2) := \left\{ \begin{pmatrix} z & -\overline{w} \\ w & \overline{z} \end{pmatrix} ; z, w \in \boldsymbol{C}, \ |z|^2 + |w|^2 = 1 \right\},$$

which can be identified with the unit 3-sphere $S^3 \subset \mathbf{R}^4 = \boldsymbol{C}^2$, where $\overline{w}, \overline{z}$ are complex conjugates of z, w. The Lie algebra of the Lie group $\mathrm{SU}(2)$ is given by

$$\mathfrak{su}(2) := \left\{ \begin{pmatrix} \sqrt{-1}t & -\overline{z} \\ z & -\sqrt{-1}t \end{pmatrix} ; t \in \boldsymbol{R}, \ z \in \boldsymbol{C} \right\},$$

which can be identified with $\mathbf{R}^3$, considering

$$e_1 := \begin{pmatrix} 0 & \sqrt{-1} \\ \sqrt{-1} & 0 \end{pmatrix}, \quad e_2 := \begin{pmatrix} 0 & -1 \\ 1 & 0 \end{pmatrix}, \quad e_3 := \begin{pmatrix} \sqrt{-1} & 0 \\ 0 & -\sqrt{-1} \end{pmatrix},$$

which is an orthonormal basis of $\boldsymbol{R}^3$ under the above identification. If we fix $a \in \mathrm{SU}(2)$, then

$$\rho(a) : \mathfrak{su}(2) \ni x \longmapsto axa^{-1} \in \mathfrak{su}(2)$$

gives an orientation preserving orthogonal transformation of $\boldsymbol{R}^3$. So we obtain a group homomorphism $\rho : S^3 \to \mathrm{SO}(3)$. We then have a double covering of $\mathrm{SO}(3)$. So we give a contact structure on S^3 such that $\pi \circ \rho : S^3 \to S^2$ gives a Legendrian fibration.

Appendix A

The Division Lemma

Here we prove the division lemma (Proposition A.1) for C^∞ functions of two variables, which is frequently used in this text.

Proposition A.1 (The Division Lemma). *Let $f(u,v)$ be a C^∞ function defined on a convex neighborhood U of the origin of $\mathbf{R}^2$. If there exists a non-negative integer k such that*

$$f(u,0) = \frac{\partial f}{\partial v}(u,0) = \frac{\partial^2 f}{\partial v^2}(u,0) = \cdots = \frac{\partial^k f}{\partial v^k}(u,0) = 0$$

holds for $(u,0) \in U$, then there exists a C^∞ function $g(u,v)$ defined on U such that

$$f(u,v) = v^{k+1} g(u,v) \quad ((u,v) \in U).$$

Proof. We prove this by induction. If $k = 0$, then since

$$f(u,v) = \left[f(u,tv) \right]_{t=0}^{t=1} = \int_0^1 \frac{df(u,tv)}{dt}\, dt$$

$$= \int_0^1 v f_v(u,tv)dt = v \int_0^1 f_v(u,tv)\, dt \quad \left(f_v := \frac{\partial f}{\partial v} \right)$$

holds, $g(u,v) := \int_0^1 f_v(u,tv)\, dt$ gives the desired function.

Next, we assume that the assertion holds for $k - 1 \geq 0$, and prove it for the case k. Using the inductive hypothesis, we apply the case of $k - 1$, so there exists a C^∞ function $h(u,v)$ defined on U such that

$$f(u,v) = v^k h(u,v) \quad ((u,v) \in U). \tag{A.1}$$

Taking the partial derivative with respect to v of (A.1) k-times, and substituting $v = 0$, we have

$$0 = \frac{\partial^k f}{\partial v^k}(u,0) = k! h(u,0).$$

Thus, $h(u, 0) = 0$. Using the case $k = 0$, we see there exists a C^∞ function $g(u, v)$ defined on U such that $h(u, v) = vg(u, v)$ $\big((u, v) \in U\big)$. Substituting this into (A.1), we see that $g(u, v)$ is the desired function. $\qquad\square$

Corollary A.2. *For a C^∞ function $f(u, v)$ defined on a convex neighborhood U of $\mathbf{R}^2$, there exist a C^∞ function $h_1(u)$ depending only on u and a C^∞ function $h_2(u, v)$ of two variables such that $f(u, v) = h_1(u) + vh_2(u, v)$.*

Proof. Considering the function $g(u, v) := f(u, v) - f(u, 0)$, then we see that $g(u, 0) = 0$. Thus g satisfies the assumption of the division lemma (Proposition A.1). Hence there exists a C^∞ function $h_2(u, v)$ such that $g(u, v) = vh_2(u, v)$. Setting $h_1(u) := f(u, 0)$, we have the assertion. $\qquad\square$

When we apply Proposition A.1 to a plane curve $\gamma(t) = (x(t), y(t))$, we need to apply it by setting $f(u, v) = x(u)$ or $f(u, v) = y(u)$. The following corollary is useful for this purpose:

Corollary A.3. *Let $f(t)$ be a C^∞ function defined on an open interval $I \subset \mathbf{R}$ containing the origin 0. If $f(0) = f'(0) = f''(0) = \cdots = f^{(k)}(0) = 0$ for a non-negative integer k, then there exists a C^∞ function g defined on I such that*

$$f(t) = t^{k+1}g(t) \qquad (t \in I).$$

Furthermore, the following "division lemma for general degree" holds.

Lemma A.4. *Let $\varphi(t)$ be a C^∞ function defined on an open interval $I \subset \mathbf{R}$ containing the origin 0. For any positive number α, the function*

$$g(t) := \frac{f(t)}{\operatorname{sgn}(t)|t|^{1+\alpha}} \qquad \left(f(t) := \int_0^t |u|^\alpha \varphi(u)\, du \right)$$

is a C^∞ function on I. Moreover, it holds that

$$g(0) = \frac{\varphi(0)}{1 + \alpha}. \tag{A.2}$$

Proof. We have

$$f(t) = \int_0^1 \frac{df(tu)}{du}\, du = \int_0^1 tf'(tu)\, du$$

$$= \int_0^1 t|tu|^\alpha \varphi(tu)\, du = \operatorname{sgn}(t)|t|^{1+\alpha} \left(\int_0^1 |u|^\alpha \varphi(tu)\, du \right).$$

Noting that $\alpha > 0$, the function

$$g(t) = \frac{f(t)}{\operatorname{sgn}(t)|t|^{1+\alpha}} = \int_0^1 |u|^\alpha \varphi(tu)\, du$$

is a C^∞ function at $t = 0$. Substituting $t = 0$ into the above equation, and calculating the definite integral on the right-hand side, we have (A.2). $\square$

Lemma A.4 can be regarded as a variant of Corollary A.3. In fact, if $\alpha = 0$, then $g(t) = f(t)/t$, and this is the $k = 0$ case of Corollary A.3.

Appendix B

Topics on Cusps

B.1. Normal Form for Cusps

In this appendix, we introduce the concept of generalized cusps, and give a useful representation formula for them.

Definition B.1.1. Let $t = c$ be a singular point of a smooth planar curve $\gamma(t)$ defined on an open interval I. Then $t = c$ is called a *generalized cusp* if $\gamma''(c) \neq \mathbf{0}$ (this condition is independent of the choice of the parameter t).

Cusps and $5/2$-cusps are typical examples of generalized cusps. Moreover, if $\gamma(t)$ is a regular curve defined on an interval containing $t = 0$, then $t \mapsto \gamma(t^2)$ gives a generalized cusp at $t = 0$. We show the following proposition.

Proposition B.1.2. *Suppose that the planar curve $\gamma(t)$ has a generalized cusp at $t = 0$. Then γ is a frontal (cf. Definition 1.5.1) on an open interval $(-\varepsilon, \varepsilon)$ for sufficiently small $\varepsilon(> 0)$. Moreover, γ is a wave front on $(-\varepsilon, \varepsilon)$ for sufficiently small $\varepsilon(> 0)$ if and only if*

$$\det\big(\gamma''(0), \gamma'''(0)\big) \neq 0, \tag{B.1}$$

that is, $t = 0$ is a cusp (cf. Theorem 1.3.2 in Chapter 1).

Proof. Without loss of generality, we may assume that $\gamma(0) = \gamma'(0) = \mathbf{0}$. By the division lemma (cf. Corollary A.3 in Appendix A), there exist C^∞ functions a and b such that $\gamma(t) = t^2(a(t), b(t))$. Then

$$\tilde{\boldsymbol{n}}(t) := (-2b(t) - tb'(t), 2a(t) + ta'(t)) \tag{B.2}$$

is perpendicular to $\gamma'(t)$ for each t. Since $t = 0$ is a generalized cusp, $\tilde{\boldsymbol{n}}(0) = (a(0), b(0))$ is not the zero vector, and so γ is a frontal at $t = 0$.

299

We next prove the second assertion. It holds that

$$\Delta := \det(\gamma''(0), \gamma'''(0)) = 12 \det \begin{pmatrix} a(0) & a'(0) \\ b(0) & b'(0) \end{pmatrix}. \tag{B.3}$$

We set $n := \tilde{n}/|\tilde{n}|$. Since

$$\tilde{n}(0) = (-2b(0), 2a(0)), \quad \tilde{n}'(0) = (-3b'(0), 3a'(0)),$$

$\tilde{n}(0)$ and $\tilde{n}'(0)$ are linearly independent if and only if $\Delta \neq 0$. So if $\Delta \neq 0$, then $n'(t)$ does not vanish at $t = 0$, and we can conclude that $\gamma(t)$ is a wave front on an open interval containing $t = 0$.

On the other hand, if $\Delta = 0$, then $\tilde{n}(0)$ and $\tilde{n}'(0)$ are linearly dependent. Then $n(0)$ and $n'(0)$ are linearly dependent. Since n is a unit vector field, $n'(0)$ must be perpendicular to $n(0)$, and so $n'(0) = \mathbf{0}$. $\qquad\square$

From the above proof, we obtain the following:

Corollary B.1.3. *The smoothly chosen unit normal vector field of $\gamma(t)$ changes from leftward (respectively, rightward) to rightward (respectively, leftward) at the generalized cusp.*

Proof. From the above proof, $n(t) := \tilde{n}(t)/|\tilde{n}(t)|$ gives the smooth unit normal vector field of $\gamma(t)$. One can easily show the identity

$$\det(\gamma'(t), \tilde{n}(t)) = t(4(a^2 + b^2) + 4t(aa' + bb') + t^2\left(a'^2 + b'^2\right)).$$

Since $a(0)^2 + b(0)^2 \neq 0$, this implies the assertion of the corollary. $\qquad\square$

Theorem B.1.4 (Normal form for generalized cusps). *Suppose the planar curve $\gamma(t)$ has a generalized cusp at $t = 0$. Then there exist a parameter t and an orientation preserving Euclidean motion (i.e., a composition of a parallel translation and a rotation around the origin) T and a C^∞ function $\psi(t)$ defined on a neighborhood of $t = 0$ such that*

$$T \circ \gamma(t) = (t^2, t^3\psi(t)). \tag{B.4}$$

Moreover, $t = 0$ is a cusp if and only if $\psi(0) \neq 0$.

The form (B.4) is called a *normal form* of a generalized cusp.

Proof. By a parallel translation in $\mathbf{R}^2$, we may assume $\gamma(0) = \mathbf{0}$. Since $t = 0$ is a generalized cusp, $\gamma''(0) \neq \mathbf{0}$ holds. By a rotation around the origin, we may assume that $\gamma''(0)$ points into the direction of the positive x-axis. We set $\gamma(t) = \big(x(t), y(t)\big)$. Then

$$x(0) = x'(0) = 0, \quad x''(0) > 0, \quad y(0) = y'(0) = y''(0) = 0.$$

By Corollary A.3, we can write

$$x(t) = t^2 a(t), \quad y(t) = t^3 b(t) \quad \big(a(0) > 0\big),$$

where $a(t)$ and $b(t)$ are C^∞ functions defined near $t = 0$. Since $a(0) > 0$, the map $t \mapsto \sqrt{a(t)}$ is C^∞ differentiable at $t = 0$. We set $s := t\sqrt{a(t)}$. Then by $ds/dt|_{t=0} = \sqrt{a(0)} > 0$, we can take s as a new parameter of the curve $\gamma(t)$, and γ is written as

$$\gamma(s) = (s^2, t(s)^3 b(t(s))),$$

using the inverse function $t = t(s)$ of the function $s = s(t)$. Since $s(0) = 0$, it holds that $t(0) = 0$. By Corollary A.3, the function $t(s)$ can be written in the form $t(s) = s\varphi(s)$, where $\varphi(s)$ is a C^∞ function defined on a neighborhood of $s = 0$. Thus γ can be written in the form

$$\gamma(s) = (s^2, s^3 \psi(s)), \tag{B.5}$$

where $\psi(s) := \varphi(s)^3 b\big(t(s)\big)$. The last assertion is obvious. $\quad\square$

By a direct calculation, the cuspidal curvature (Definition 1.3.9 in Chapter 1) of the normal form $\gamma(t) = \big(t^2, t^3 \psi(t)\big)$ (cf. (B.4)) is $\mu = 3\psi(0)/\sqrt{2}$. We consider the case that γ has a cusp at $t = 0$. Since the cycloid having the given cuspidal curvature is uniquely determined, up to a rotation and a parallel translation, as in Example 1.3.13, we have the following corollary.

Corollary B.1.5. *Let a planar curve $\gamma(t)$ have a cusp at $t = c$. Then the absolute value of the inverse of the cuspidal curvature of γ at $t = c$ coincides with the square of the radius of the cycloid which gives the best approximation of that cusp at a cusp of the cycloid.*

At a cusp, we have a normal vector. We call the line passing through the cusp and perpendicular to the normal vector the *center line* of the cusp (see Fig. B.1).

The center line of the standard cusp $\gamma_0(t) = (t^2, t^3)$ is the x-axis, and the center line of the cycloid (1.1) in Chapter 1 at the origin is the y-axis.

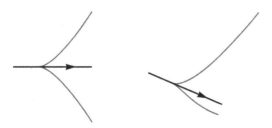

Fig. B.1. Center lines of cusps.

We can give an orientation to the center line so that line points into the inner region of the cusp.

Proposition B.1.6. *If a planar curve $\gamma(t)$ has a cusp at $t = 0$, then the center line divides the curve into two parts, one on each side.*

Proof. The center line of the normal form of the cusp (B.4) is the x-axis. If the cusp is zig (i.e., $\mu > 0$, see Section 1.3 in Chapter 1), then since $\psi(0) > 0$, it follows that the curve moves from the lower region of the x-axis into the upper region of the axis x-axis. Thus the center line divides the curve at the singular point. One can show the case that the cusp is zag by the same argument.   □

In the case of a 5/2-cusp, Proposition B.1.6 does not hold in general. There exists an example where the image lies in one half-plane bounded by the center line at the singular point (Exercise **2** in this section).

B.2.   Half-arc-length Parameters for Generalized Cusps

For regular curves in $\boldsymbol{R}^2$, the "fundamental theorem for planar curves" (Theorem 1.2.3) holds. We would like to generalize this theorem to planar curves with generalized cusps. The arc-length parameter is not differentiable at generalized cusps, but we can show the square root of the arc-length parameter is C^∞ by using Lemma A.4 in Appendix A. Moreover, we give a generalization of the fundamental theorem for planar curves using this special parametrization.

Proposition B.2.1 (**[75]**). *Suppose that a planar curve $\gamma(t)$ has a generalized cusp at $t = 0$. Then the square root of the arc-length function*

$$\tau(t) := \operatorname{sgn}(t)\sqrt{|s(t)|}, \quad s(t) := \int_0^t |\gamma'(u)|\, du, \qquad \text{(B.6)}$$

from $t = 0$ gives a C^∞ function satisfying $d\tau(t)/dt > 0$. Thus, τ can be considered as a parameter of γ, and $\tau = 0$ corresponds to the generalized cusp.

We call τ the *half-arc-length parameter* of γ at the generalized cusp.

Proof. Since $\gamma'(0) = \boldsymbol{0}$, by Corollary A.3 in Appendix A,

$$\boldsymbol{v}(t) := \frac{\gamma'(t)}{t} \qquad \text{(B.7)}$$

is a vector-valued C^∞ function on a neighborhood of $t = 0$, and satisfies $v(0) = \gamma''(0) \neq \mathbf{0}$. Thus, the function $\varphi(t) := |v(t)|$ is a C^∞ function defined on a neighborhood of $t = 0$. Then we have $s(t) = \int_0^t |u|\varphi(u)\, du$. By the division lemma for general degree (Lemma A.4 in Appendix A),

$$\Psi(t) := \frac{s(t)}{\operatorname{sgn}(t)t^2} = \frac{\operatorname{sgn}(t)s(t)}{t^2} = \frac{|s(t)|}{t^2} \tag{B.8}$$

is a C^∞ function and, by (1.30) in Chapter 1 and the property $\gamma''(0) \neq \mathbf{0}$, we have $\Psi(0) > 0$. Thus $\sqrt{\Psi(t)}$ is a C^∞ function defined on a neighborhood of $t = 0$. Since

$$\tau(t) = \operatorname{sgn}(t)\sqrt{|s(t)|} = \operatorname{sgn}(t)|t|\sqrt{\Psi(t)} = t\sqrt{\Psi(t)},$$

$\tau(t)$ is also a C^∞ function. Moreover, since

$$\left.\frac{d\tau(t)}{dt}\right|_{t=0} = \sqrt{\Psi(0)} > 0,$$

$\tau(t)$ is strictly monotone increasing near $t = 0$. □

In the above situation, we set

$$\mu(t) := \begin{cases} 2\sqrt{2}\,\kappa(t)\sqrt{|s(t)|} & (\text{if } t \neq 0), \\ \mu_0 & (\text{if } t = 0), \end{cases} \tag{B.9}$$

where μ_0 is the cuspidal curvature of $\gamma(t)$ at $t = 0$ (see Definition 1.3.9 in Chapter 1) and $\kappa(t)$ is the curvature function of $\gamma(t)$ with respect to the leftward unit normal vector field $n_L(t)$ of $\gamma(t)$ for $t \neq 0$. By Theorem 1.3.8 in Chapter 1, $\mu(t)$ is continuous at $t = 0$. We call $\mu(t)$ the *normalized curvature function* or *extended cuspidal curvature function*. We can prove the following assertion:

Proposition B.2.2 ([75]). *Let a planar curve $\gamma(t)$ have a generalized cusp at $t = 0$. Then the normalized curvature function $\mu(t)$ is a C^∞ function near $t = 0$ and coincides with the cuspidal curvature[1] of $\gamma(t)$ at $t = 0$.*

Proof. We may assume that t is a half-arc-length parameter of γ and denote it by τ. We may also assume $\gamma(\tau)$ is defined on $(-\varepsilon, \varepsilon)$, where $\varepsilon > 0$. Since a generalized cusp is a frontal (cf. Proposition B.1.2), we can take a smooth unit normal vector field $n(\tau)$. By Corollary B.1.3, it holds that

$$n(\tau) := \begin{cases} n_L(\tau) & (\tau \geq 0), \\ -n_L(\tau) & (\tau < 0). \end{cases}$$

---

[1]If $t = 0$ is a generalized cusp but not a cusp, then we define the cuspidal curvature to be zero, since $\mu(0)$ is zero.

By (B.6), we have

$$s = \operatorname{sgn}(s)|s| = \operatorname{sgn}(s)(\sqrt{|s|})^2 = \operatorname{sgn}(s)\tau^2 = \operatorname{sgn}(\tau)\tau^2$$

and $ds/d\tau = 2\operatorname{sgn}(\tau)\tau$. We let $e(\tau)$ be the smooth unit tangent vector field of $\gamma(\tau)$ which is obtained by the counterclockwise 90°-rotation of the smooth vector field $n(\tau)$. Then we have

$$\det(e(\tau), n(\tau)) = 1. \tag{B.10}$$

In particular, if s is the arc-length parameter of γ, we have

$$e(s) = \begin{cases} d\gamma/ds & \text{if } s > 0, \\ -d\gamma/ds & \text{if } s > 0. \end{cases}$$

Moreover, it holds that

$$\frac{d\gamma}{d\tau} = \frac{ds}{d\tau}\frac{d\gamma}{ds} = (2\operatorname{sgn}(\tau)\tau)(\operatorname{sgn}(\tau)e(\tau)) = 2\tau e(\tau) \tag{B.11}$$

and the formula $d^2\gamma/ds^2 = \kappa(s)n_L(s)$ yields that

$$\frac{de(s)}{ds} = \operatorname{sgn}(s)\kappa(s)n_L(s).$$

In particular, we have

$$\frac{de}{d\tau} = \frac{ds}{d\tau}\frac{de}{ds} = (2\operatorname{sgn}(\tau)\tau)(\operatorname{sgn}(\tau)\kappa(\tau)n_L(\tau)) = 2\tau\operatorname{sgn}(\tau)\kappa(\tau)n(\tau)$$

for $\tau \neq 0$. Since $2\sqrt{2}\tau\operatorname{sgn}(\tau)\kappa(\tau) = \mu(\tau)$ (see (B.9)), the continuity of μ at $t = 0$ yields that we have

$$\frac{de(\tau)}{d\tau} = \frac{\mu(\tau)}{\sqrt{2}}n(\tau) \tag{B.12}$$

holds for $t \in (-\varepsilon, \varepsilon)$, which implies that $\mu(\tau)$ is a smooth function.   □

What is important here is that when we choose τ as a parameter of the curve γ, the function $\mu(\tau)$ is invariant under rotations and parallel translations of the curve. Moreover, by Theorem 1.3.8 in Chapter 1, $\mu(0)$ coincides with the cuspidal curvature defined in (1.33) in Chapter 1. The following theorem can be considered as the cusp version of the fundamental theorem for planar curves (cf. Theorem 1.2.3).

Theorem B.2.3 ([13, 75]). *Let $\mu(\tau)$ be a C^∞ function defined on a neighborhood of $\tau = 0$. Then there exists a planar curve such that τ is a half-arc-length parameter, $\mu(\tau)$ is the normalized curvature function and*

$\tau = 0$ *is a generalized cusp. Moreover, after a suitable parallel translation and rotation, such a curve coincides with*

$$\gamma(\tau) := 2 \int_0^\tau \left(u \cos \theta(u), u \sin \theta(u) \right) du, \qquad \text{(B.13)}$$

$$\left(\theta(\tau) := \frac{1}{\sqrt{2}} \int_0^\tau \mu(u)\, du \right).$$

The unit normal vector of $\gamma(\tau)$ *is given by* $\boldsymbol{n}(\tau) := (-\sin\theta(\tau), \cos\theta(\tau))$. *In this situation,* γ *gives a cusp at* $\tau = 0$ *if and only if* $\mu(0) \neq 0$.

Remark B.2.4. Kossowski [13] showed a formula similar to (B.13) using differential forms, where the cuspidal curvature and normalized curvature do not appear in his formula. The above formula using the normalized curvature is given in Shiba–Umehara [75].

Proof. If we set $\boldsymbol{e}(\tau) = (a(\tau), b(\tau))$ then (cf. (B.10))

$$\boldsymbol{n}(\tau) = (-b(\tau), a(\tau)).$$

So, (B.12) implies

$$\frac{d\boldsymbol{n}(\tau)}{d\tau} = -\frac{-\mu(\tau)}{\sqrt{2}} \boldsymbol{e}(\tau). \qquad \text{(B.14)}$$

Since (B.11), (B.12) and (B.14) can be considered as a system of ordinary differential equations with unknown vector valued functions $\gamma, \boldsymbol{e}$ and $\boldsymbol{n}$, we obtain the uniqueness of the curve determined by the function μ.

On the other hand, it can be easily checked that (B.13) with

$$\boldsymbol{e}(\tau) = (\cos\theta(\tau), \sin\theta(\tau)), \quad \boldsymbol{n}(\tau) = (-\sin\theta(\tau), \cos\theta(\tau))$$

satisfy (B.11), (B.12) and (B.14), and we obtain the assertion.

We have not yet used the assumption $\mu(0) \neq 0$, and this condition implies that $\gamma(\tau)$ has a cusp at $\tau = 0$. $\qquad \square$

By the formula (B.13), we have

$$\mu(\tau) = \sqrt{2} \frac{\det(\dot{\gamma}(\tau), \ddot{\gamma}(\tau))}{|\dot{\gamma}(\tau)|^2},$$

where the dot stands for the derivative with respect to the half-arc-length parameter τ.

Corollary B.2.5 ([75]). *Let* $\gamma(t)$ *be a smooth curve in* $\boldsymbol{R}^2$ *such that* $t = 0$ *is a generalized cusp. Then* t *is a half-arc-length parameter if and only if*

$$|\gamma'(t)| = 2|t|. \qquad \text{(B.15)}$$

Proof. If t is the half-arc-length parameter, then we can prove (B.15) by differentiating (B.13). Conversely, we suppose that $\gamma(t)$ satisfies (B.15). Then by (B.6),

$$s(t) = \int_0^t |\gamma'(u)|\, du = \int_0^t 2|t|\, du = \operatorname{sgn}(t) t^2.$$

Thus, we have

$$\tau(t) = \operatorname{sgn}(t)\sqrt{|s(t)|} = t,$$

proving the assertion.   □

Remark B.2.6. Let $\gamma(t)$ be a smooth curve in $\mathbf{R}^2$ such that $t = 0$ is a singular point. If (B.15) holds, then $t = 0$ must be a generalized cusp. In other words, the existence of the half-arc-length parameter characterizes generalized cusps.

Remark B.2.7. If one wishes to avoid the constants 2 and $1/\sqrt{2}$, it might be better to use the parameter

$$v := \sqrt{2}\tau,$$

which was introduced by Fukui [16]. We call this the *normalized half-arc-length parameter*. If we use this parametrization, then $\gamma(v)$ satisfies $|d\gamma/dv| = |v|$, and the representation formula (B.13) can be rewritten as

$$\gamma(v) := \int_0^v (u\cos\theta(u), u\sin\theta(u))\, du, \quad \left(\theta(v) := \int_0^v \hat{\mu}(u)\, du\right), \quad \text{(B.16)}$$

where

$$\hat{\mu}(v) = \frac{\mu(v/\sqrt{2})}{2}$$

is called the *extended half-cuspidal curvature function* (cf. [30]). By (B.16), we have

$$\hat{\mu}(v) = \frac{\det(\gamma_v(v), \gamma_{vv}(v))}{|\gamma_v(v)|^2},$$

where the subscript v stands for the derivative with respect to the normalized half-arc-length parameter v.

Corollary B.2.8. *Suppose that $\gamma_1(t)$ (respectively, $\gamma_2(s)$) satisfies the condition in the criteria for a cusp (1.20) at $t = 0$ (respectively, $s = 0$). Then the following two conditions are equivalent:*

(1) the image of one curve is congruent to the image of the other curve via an orientation preserving isometry of $\mathbf{R}^2$,

(2) the two normalized curvatures (cf. (B.9)) of the two curves coincide.

Proof. Without loss of generality, we may assume that the two curves are parametrized by the common half-arc-length parameter τ. Then the fact that (2) implies (1) is a consequence of Theorem B.2.3.

Thus, to show the corollary here, it is enough to show that (1) implies (2). Let $\mu_i(\tau)$ ($i = 1, 2$) be the normalized curvatures of the two curves $\gamma_i(\tau)$. If these two functions do not coincide, then there exists a sufficiently small τ_0 such that $\mu_1(\tau_0) \neq \mu_2(\tau_0)$. If $\tau_0 = 0$, then the two cusps have different cuspidal curvature, and this is a contradiction. Thus, we may assume $\tau_0 \neq 0$. Since τ is the half-arc-length parameter, the formula (B.9) tells us that the two curvatures at the points where the arc-lengths are distance τ_0^2 from the singular points are different. This is also a contradiction. □

Example B.2.9. Since the arc-length parameter of the curve $\gamma(t) = a(t^2, t^3)$ ($a > 0$) is

$$s(t) = \frac{a}{27}(\varphi(t) - 8) \qquad \left(\varphi(t) := \left(9t^2 + 4\right)^{3/2}\right),$$

the half-arc-length parameter is $\tau = \mathrm{sgn}(t)\sqrt{a}\sqrt{\varphi(t) - 8}/(3\sqrt{3})$. The normalized curvature is

$$\mu(t) = \frac{2\,\mathrm{sgn}(t)\sqrt{\varphi(t) - 8}}{\sqrt{3a}\,t\,\varphi(t)},$$

and this is a C^∞ function.

Example B.2.10. The curve

$$\sigma_a(\tau) = \frac{1}{2a^2}\left(2a\tau \sin(2a\tau) + \cos(2a\tau), \sin(2a\tau) - 2a\tau \cos(2a\tau)\right) \quad (a > 0)$$

is parametrized by a half-arc-length parameter, and the normalized curvature is the constant $2\sqrt{2}a$ (cf. Fig. B.2).

The cuspidal curvature, the half-arc-length parameter and the normalized curvature can be defined for curves in general two-dimensional manifolds. There are also affine geometric analogies of Theorem B.2.3 for cusps and inflection points; see [71, 75] for details.

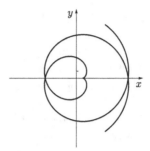

Fig. B.2.   Example B.2.10 (the case of $a = 1$).

Exercises B

1 Show the normalized curvature of the cycloid ((1.1) in Chapter 1) parametrized by half-arc-length at the origin is

$$\mu(\tau) = -\frac{2\sqrt{2}}{\sqrt{a(1 - \tau^2/(8a))}} = -\frac{1}{\sqrt{a}} - \frac{\tau^2}{16\sqrt{a}^3} + O\left(\tau^3\right).$$

In particular, show that the cuspidal curvature is given by $-1/\sqrt{a}$. Confirm that the half-arc-length is $\tau(t) = 2\sqrt{a}\,\mathrm{sgn}(t)\sqrt{1 - \cos(t/2)}$, and the normalized curvature is $-\dfrac{1}{2\sqrt{a}\cos(t/4)}$.

2 Show $\gamma(t) = (t^2, t^4 + t^5)$ has a 5/2-cusp at $t = 0$ (by using (1.21) in Chapter 1), and see that the image of γ near $t = 0$ lies in one half-plane determined by the center line of γ at $t = 0$.

3 Regarding the integral formula (B.13) in Theorem B.2.3, show that $\gamma(\tau)$ at $\tau = 0$ is a 5/2-cusp if and only if $\mu(0) = 0$, $\mu''(0) \neq 0$, where τ is the half-arc-length parameter.

Appendix C

A Criterion for 4/3-Cusps

Here, we show that a necessary and sufficient condition for a map-germ $\gamma : (\boldsymbol{R}, 0) \to (\boldsymbol{R}^2, 0)$ to be right–left equivalent to $t \mapsto (t^3, t^4)$ is

$$\gamma'(0) = \gamma''(0) = \boldsymbol{0}, \quad \det(\gamma'''(0), \gamma^{(4)}(0)) \neq 0 \qquad \text{(C.1)}$$

(Theorem 1.3.4 in Chapter 1). Clearly the standard 4/3-cusp $t \mapsto (t^3, t^4)$ satisfies (C.1), and one can easily check that the condition does not depend on the choice of coordinate systems on the source and target spaces. Thus it is enough to show that a map-germ satisfying the above condition is right-left equivalent to the standard 4/3-cusp. Since $\gamma'(0) = \gamma''(0) = \boldsymbol{0}$, γ has the form

$$\gamma(t) = (a_3 t^3 + a_4 t^4 + a_5 t^5, b_3 t^3 + b_4 t^4 + b_5 t^5) + h_6(t),$$

where $h_6(t)$ is a $\boldsymbol{R}^2$-valued C^∞ function satisfying

$$h_6(0) = h_6'(0) = \cdots = h_6^{(5)}(0) = \boldsymbol{0}. \qquad \text{(C.2)}$$

With the condition $\det(\gamma'''(0), \gamma^{(4)}(0)) \neq 0$, we may assume that γ has the form

$$\gamma(t) = (t^3 + \hat{a}_5 t^5, t^4 + \hat{b}_5 t^5) + \hat{h}_6(t), \qquad \text{(C.3)}$$

by considering a composition of γ and a linear coordinate transformation in the target space $\boldsymbol{R}^2$, where $\hat{h}_6(t)$ is a $\boldsymbol{R}^2$-valued C^∞-function satisfying the same condition as (C.2). We set

$$c_1 := -\frac{\hat{b}_5}{4}, \quad c_2 := -\frac{16\hat{a}_5 + 3\hat{b}_5^2}{48},$$

and take $\tilde{t}$ satisfying

$$t = \tilde{t} + c_1 \tilde{t}^2 + c_2 \tilde{t}^3. \qquad \text{(C.4)}$$

Substituting (C.4) into (C.3), and looking at the terms with degrees less than or equal to 5 with respect to $\tilde{t}$, we have

$$\gamma(\tilde{t}) = \left(\tilde{t}^3 - \frac{3b_5\tilde{t}^4}{4}, \tilde{t}^4\right) + \check{h}_6(\tilde{t}),$$

where $\check{h}_6(\tilde{t})$ is a $\boldsymbol{R}^2$-valued C^∞ function satisfying the condition (C.2). By considering a linear coordinate transformation of $\boldsymbol{R}^3$, we see that γ is right–left equivalent to $(\tilde{t}^3, \tilde{t}^4) + \check{h}_6(\tilde{t})$. Renaming the terms involved, we may assume

$$\gamma(t) = (t^3, t^4) + h(t), \tag{C.5}$$

where $h(t)$ is a $\boldsymbol{R}^2$-valued C^∞ function satisfying the condition (C.2).

As we saw in the proof of Theorem 1.3.2 in Chapter 1, by Theorem 3.1.7 and Corollary 3.1.12 in Chapter 3, for any smooth function $f(t)$, there exist C^∞ functions f_1, f_2 such that $f(t) = f_1(t^2) + tf_2(t^2)$, and applying the same argument to f_1 and f_2, we see there exist C^∞ functions $f_{11}, f_{12}, f_{21}, f_{22}$ such that

$$f(t) = f_{11}(t^4) + tf_{21}(t^4) + t^2 f_{12}(t^4) + t^3 f_{22}(t^4). \tag{C.6}$$

By setting $h = (h_1, h_2)$ (cf. (C.5)), and by applying the formula (C.6) for $f = h_i$ ($i = 1, 2$), there exist $\boldsymbol{R}^2$-valued C^∞ functions g_1, g_2, g_3, g_4 such that

$$h(t) = g_1(t^4) + tg_2(t^4) + t^2 g_3(t^4) + t^3 g_4(t^4).$$

Since $h(t)$ satisfies the condition (C.2), we have

$$g_1(0) = g_2(0) = g_3(0) = g_4(0) = 0, \quad g_1'(0) = g_2'(0) = 0.$$

Thus, by the division lemma (Lemma A.3), there exist $\boldsymbol{R}^2$-valued C^∞ functions $\tilde{g}_1, \tilde{g}_2, \tilde{g}_3, \tilde{g}_4$ such that

$$g_1(x) = x^2\tilde{g}_1(x), \quad g_2(x) = x^2\tilde{g}_2(x), \quad g_3(x) = x\tilde{g}_3(x), \quad g_4(x) = x\tilde{g}_4(x).$$

Hence $h(t)$ can be written as

$$h(t) = t^8\tilde{g}_1(t^4) + t^9\tilde{g}_2(t^4) + t^6\tilde{g}_3(t^4) + t^7\tilde{g}_4(t^4).$$

We set

$$\varphi(X, Y) := Y^2\tilde{g}_1(Y) + X^3\tilde{g}_2(Y) + X^2\tilde{g}_3(Y) + XY\tilde{g}_4(Y).$$

Then

$$\varphi(t^3, t^4) = h(t). \tag{C.7}$$

We consider a coordinate transformation $\Psi(X, Y) := (X, Y) + \varphi(X, Y)$. By the definition of φ,

$$\Psi : (\boldsymbol{R}^2, \boldsymbol{0}) \to (\boldsymbol{R}^2, \boldsymbol{0})$$

is a local diffeomorphism. By (C.7), it holds that $\Psi(t^3, t^4) = \gamma(t)$. Thus, $\gamma(t)$ is right–left equivalent to (t^3, t^4).

Proof of the Criterion for Whitney Cusps

Here, we give a proof of the criterion for Whitney cusps (Theorem 4.4.2) introduced in Chapter 4.

The criterion Theorem 4.4.2 we stated in Chapter 4 does not depend on the choice of coordinate system on the domain, nor that on the target, nor on the identifier of singularities and extended null vector field. To show the necessity of the criterion, it is enough to show that the standard Whitney cusp satisfies the condition of Theorem 4.4.2 in Chapter 4. We already saw this in Example 4.4.4 in Chapter 4. Thus, here we need to show only the sufficiency of the criterion.

Let U be a domain in $\boldsymbol{R}^2$ containing the origin. In the following discussions, we consider maps from U to $\boldsymbol{R}^2$, and we denote by (x, y) the standard coordinate system of $\boldsymbol{R}^2$ as the target space, unless otherwise stated. Also, we denote by (u, v) the local coordinate system on U near the origin.

In this appendix, we prove the following theorem:

Theorem D (Sufficiency of Theorem 4.4.2). *Let* $f \colon (\boldsymbol{R}^2, 0) \to (\boldsymbol{R}^2, 0)$ *be a* C^∞ *germ satisfying the following conditions:*

(i) *the origin is a non-degenerate singular point (cf. Section 4.4),*

(ii) *an identifier of singularities (cf. Section 4.4)* Λ *satisfies* $\Lambda_\eta(0) = 0$ *and* $\Lambda_{\eta\eta}(0) \neq 0$, *where* $\tilde{\eta}$ *is an extended null vector field and the subscript* η *stands for differentiation with respect to* $\tilde{\eta}$.

Then there exist a local coordinate system $(\tilde{u}, \tilde{v})$ *defined on a neighborhood of the origin and a local diffeomorphism* Φ *at the origin of the target space such that*

$$\Phi \circ f(\tilde{u}, \tilde{v}) = (\tilde{u}\tilde{v} - \tilde{u}^3, \tilde{v}). \tag{D.1}$$

In Example 4.4.4 in Chapter 4, we considered the standard Whitney cusp $f(u, v) = (u^3 - 3uv, v)$. With the coordinate transformation $(u, v) \mapsto (-u, v/3)$ and the diffeomorphism $\Phi : (x, y) \mapsto (x, 3y)$ on the target, one can change f to the form (D.1), and so if Theorem D is shown, then the criterion for Whitney cusps (Theorem 4.4.2 in Chapter 4) is shown.

The following proof is due to Whitney [91]. Although a bit complicated, the method is basically the same as that of the proofs of the criteria for cusps, cross caps and cuspidal edges given in Chapter 3. We show that for a map-germ f satisfying the condition, we can find a suitable coordinate system on the domain and diffeomorphism on the target space such that f is changed into the right-hand side of (D.1).

D.1. Introducing Notations

In the proof of Theorem D, to represent the higher order terms, we introduce the following notation: Let $h(u, v)$ be a C^∞ function of two variables defined on a neighborhood of $(u, v) = (0, 0)$. The function $h(u, v)$ *belongs to the set of functions* $\mathcal{O}(d)$ if

$$\frac{|h(u, v)|}{(u^2 + v^2)^{d/2}} \quad (d \text{ is a non-negative integer})$$

is bounded near $(0, 0)$. By definition, $\mathcal{O}(d)$ is closed under the operations of summation and scalar multiplication. Moreover, if $h_1 \in \mathcal{O}(d_1)$ and $h_2 \in \mathcal{O}(d_2)$, then the product $h_1 h_2$ belongs to $\mathcal{O}(d_1 + d_2)$. If $h \in \mathcal{O}(d)$, then at most $(d - 1)$-times derivatives of $h(u, v)$ vanish at the origin. For example, $h \in \mathcal{O}(3)$ implies

$$h(0, 0) = h_u(0, 0) = h_v(0, 0) = h_{uu}(0, 0) = h_{uv}(0, 0) = h_{vv}(0, 0) = 0.$$

Furthermore, if C^∞-functions $g(u, v)$ and $h(u, v)$ defined on a neighborhood of the origin satisfy $h - g \in \mathcal{O}(d + 1)$ for $d(\geq 0)$, we denote this by

$$g(u, v) \equiv_d h(u, v). \tag{D.2}$$

For a non-negative integer d, (D.2) is equivalent to having all of the derivatives of g and h of degree at most d vanish at the origin. For example,

$$u^2 + uv + v^3 \equiv_2 u^2 + uv.$$

The relation $\equiv_d$ is independent of coordinate transformations that map the origin to the origin.

When there is no possibility of confusion, we use the same notation for C^∞-functions $\varphi(t)$ of one variable. Namely, $\varphi \in \mathcal{O}(d)$ implies that $|\varphi(t)|/|t|^d$ is bounded on a neighborhood of $t = 0$.

D.2. Coordinate Change, First Step

In this section, we show the following proposition.

Proposition D.2.1. *Let U be a neighborhood of the origin in $\mathbf{R}^2$, and let $f : (U, \mathbf{0}) \to (\mathbf{R}^2, \mathbf{0})$ be a C^∞ map satisfying the assumptions of Theorem D. Then there exist a local coordinate system $(\tilde{u}, \tilde{v})$ on the domain and a local diffeomorphism Φ of the target $\mathbf{R}^2$ near the origin such that*

$$\Phi \circ f = (k(\tilde{u}, \tilde{v}), \tilde{v}), \quad k(\tilde{u}, \tilde{v}) \equiv_3 \tilde{u}\tilde{v} - \tilde{u}^3. \tag{D.3}$$

Furthermore, with this form, $\partial_{\tilde{u}}(= \partial/\partial\tilde{u})$ becomes an extended null vector field (see Remark D.2.2 below).

Remark D.2.2. In general, for a C^∞ map of the form $f(u, v) = (k(u, v), v)$, the vector field ∂_u gives a null vector field on the set of singular points. Namely, ∂_u can be taken as an extended null vector field. In fact, taking an identifier of singularities Λ as the determinant of the Jacobi matrix J_f of f, then, since

$$\Lambda = J_f = \det(f_u, f_v) = \det \begin{pmatrix} k_u & k_v \\ 0 & 1 \end{pmatrix} = k_u,$$

the set of singular points is $\{(u, v) \, ; \, k_u(u, v) = 0\}$. Thus at a singular point,

$$df(\partial_u) = f_u = (k_u, 0) = \mathbf{0}$$

holds. Hence ∂_u gives a null vector field.

We show Proposition D.2.1, by taking coordinate changes step by step for a given f.

Lemma D.2.3. *Under the assumption of Proposition D.2.1, there exist a local coordinate system $(\tilde{u}, \tilde{v})$ on the domain and a local diffeomorphism Φ of the target $\mathbf{R}^2$ near the origin such that*

$$\Phi \circ f = (\tilde{u}(\tilde{v} + s(\tilde{u}, \tilde{v})), \tilde{v}), \quad s \in \mathcal{O}(2). \tag{D.4}$$

Proof. By the assumption (i) in Theorem D, the origin is a non-degenerate singular point of f, and one of $f_u(0, 0)$ or $f_v(0, 0)$ does not vanish, and switching u and v if necessary, we may assume $f_v(0, 0) \neq \mathbf{0}$. Moreover, again changing the coordinate system of $\boldsymbol{R}^2$ if necessary, we may assume $\bar{y}_v(0,0) \neq 0$, where we set $f = (\bar{x}(u,v), \bar{y}(u,v))$.

Setting $\tilde{u} = u$ and $\tilde{v} = \bar{y}(u,v)$, then $(u,v) \mapsto (\tilde{u}, \tilde{v})$ gives a coordinate transformation near the origin in U. For this local coordinate system, f is written as $f(\tilde{u}, \tilde{v}) = (\bar{x}(\tilde{u}, \tilde{v}), \tilde{v})$. Setting $\Phi(x,y) := (x - \bar{x}(0,y), y)$, we have a local diffeomorphism of $\boldsymbol{R}^2$ near the origin. By setting $\tilde{k}(u,v) := \bar{x}(u,v) - \bar{x}(0,v)$, we have

$$\Phi \circ f(\tilde{u}, \tilde{v}) = (\tilde{k}(\tilde{u}, \tilde{v}), \tilde{v}), \quad \tilde{k}(0, \tilde{v}) = 0.$$

By the division lemma (Proposition A.1), there exists a C^∞ function h such that $\tilde{k}(\tilde{u}, \tilde{v}) = \tilde{u} h(\tilde{u}, \tilde{v})$. Since the origin is a singular point of f, it holds that $h(0,0) = 0$.

We then replace $\Phi \circ f$ by f notationally, and set the identifier of singularities Λ to be the Jacobian $J_f := \det(f_{\tilde{u}}, f_{\tilde{v}})$. By Remark D.2.2, $\tilde{\eta} := \partial_{\tilde{u}}$ gives an extended null vector field, and $\Lambda = h + \tilde{u} h_{\tilde{u}}$. By the assumption (i) of (D.1),

$$0 = \Lambda_{\tilde{\eta}}(0,0) = \Lambda_{\tilde{u}}(0,0) = (h + \tilde{u} h_{\tilde{u}})_{\tilde{u}}\big|_{(\tilde{u},\tilde{v})=(0,0)} = 2h_{\tilde{u}}(0,0).$$

On the other hand, by non-degeneracy and condition (D.1),

$$\Lambda_{\tilde{v}}(0,0) = (h + \tilde{u} h_{\tilde{u}})_{\tilde{v}}\big|_{(\tilde{u},\tilde{v})=(0,0)} = h_{\tilde{v}}(0,0) \neq 0.$$

Thus, setting $b := h_{\tilde{v}}(0,0)$, we can write

$$h(\tilde{u}, \tilde{v}) = b\tilde{v} + s(\tilde{u}, \tilde{v}) \quad (b \neq 0, \ s \in \mathcal{O}(2)).$$

Then setting $\Phi(x,y) := (x/b, y)$, we have

$$\Phi \circ f(\tilde{u}, \tilde{v}) = (\tilde{u}(\tilde{v} + s(\tilde{u}, \tilde{v})/b), \tilde{v}).$$

Rewriting $s(\tilde{u}, \tilde{v})/b$ as $s(\tilde{u}, \tilde{v})$, we have the assertion. □

The following lemma is a refinement of Lemma D.2.3.

Lemma D.2.4. *Under the assumption of Proposition D.2.1, there exist a local coordinate system $(\tilde{u}, \tilde{v})$ on the domain and a local diffeomorphism Φ of the target $\boldsymbol{R}^2$ near the origin and a C^∞ function $k(\tilde{u}, \tilde{v})$ defined on a coordinate neighborhood $(\tilde{u}, \tilde{v})$ such that*

$$\Phi \circ f = (k(\tilde{u}, \tilde{v}), \tilde{v}), \quad k(\tilde{u}, \tilde{v}) \equiv_3 \tilde{u}(\tilde{v} + a_{30}\tilde{u}^2),$$

where a_{30} is a constant.

Proof. By Lemma D.2.3, we may assume that f is written as

$$f(u,v) = \big(x(u,v), v\big), \quad x(u,v) := u(v + s(u,v)), \quad s \in \mathcal{O}(2). \quad \text{(D.5)}$$

Since $u\, s(u,v) \in \mathcal{O}(3)$, there exist $a_{30}, a_{21}, a_{12} \in \mathbf{R}$ such that

$$us(u,v) \equiv_3 a_{30}u^3 + a_{21}u^2v + a_{12}uv^2.$$

Then we have

$$k(u,v) := x(u,v) - a_{12}x(u,v)v \equiv_3 uv + a_{30}u^3 + a_{21}u^2v.$$

Since $\Phi : (x,y) \mapsto (x - a_{12}xy, y)$ is a local diffeomorphism at the origin, we can write $\Phi \circ f(u,v) = (k(u,v), v)$ and

$$k(u,v) \equiv_3 uv + a_{30}u^3 + a_{21}u^2v \equiv_3 v(u + a_{21}u^2) + a_{30}(u + a_{21}u^2)^3.$$

Setting a new coordinate system $(\tilde{u}, \tilde{v})$ on the domain by

$$\tilde{u} := u + a_{21}u^2, \quad \tilde{v} := v,$$

we obtain the assertion. $\qquad\square$

Proof of Proposition D.2.1. By Lemma D.2.4, we may assume that f is written in the form

$$f(u,v) = (x(u,v), v), \quad x(u,v) := uv + a_{30}u^3 + r(u,v), \quad r \in \mathcal{O}(4).$$

Taking an identifier of singularities Λ as the determinant of the Jacobi matrix, then $\Lambda = v + 3a_{30}u^2 + r_u$. By Remark D.2.2, ∂_u is an extended null vector field. By condition (ii) of Theorem D and $r_u \in \mathcal{O}(3)$,

$$0 \neq \Lambda_{\eta\eta}(0,0) = \Lambda_{uu}(0,0) = 6a_{30}$$

holds. Define a coordinate system $(\tilde{u}, \tilde{v})$ on the domain by $u := \tilde{u}/\alpha$ and $v := \alpha v$, where α is a real number satisfying $\alpha^3 = a_{30}$. Then we have $x(u,v) \equiv_3 \tilde{u}\tilde{v} + \tilde{u}^3$. So define a coordinate transformation Φ on the target by $\Phi(x,y) := \left(x, \frac{y}{\alpha}\right)$, and then $\Phi \circ f(\tilde{u}, \tilde{v})$ has the form (D.3). $\qquad\square$

D.3. Coordinate Change, Second Step

By Proposition D.2.1, a map f which satisfies the assumption of Theorem D can be written in the form

$$f(u,v) = \big(uv - u^3 + r(u,v), v\big), \quad r \in \mathcal{O}(4), \quad \text{(D.6)}$$

by coordinate changes on the domain and the target. Here, we show that we may assume $r \in \mathcal{O}(7)$, namely, we show the following proposition.

Proposition D.3.1. *We assume that a C^∞ map $f: (U, \mathbf{0}) \to (\mathbf{R}^2, \mathbf{0})$ satisfies the assumptions of Theorem D. Then there exist a local coordinate system $(\tilde{u}, \tilde{v})$ on the domain and a local diffeomorphism Φ of the target $\mathbf{R}^2$*

near the origin such that

$$\Phi \circ f(\tilde{u}, \tilde{v}) = \left(\tilde{u}\tilde{v} - \tilde{u}^3 + r(\tilde{u}, \tilde{v}), \tilde{v}\right), \quad r \in \mathcal{O}(7). \tag{D.7}$$

We show the following lemma as a preparation.

Lemma D.3.2. *For an integer d $(d \geq 4)$, and for a given f of the form*

$$f(u, v) = \left(uv - u^3 + r(u, v), v\right), \quad r \in \mathcal{O}(d),$$

there exists a local coordinate system $(\tilde{u}, \tilde{v})$ on the domain such that f has the form

$$f(\tilde{u}, \tilde{v}) = \left(\tilde{u}\tilde{v} - \tilde{u}^3 + a\tilde{u}^d + r_1(\tilde{u}, \tilde{v}), \tilde{v}\right), \quad r_1 \in \mathcal{O}(d+1),$$

where a is a constant.

Proof. Separating the function $r(u, v)$ into the u^d-term, the other dth degree terms and the $(d+1)$-st and higher-order terms, we have

$$r(u, v) = au^d + vs(u, v) + t(u, v),$$

where $a \in \mathbf{R}$, $s \in \mathcal{O}(d-1)$ and $t \in \mathcal{O}(d+1)$. By setting

$$\tilde{u} := u + s(u, v), \qquad \tilde{v} := v,$$

$(\tilde{u}, \tilde{v})$ is a new coordinate system on the domain near the origin. Then it can be easily checked that

$$\tilde{u}\tilde{v} - \tilde{u}^3 + a\tilde{u}^d \equiv_d uv - u^3 + au^d + vs(u, v) \equiv_d uv - u^3 + r(u, v)$$

holds. So we obtain the assertion. □

We show in the formula (D.6) that we can refine the statement from $r \in \mathcal{O}(4)$ to $r \in \mathcal{O}(5)$, by using Lemma D.3.2.

Proposition D.3.3. *For the map f of the form (D.6), there exist a local coordinate system $(\tilde{u}, \tilde{v})$ on the domain and a local diffeomorphism Φ of the target $\mathbf{R}^2$ near the origin such that*

$$\Phi \circ f = \left(\tilde{u}\tilde{v} - \tilde{u}^3 + r(\tilde{u}, \tilde{v}), \tilde{v}\right), \quad r \in \mathcal{O}(5).$$

Proof. Applying Lemma D.3.2 to the map f of the form (D.6), we may assume that f is written in the form (D.7) with $d = 4$. Setting $\alpha := a/2$, we may assume

$$f(u, v) = \left(x(u, v), v\right), \quad x(u, v) := uv - u^3 + 2\alpha u^4 + r(u, v), \quad r \in \mathcal{O}(5).$$

Setting a new coordinate system $(\tilde{u}, \tilde{v})$ by

$$\begin{cases} \tilde{u} := u/(1 + \alpha u), \\ \tilde{v} := (1 + \alpha u)v - \alpha(u^3 - 2\alpha u^4 - r(u, v)). \end{cases}$$

Then by

$$\tilde{u}\tilde{v} \equiv_4 uv - \frac{\alpha u}{1 + \alpha u}(u^3 - 2\alpha u^4 - r(u, v)) \equiv_4 uv - \alpha u^4,$$

$$\tilde{u}^3 \equiv_4 \frac{u^3}{(1 + \alpha u)^3} \equiv_4 u^3(1 - \alpha u + \alpha^2 u^2 + \cdots)^3 \equiv_4 u^3 - 3\alpha u^4,$$

we have

$$\tilde{v} = v + \alpha x(u, v),$$

$$\tilde{u}\tilde{v} - \tilde{u}^3 \equiv_4 uv - u^3 + 2\alpha u^4 \equiv_4 x(u, v).$$

Thus, by the coordinate change Φ on $\mathbf{R}^2$ defined by $\Phi : (x, y) \mapsto (x, y + \alpha x)$, f has the form $\Phi \circ f(u, v) \equiv_4 (\tilde{u}\tilde{v} - \tilde{u}^3, \tilde{v})$, proving the assertion. □

We next show the following.

Proposition D.3.4. *For the map f of the form*

$$f(u, v) = (uv - u^3 + r(u, v), v), \quad r \in \mathcal{O}(5), \tag{D.8}$$

there exist a local coordinate system $(\tilde{u}, \tilde{v})$ on the domain and a local diffeomorphism Φ of the target $\mathbf{R}^2$ near the origin such that

$$\Phi \circ f(\tilde{u}, \tilde{v}) = (\tilde{u}\tilde{v} - \tilde{u}^3 + r_1(\tilde{u}, \tilde{v}), \tilde{v}), \quad r_1 \in \mathcal{O}(6). \tag{D.9}$$

Proof. Applying Lemma D.3.2 to the map f of the form (D.8), we may assume that f is written in the form

$$f = (x(u, v), v), \quad x(u, v) := uv - u^3 + au^5 + s(u, v), \quad s \in \mathcal{O}(6).$$

Setting a new coordinate system $(\tilde{u}, \tilde{v})$ by

$$\begin{cases} \tilde{u} := u - \dfrac{a}{6}uv - \dfrac{a}{3}u^3 + \dfrac{a^2}{12}u^3 v \\ \tilde{v} := v, \end{cases}$$

it can be easily checked that

$$\tilde{u}^3 \equiv_5 u^3 - \frac{au^3 v}{2} - au^5 + \frac{a^2 u^3 v^2}{12}.$$

In particular, we have

$$\tilde{u}\tilde{v} - \tilde{u}^3 \equiv_5 x(u, v) - \frac{avx(u, v)}{6}.$$

Thus, by the coordinate change Φ on $\mathbf{R}^2$ defined by

$$\Phi : (x, y) \mapsto \left(x - \frac{axy}{6}, y\right),$$

we have the form (D.9). □

Proof of Proposition D.3.1. It is enough to show that in (D.9), we can change the assertion from $r_1 \in \mathcal{O}(6)$ to $r_1 \in \mathcal{O}(7)$.

We assume that f is written as in the form (D.9). Moreover, applying Lemma D.3.2 and writing $a := 4\alpha$, we may assume that f has the form

$$f = \big(x(u,v), v\big),$$

where

$$x(u,v) := uv - u^3 + 4\alpha u^6 + s(u,v), \quad s \in \mathcal{O}(7).$$

Setting a new coordinate system $(\tilde{u}, \tilde{v})$ by

$$\begin{cases} \tilde{u} := u - \alpha u^2 v - \alpha u^4, \\ \tilde{v} := v, \end{cases}$$

then

$$\tilde{u}\tilde{v} - \tilde{u}^3 \equiv_6 uv - \alpha u^2 v^2 + 2\alpha u^4 v - u^3 + 3\alpha u^6$$

$$\equiv_6 uv - u^3 + 4\alpha u^6 - \alpha(uv - u^3 + 4\alpha u^6)^2.$$

So, we have

$$x(u,v) - \alpha x(u,v)^2 \equiv_6 \tilde{u}\tilde{v} - \tilde{u}^3.$$

Thus, by the coordinate change $\Phi : (x,y) \mapsto (x - \alpha x^2, y)$ on $\mathbf{R}^2$, we see that the map f can be written in the form (D.7). $\qquad\square$

D.4.   Normalizing the Set of Singular Points

By Proposition D.3.1, a map f satisfying the assumption of Theorem D can be written in the form (D.7) by local diffeomorphisms on U and the target $\mathbf{R}^2$. Using this fact, next we show that the set of singular points of f can be expressed as $\{v = 3u^2\}$, which is the singular set of the standard cusp. For this purpose, we prepare the following proposition.

Proposition D.4.1. *Let* $f : (U, 0) \to (\mathbf{R}^2, 0)$ *be a* C^∞ *map satisfying the assumptions of Theorem D. Then there exist a local coordinate system* $(\tilde{u}, \tilde{v})$ *on the domain and a local diffeomorphism* Φ *of the target* $\mathbf{R}^2$ *near the origin such that*

$$f(\tilde{u}, \tilde{v}) = \big(\tilde{u}\tilde{v} - \tilde{u}^3 + r(\tilde{u}, \tilde{v}), \tilde{v}\big), \quad r \in \mathcal{O}(6),$$

$$r_{\tilde{u}}(\tilde{u}, 3\tilde{u}^2) = 0.$$

If f *is given in the above form, then* $\{\tilde{v} = 3\tilde{u}^2\}$ *is the singular set.*

Proof. By Proposition D.3.1, we may assume that f is written in the form

$$f(u, v) = (uv - u^3 + r(u, v), v), \quad r \in \mathcal{O}(7). \tag{D.10}$$

In particular, $\Lambda(u, v) = v - 3u^2 + r_u(u, v)$ is the determinant of its Jacobian matrix. Since we can take ∂_u as an extended null vector field (Remark D.2.2), and by the non-degeneracy (i) in Theorem D, it holds that $\Lambda_v(0, 0) \neq 0$. Thus, by the implicit function theorem, there exists a C^∞-function $\varphi(u)$ defined on a neighborhood of $u = 0$ such that

$$\Lambda(u, \varphi(u)) = \varphi(u) - 3u^2 + r_u(u, \varphi(u)) = 0, \quad \varphi(0) = 0. \tag{D.11}$$

Note that the set of singular points of f is $\{v = \varphi(u)\}$. By (D.10), it holds that $r_u \in \mathcal{O}(6)$. Thus, by differentiating (D.11),

$$\varphi'(u) - 6u \in \mathcal{O}(5), \quad \varphi''(u) - 6 \in \mathcal{O}(4), \quad (' = d/du) \tag{D.12}$$

hold, so $\varphi(0) = \varphi'(0) = 0$ and $\varphi''(0) = 6$. Thus, by the division lemma (Proposition A.1 in Appendix A), there exists a C^∞-function $\psi(u)$ such that

$$\varphi(u) = 3u^2 \psi(u), \quad \psi(0) = 1.$$

In particular, setting $s(u) := \sqrt{\psi(u)}$, then s is a C^∞-function, and satisfies

$$\varphi(u) = 3u^2 s(u)^2, \quad s(0) = 1.$$

By (D.12), we have $\varphi^{(3)}(0) = \varphi^{(4)}(0) = \varphi^{(5)}(0) = 0$. This implies that $\psi'(0) = \psi''(0) = \psi^{(3)}(0) = 0$. Thus, differentiating $s(u) = \sqrt{\psi(u)}$, we see

$$s(0) = 1, \quad s'(0) = s''(0) = s^{(3)}(0) = 0.$$

Hence there exists a C^∞-function s_1 such that

$$us(u) = u + s_1(u), \quad s_1 \in \mathcal{O}(5).$$

In particular $(\tilde{u}, \tilde{v}) = (u + s_1(u), v)$ gives new local coordinates, and its inverse map is written as $u = \tilde{u} + s_2(\tilde{u})$, where $s_2 \in \mathcal{O}(5)$. Since

$$uv - u^3 \equiv_7 \tilde{u}\tilde{v} - \tilde{u}^3 + s_2 v \equiv_6 \tilde{u}\tilde{v} - \tilde{u}^3.$$

By (D.10), we can write

$$f(\tilde{u}, \tilde{v}) = (k(\tilde{u}, \tilde{v}), \tilde{v}), \quad k(\tilde{u}, \tilde{v}) := \tilde{u}\tilde{v} - \tilde{u}^3 + r_1(\tilde{u}, \tilde{v}) \quad r_1 \in \mathcal{O}(6).$$

Since $\varphi(u) = 3(us(u))^2 = 3(u + s_1(u))^2 = 3\tilde{u}^2$, the set of singular points of f is $\{(\tilde{u}, \tilde{v}); \tilde{v} = 3\tilde{u}^2\}$, which is the last assertion. Substituting $\tilde{v} = 3\tilde{u}^2$,

$$0 = k_{\tilde{u}} = \tilde{v} - 3\tilde{u}^2 + (r_1)_{\tilde{u}}(\tilde{u}, \tilde{v}) = (r_1)_{\tilde{u}}(\tilde{u}, 3\tilde{u}^2)$$

holds, and all the assertions are obtained. $\qquad\square$

D.5.   Normalizing the Singular Set Image

We show the following proposition.

Proposition D.5.1. *Let $f : (U, 0) \to (\mathbf{R}^2, 0)$ be a C^∞ map satisfying the assumptions of Theorem D. Then there exist a local coordinate system $(\tilde{u}, \tilde{v})$ on the domain and a local diffeomorphism Φ of the target $\mathbf{R}^2$ near the origin such that*

$$f(\tilde{u}, \tilde{v}) = \left(\tilde{u}\tilde{v} - \tilde{u}^3 + r(\tilde{u}, \tilde{v}), \tilde{v}\right), \quad r \in \mathcal{O}(3), \tag{D.13}$$

$\{\tilde{v} = 3\tilde{u}^2\}$ is the set of singular points, and $\{27x^2 = 4y^3\}$ is the image of the set of singular points. Moreover, we may assume that r satisfies

$$r(\tilde{u}, 3\tilde{u}^2) = r_{\tilde{u}}(\tilde{u}, 3\tilde{u}^2) = r_{\tilde{v}}(\tilde{u}, 3\tilde{u}^2) = 0. \tag{D.14}$$

Remark D.5.2. If a map f given in the form (D.13) satisfies (D.14), then $f_u(u, 3u^2) = \mathbf{0}$, the set of singular points is $\{v = 3u^2\}$, and $f(u, 3u^2) = (2u^3, 3u^2)$ holds. Thus, the image of the set of singular points is $\{27x^2 = 4y^3\}$. This coincides with the set of singular points and the image of the map given by (D.1).

Proof. By Proposition D.4.1, we may assume f is given by

$$f(u, v) = \left(uv - u^3 + r(u, v), v\right), \quad r_u(u, 3u^2) = 0, \quad r \in \mathcal{O}(6). \tag{D.15}$$

Noting that $r(u, v) \in \mathcal{O}(6)$, by the division lemma (Proposition A.1), there exists a C^∞ function ρ of one variable such that

$$r(u, 3u^2) = u^2 \rho(u), \quad \rho(u) \in \mathcal{O}(4).$$

We divide $\rho(u)$ into an even function and an odd function as follows:

$$\rho(u) = \rho_1(u) + \rho_2(u)$$
$$\left(\rho_1(u) := \frac{\rho(u) + \rho(-u)}{2}, \quad \rho_2(u) := \frac{\rho(u) - \rho(-u)}{2}\right).$$

Then by the Whitney lemma (Theorem 3.1.7 in Chapter 3), there exist C^∞ functions s_1, s_2 of one variable such that $\rho_1(u) = s_1(u^2)$ and $\rho_2(u) = us_2(u^2)$. In particular, by $\rho \in \mathcal{O}(4)$, we have $s_1(t) \in \mathcal{O}(2)$ and $s_2(t) \in \mathcal{O}(1)$. We introduce two C^∞ functions σ_1, σ_2 of one variable:

$$\sigma_1(t) := -\frac{2ts_1(t/3)}{3(2 + s_2(t/3))}, \quad \sigma_2(t) := -\frac{s_2(t/3)}{2 + s_2(t/3)}.$$

Then we see that $\sigma_1 \in \mathcal{O}(3)$ and $\sigma_2 \in \mathcal{O}(1)$.

By the definition of ρ, since $r(u, 3u^2) = u^2\big(s_1(u^2) + us_2(u^2)\big)$, setting

$$R := \sigma_1(3u^2) + \big(2u^3 + r(u, 3u^2)\big)\sigma_2(3u^2), \tag{D.16}$$

a direct calculation yields

$$R = -u^2\big(s_1(u^2) + us_2(u^2)\big) = -r(u, 3u^2). \tag{D.17}$$

Consider the coordinate change $\Phi : (x, y) \mapsto \big(x\big(1 + \sigma_2(y)\big) + \sigma_1(y), y\big)$, and defining a C^∞ function $k(u, v)$ by

$$\Phi \circ f(u, v) = \big(k(u, v), v\big), \tag{D.18}$$

we have $k(u, v) = (uv - u^3 + r(u, v))(1 + \sigma_2(v)) + \sigma_1(v)$. Since (D.16) and (D.17) are equal, it holds that

$$k(u, 3u^2) = 2u^3 + r(u, 3u^2) + R = 2u^3.$$

Let us set $r_1(u, v)$ to be

$$r_1(u, v) := k(u, v) - (uv - u^3). \tag{D.19}$$

Then, since $r \in \mathcal{O}(6)$, $\sigma_1 \in \mathcal{O}(3)$ and $\sigma_2 \in \mathcal{O}(1)$, it holds that

$$r_1(u, v) = r(u, v) + (uv - u^3 + r(u, v))\sigma_2(v) + \sigma_1(v) \in \mathcal{O}(3).$$

Hence if we show

$$r_1(u, 3u^2) = (r_1)_u(u, 3u^2) = (r_1)_v(u, 3u^2) = 0, \tag{D.20}$$

then rewriting r_1 to r, the assertion is shown. In what follows, we show (D.20). By the definition of k and the equality of (D.16) and (D.17), we have

$$r_1(u, 3u^2) = r(u, 3u^2) + R = 0.$$

Next we show $(r_1)_u(u, 3u^2) = 0$. The set of singular points of f in (D.15) is $\{v = 3u^2\}$, and since ∂_u gives an extended null vector field (cf. Remark D.2.2), we see $f_u(u, 3u^2) = 0$. Thus, by (D.18), it holds that $k_u(u, 3u^2) = 0$. By (D.19), $0 = k_u(u, 3u^2) = (r_1)_u(u, 3u^2)$ holds. Thus, we get $(r_1)_u(u, 3u^2) = 0$. Differentiating $r_1(u, 3u^2) = 0$ by u, we have

$$0 = \frac{d}{du}r_1(u, 3u^2) = (r_1)_u(u, 3u^2) + 6u(r_1)_v(u, 3u^2) = 6u(r_1)_v(u, 3u^2).$$

Thus, $(r_1)_v(u, 3u^2) = 0$ holds when $u \neq 0$. By continuity, we have the same formula for the case of $u = 0$. $\qquad\square$

D.6.   Proof of Theorem D

By Proposition D.5.1, we may assume $f(\tilde{u}, \tilde{v})$ is given by (D.13) with (D.14). We then set $\tilde{u} := u$ and $\tilde{v} := v$. Defining $\tilde{r}(u, t) := r(u, t + 3u^2)$, and by (D.14), it holds that

$$\tilde{r}(u, 0) = r(u, 3u^2) = 0, \quad \tilde{r}_t(u, 0) = r_v(u, 3u^2) = 0.$$

Thus, by the division lemma (Lemma A.1 in Appendix A), there exists a C^∞ function ρ such that $\tilde{r}(u, t) = t^2 \rho(u, t)$. Substituting $v - 3u^2$ into t in this formula, we have

$$r(u, v) = (v - 3u^2)^2 r_1(u, v),$$

where $r_1(u, v) := \rho(u, v - 3u^2)$. Since $r \in \mathcal{O}(3)$, it holds that $r_1 \in \mathcal{O}(1)$. Defining a C^∞ function H of the variables u, v, w by

$$H(u, v, w) := (1 - 3uw - (v - 3u^2)w^2)w - r_1(u, v),$$

then $H(0, 0, 0) = 0$ and $H_w(0, 0, 0) = 1$. Thus, by the implicit function theorem, there exists a C^∞ function $(w =)a(u, v)$ such that

$$H(u, v, a(u, v)) = 0, \quad a(0, 0) = 0.$$

By the inverse function theorem, we may set a new coordinate system on a neighborhood of the origin by $(\tilde{u}, \tilde{v}) := (u + (v - 3u^2)a(u, v), v)$. Since $r = (v - 3u^2)^2 r_1$,

$$\tilde{u}\tilde{v} - \tilde{u}^3 - (uv - u^3 + r(u, v))$$

$$= (v - 3u^2)^2 H(u, v, a(u, v)) = 0.$$

Thus, we obtain $f(\tilde{u}, \tilde{v}) = (\tilde{u}\tilde{v} - \tilde{u}^3, \tilde{v})$. $\square$

Finally, we remark that the criterion of Whitney cusps is given [91], and the discussions in this appendix are in [91]. See [22, Chapter 7] for a modern proof of Whitney cusps using an algebraic version of the Malgrange preparation theorem.

Appendix E

A Zakalyukin-Type Lemma

This appendix is prepared for the discussions in Chapter 8.

E.1. Preliminaries

We fix a positive integer n. For each $P = (y_1, \ldots, y_n) \in \mathbf{R}^n$ $(n \geq 1)$, we set

$$|P| := \sqrt{\sum_{i=1}^{n} y_i^2}.$$

Let

$$B^n(P, r) := \{ Q \in \mathbf{R}^n \, ; \, |Q - P| < r \}$$

be the ball of radius $r(> 0)$ centered at $P \in \mathbf{R}^n$. We let m be a positive integer satisfying $(m \leq n)$ and consider a non-empty open subset U of $\mathbf{R}^m$. A continuous map $f : U \to \mathbf{R}^n$ is called a *proper map* if $f^{-1}(K)$ is a compact subset of U for each compact subset $K(\subset \mathbf{R}^m)$. We now localize this concept as follows:

Definition E.1.1 ([32]). Let $V(\subset U)$ be a neighborhood of a point $p \in U$. We denote by $f|_V$ the restriction of the map f to the subset V. A continuous map $f : U \to \mathbf{R}^n$ is said to be *V-proper at a point* $p \in U$ if there exists a positive number r_0 such that $(f|_V)^{-1}(\overline{B^n(f(p), r)})$ is a compact subset of V for each $r \in (0, r_0)$, where $\overline{B^n(f(p), r)}$ is the closure of $B^n(f(p), r)$ in $\mathbf{R}^n$.

Moreover, f is called *strongly V-proper at* $p \in V$, if, for each neighborhood $W(\subset V)$ of p, there exists $r_0(> 0)$ such that $(f|_V)^{-1}(\overline{B^n(f(p), r)})$ is a compact subset of W for $r \in (0, r_0)$.

We then give the following definition:

Definition E.1.2 ([32]). A continuous map $f : U \to \mathbf{R}^n$ is called *proper at a point* $p \in U$ if there exists a neighborhood $V(\subset U)$ of p such that f is strongly V-proper at p.

We give here a few examples:

Example E.1.3. We consider a function $f : \mathbf{R} \to \mathbf{R}$ defined by $f(x) := xe^{-x^2}$. This function itself is not a proper map, but for any positive $\varepsilon > 0$, the restriction of f to $U := (-\varepsilon, \varepsilon)$ is strongly U-proper at $x = 0$.

Example E.1.4. Consider a continuous function $f : \mathbf{R} \to \mathbf{R}$ such that $f(x) = x(1 - |x|^{-1})$ if $|x| > 1$ and $f(x) = 0$ if $|x| \le 1$. Obviously, f is a proper map, but not proper at $x = 0$. In fact, if we set $U := (-\varepsilon, \varepsilon)$ $(0 < \varepsilon < 1)$, then for each $r \in (0, \varepsilon)$

$$(f|_U)^{-1}([-r, r]) = (f|_U)^{-1}(\{0\}) = U$$

and so f cannot be U-proper at $x = 0$.

Example E.1.5. Consider a continuous function

$$f(x) := x \sin \frac{1}{x}, \quad (x \in [-1, 1]).$$

Then f is a proper map because $[-1, 1]$ is compact. Moreover, one can easily check that f is U-proper at $x = 0$ for each choice of an open interval $U := (-\varepsilon, \varepsilon)$ $(0 < \varepsilon < 1)$. However, f is not strongly U-proper at $x = 0$. In fact, we set

$$V_k := (-a_k, a_k), \quad a_k := \frac{1}{k\pi}.$$

Then we have $V_k \subset U$ for sufficiently large positive integer k. We fix such a V_k and set $K_r := [-r, r]$ $(r > 0)$. Since 0 is an interior point of $f^{-1}(K_r)$ and $f(\pm a_k) = 0$, there exists $\delta(> 0)$ depending on r satisfying $f((a_k - \delta, a_k)) \subset K_r$, which implies that $f|_{V_k}^{-1}(K_r)$ cannot be a compact subset of V_k for any choice of r.

We prepare the following:

Proposition E.1.6 ([32]). *Let* $U(\subset \mathbf{R}^m)$ *be a neighborhood of* $p \in \mathbf{R}^m$ *and* $f : U \to \mathbf{R}^n$ *a continuous map. If* f *is strongly* U-proper at p, *then for each neighborhood* $V(\subset U)$ *of* p, *the map* f *is strongly* V-proper at p.

Proof. We set $g := f|_V$. Let $O(\subset V)$ be an arbitrarily fixed neighborhood of p. Since f is strongly U-proper at p, there exists $r_0(> 0)$ such that $K_r := f^{-1}(\overline{B^n(f(p), r)})$ is a compact subset of O for $r \in (0, r_0)$. Since $O \subset V$, $K_r \subset g^{-1}(\overline{B^n(f(p), r)})$ holds. On the other hand, $g^{-1}(\overline{B^n(f(p), r)}) \subset K_r$ is clear. So we have $K_r = g^{-1}(\overline{B^n(f(p), r)})$. $\square$

We can prove the following:

Proposition E.1.7 ([32]). *Let $U(\subset \boldsymbol{R}^m)$ be a neighborhood of $p \in \boldsymbol{R}^m$ and $f : U \to \boldsymbol{R}^n$ a continuous map. For a neighborhood $V(\subset U)$ of p, if f is V-proper at p and $(f|_V)^{-1}(f(p)) = \{p\}$, then f is strongly V-proper at p.*

Proof. We fix a neighborhood $W(\subset V)$ of p. It is sufficient to show that

$$(K :=)(f|_V)^{-1}(\overline{B^n(f(p), r)}) \subset W$$

for sufficiently small $r(> 0)$. (In fact, we may assume that K is compact, because f is a V-proper at p.) If this fails, then, for each positive integer k satisfying $1/k < r$, there exists $q_k \in (f|_V)^{-1}(\overline{B^n(f|_V(p), 1/k)})(\subset K)$ which is not belonging to W. Since K is compact, the sequence $\{q_k\}_{k=1}^\infty$ has an accumulation point $q_\infty \in K$. In particular $q_\infty \in V$ and $f(q_\infty) = f(p)$. Since $(f|_V)^{-1}(f(p)) = \{p\}$, we have $q_\infty = p$. On the other hand, since $q_k \in K \backslash V$, we have $q_\infty \in K \setminus V$, contradicting the fact $q_\infty = p$. $\square$

Corollary E.1.8. *Let $f : U \to \boldsymbol{R}^n$ be a proper map. If $f^{-1}(f(p)) = \{p\}$ holds, then f is strongly V-proper at p for each open neighborhood $V(\subset U)$ of p.*

Proof. Since f is a proper map, f is U-proper at p. Since $f^{-1}(f(p)) = \{p\}$, Proposition E.1.7 implies that f is strongly U-proper at p. In particular, f is strongly V-proper at p by Proposition E.1.6. $\square$

We next prove the following:

Lemma E.1.9 ([32]). *Let $f : U(\subset \boldsymbol{R}^m) \to \boldsymbol{R}^n$ be a continuous map and $V(\subset U)$ a neighborhood of $p \in U$. If f is strongly V-proper at p, then $(f|_V)^{-1}(f(p))$ consists of the single point p.*

Proof. We set $g := f|_V$ and suppose that $g^{-1}(f(p))$ has a point q other than p. Since U is a Hausdorff space, there exists a pair (W_1, W_2) of disjoint open subsets of U such that $p \in W_1$ and $q \in W_2$.

Since f is strongly V-proper at p, there exists $\varepsilon(> 0)$ such that

$$g^{-1}(\overline{B^n(f(p), \varepsilon)}) \subset W_1.$$

Then we have that

$$q \in g^{-1}(f(p)) \subset g^{-1}(\overline{B^n(f(p), \varepsilon)}) \subset W_1,$$

which contradicts the fact that $q \in W_2$. □

We next prepare the following assertion.

Proposition E.1.10 ([45]). *Let $U(\subset \mathbf{R}^m)$ be a neighborhood of $p \in \mathbf{R}^m$ and $f : U \to \mathbf{R}^n$ a continuous map such that $f^{-1}(\mathrm{P})$ ($\mathrm{P} := f(p)$) is a finite point set. Then there exist $\delta(> 0)$ such that the connected component V_r of $f^{-1}(B^n(\mathrm{P}, r))$ containing p satisfies $\overline{V_r} \subset U$ for each $r \in (0, \delta]$. Moreover, V_r is a relatively compact open neighborhood of p and f is V_r-proper at p.*

Proof. We can take a relatively compact neighborhood W of p such that $\overline{W}$ is contained in U. Since $f^{-1}(\mathrm{P})$ is a finite point set, we may assume that

$$f^{-1}(\mathrm{P}) \cap W = \{p\} \tag{E.1}$$

holds. As in the statement of the proposition, let V_r be the connected component of $f^{-1}(B^n(p, r))$ containing p for each $r > 0$. Since $V_r \subset V_s$ for $r < s$, it is sufficient to show that $\overline{V_{1/k}} \subset W$ for sufficiently large integer $k > 0$. If not, there exists a point not in W but lying in $\overline{V_{1/k}}$. Amongst them, we would like to show the existence of a point $p_k \in \partial W \cap \overline{V_{1/k}}$, where $\partial W := \overline{W} \setminus W$. If not, we have the following decomposition:

$$\overline{V_{1/k}} = (W \cap \overline{V_{1/k}}) \cup ((U \setminus \overline{W}) \cap \overline{V_{1/k}}).$$

The right-hand side is the union of two non-empty open subsets of $\overline{V_{1/k}}$, which is a contradiction, because $\overline{V_{1/k}}$ is connected. Thus, we can find a point $p_k \in \overline{V_{1/k}} \cap \partial W$ for each k. Since f is continuous, we have

$$f(\overline{V_{1/k}}) \subset \overline{f(V_{1/k})} \subset \overline{B^n(\mathrm{P}, 1/k)}.$$

By our construction, the sequence $\{p_k\}_{k=1}^{\infty}$ consists of infinitely many points. Since ∂W is compact, the sequence has an accumulation point $p_{\infty} \in \partial W$. Since $f(p_k) \in B^n(\mathrm{P}, 1/k)$, $\{f(p_k)\}_{k=1}^{\infty}$ converges to a point P. So we have $f(p_{\infty}) = \mathrm{P}$, which contradicts (E.1). So we have shown that $\overline{V_{1/k}} \subset W$ for sufficiently large integer $k > 0$. Since $\overline{W}$ is compact, $\overline{V_{1/k}}$ is also compact.

We fix such an integer k and set $r_0 := 1/k$. We next prove that f is V_r-proper at p under the assumption that $r < 1/k$. We have shown that $\overline{V_r}(\subset U)$ is compact. We set $g := f|_{V_r}$ and $K := \overline{B^n(\mathrm{P}, s)}(s < r)$, which is a compact subset of $B^n(P, r)$. Suppose that $g^{-1}(K)$ is not compact. Then there exists a sequence $\{x_k\}_{k=1}^{\infty}$ in $g^{-1}(K)$ which does not accumulate to any point in V_r. Since $\overline{V_r}$ is compact, $\{x_k\}_{k=1}^{\infty}$ must have an accumulation point $x_\infty \in \partial V_r$. Since $f(x_k) \in K$, we have

$$f(x_\infty) \in K \subset B^n(P, r).$$

In particular, there exists a neighborhood O of x_∞ such that $f(O) \subset B^n(P, r)$, which implies

$$f(V_r \cup O) \subset B^n(P, r).$$

Since $x_\infty \in V_r \cap O$, the union $V_r \cup O$ is connected. Since V_r is the connected component of $f_U^{-1}(B^n(P, r))$, we have $V_r \cup O = V_r$, contradicting the fact that $x_\infty \in \partial V_r$. Thus, $g^{-1}(K)$ is a compact subset of V_r. $\qquad\square$

Theorem E.1.11 ([32]). *Let $U(\subset \mathbf{R}^m)$ be a neighborhood of $p \in \mathbf{R}^m$ and $f : U \to \mathbf{R}^n$ a continuous map. Then the following three conditions are equivalent to each other:*

(1) *f is proper at p.*
(2) *$f^{-1}(f(p))$ is a finite point set.*
(3) *There exists a neighborhood $V(\subset U)$ of p such that $(f|_V)^{-1}(f(p)) = \{p\}$, and f is V-proper at p.*

In particular, (2) can be considered as a useful criterion for the point-wise properness of continuous maps.

Proof. By Lemma E.1.9, (1) implies (2). By Proposition E.1.10, (2) implies that f is V-proper at p for a sufficiently small neighborhood V of p. We set $g := f|_V$. Since $f^{-1}(f(p))$ is a finite point set, we may assume that $g^{-1}(f(p)) = \{p\}$. Then (3) is obtained. Finally, (3) implies (1) by Proposition E.1.7. $\qquad\square$

Corollary E.1.12. *Let U be a non-empty open subset of $\mathbf{R}^m$ and $f : U(\subset \mathbf{R}^m) \to \mathbf{R}^n$ ($m \leq n$) a C^∞ immersion. Then f is proper at each point of U.*

Proof. Since f is an immersion, it is a local injective map. So we obtain the assertion. $\qquad\square$

The standard cusp, the standard cuspidal edge and the standard swallowtail are defined by

$$\gamma_C(u) = (u^2, u^3), \qquad f_C(u, v) = (u^2, u^3, v),$$
$$f_S(u, v) = (3u^4 + u^2 v, 4u^3 + 2uv, v) \tag{E.2}$$

as a map from $\boldsymbol{R}$ into $\boldsymbol{R}^2$ and maps from $\boldsymbol{R}^2$ into $\boldsymbol{R}^3$, respectively. Using these expressions, we can prove the following as a corollary of Proposition E.1.11.

Proposition E.1.13. *The standard cusp γ_C, the standard cuspidal edge f_C and the standard swallowtail f_S are U-proper at their singular point $o := (0,0)$ (which is the origin of the domain) for any choice of an open neighborhood of o. Moreover, their inverse images coincide with $\{o\}$.*

Proof. The last assertion is obvious. By Corollary E.1.8, it is sufficient to show that γ_C, f_C and f_S are proper maps. Here we only show that f_S is a proper map. (The properness of the other two maps can be proved using the same argument.) We let K be a compact subset of $\boldsymbol{R}^2$, and let $\{(u_k, v_k)\}_{k=1}^{\infty}$ be a sequence in $f_S^{-1}(K)$. We set $f_S(u_k, v_k) = (a_k, b_k, c_k)$, then $\{a_k\}_{k=1}^{\infty}$, $\{b_k\}_{k=1}^{\infty}$ and $\{c_k\}_{k=1}^{\infty}$ are bounded sequences in $\boldsymbol{R}$ because of the compactness of K. Since $v_k = c_k$, $\{v_k\}_{k=1}^{\infty}$ is bounded. Moreover, the first component of $f_S(u_k, v_k)$ satisfies $3u_k^4 + u_k^2 v_k = a_k$, that is, u_k is a solution of the equation $3t^4 + c_k t^2 = a_k$. Since c_k, a_k are bounded, we can conclude that $\{u_k\}_{k=1}^{\infty}$ is also bounded. Thus $\{(u_k, v_k)\}_{k=1}^{\infty}$ contains a convergent subsequence. So $f_S^{-1}(K)$ is compact. $\square$

Corollary E.1.14. *Let $U(\subset \boldsymbol{R}^m)$ be a neighborhood of $p \in \boldsymbol{R}^m$ and $f : U \to \boldsymbol{R}^{m+1}$ a smooth map which is right–left equivalent to the standard cusp, cuspidal edge, or swallowtail as a map-germ. Then f is proper at p.*

Proof. It is sufficient to show that $(f|_V)^{-1}(f(p)) = \{p\}$ for sufficiently small neighborhood V of p. However, it is obvious because the standard cusp, cuspidal edge and swallowtail have such a property. $\square$

We next prove the following:

Proposition E.1.15 ([32]). *Let $U_i(\subset \boldsymbol{R}^m)$ $(i = 1, 2)$ be a neighborhood of $p_i \in \boldsymbol{R}^m$ and $f_i : U_i \to \boldsymbol{R}^n$ a continuous map. Suppose that f_2 is U_2-proper at p_2 and $f_2^{-1}(f_2(p_2)) = \{p_2\}$. Then the following three conditions are equivalent each other:*

(1) *There exists a neighborhood $V_i(\subset U_i)$ of p_i for each $i = 1, 2$ such that $f_1(V_1) \subset f_2(V_2)$.*

(2) *There exist $r > 0$ and a neighborhood $V_i(\subset U_i)$ of p_i for each $i = 1, 2$ such that $f_1(V_1) \cap B^n(\mathrm{P}, r) \subset f_2(V_2) \cap B^n(\mathrm{P}, r)$, where $\mathrm{P} := f_1(p_1) = f_2(p_2)$.*

(3) *For each neighborhood $V_i (\subset U_i)$ of p_i $(i = 1, 2)$, there exists a relatively compact neighborhood W_i of p_i such that $f_1(W_1) \subset f_2(W_2)$ and $\overline{W_i} \subset V_i$.*

Proof. Obviously (1) implies (2), and also (3) implies (1). So it is sufficient to show that (2) implies (3). So we assume (2). We set $g_i := f_i|_{V_i}$ $(i = 1, 2)$. Since f_2 is U_2-proper at p_2 and $g_2^{-1}(f_2(p_2)) = \{p_2\}$, Lemma E.1.7 implies that f_2 is strongly U_2-proper at p_2. Hence, f_2 is strongly V_2-proper at p_2 by Proposition E.1.6. As a consequence,

$$K_2 := g_2^{-1}(\overline{B^n(\mathrm{P}, r)})$$

is a compact subset of V_2 for sufficiently small $r (> 0)$. We set

$$W_2 := g_2^{-1}(B^n(\mathrm{P}, r)),$$

then $\overline{W_2} \subset K_2 \subset V_2$. Since V_1 is locally compact, there exists a relatively compact neighborhood W_1 of p_1 such that $\overline{W_1} \subset V_1$. Moreover, since f_1 is continuous, we may assume $f_1(W_1) \subset B^n(\mathrm{P}, r)$. Then we have

$$W_1 \subset g_1^{-1}(B^n(\mathrm{P}, r) \cap g_1(V_1)),$$

and we have

$$f_1(W_1) = g_1(W_1) \subset B^n(\mathrm{P}, r) \cap g_1(V_1)$$
$$\subset B^n(\mathrm{P}, r) \cap g_2(V_2) = g_2(W_2) = f_2(W_2).$$

Since $\overline{W_i} \subset V_i$ $(i = 1, 2)$, we obtain (3). $\square$

Example E.1.16. In Proposition E.1.15, the assumption that $f_2 : U_2 \to \mathbf{R}^n$ is proper at p_2 cannot be removed. In fact, we set $f_1(x) := x$ $(x \in \mathbf{R})$ and let $f_2(x)$ be the function on $\mathbf{R}$ defined in Example E.1.4. Then $f_1(\mathbf{R}) = f_2(\mathbf{R}) = \mathbf{R}$ holds. However, if we choose $V_1 = V_2 = (-1, 1)$, then $f_1(\overline{W_1}) \subset f_2(\overline{W_2})$ never holds for any choice of a pair of open intervals (W_1, W_2) containing 0 in $(-1, 1)$. In this case, $f_2(x)$ is not proper at $x = 0$ as shown in Example E.1.4.

E.2. A Zakalyukin-Type Lemma

Zakalyukin [92] pointed out that two given fronts f_1 and f_2 whose images are locally diffeomorphic satisfying a suitable genericity assumption are right–left equivalent. We now explain this: Let $f_i : (U_i, p_i) \to (\mathbf{R}^{m+1}, \mathrm{P})$ $(i = 1, 2)$ be a front defined on a domain $U_i \subset \mathbf{R}^m$ containing p_i, and let $\nu_i : U_i \to S^m$ be its Gauss map (i.e., the unit normal vector field), where S^m

is the unit sphere centered at the origin in $\boldsymbol{R}^{m+1}$. Let us consider following conditions:

(1) $p_i \in U_i$ $(i = 1, 2)$ is a singular point of f_i,
(2) the set of regular points of f_i $(i = 1, 2)$ is a dense subset of U_i.

The following assertion is a Zakalyukin-type lemma:

Theorem E.2.1 (Theorem A in [32]). *Let $f_i \colon U_i \to \boldsymbol{R}^{m+1}$ $(i = 1, 2)$ be two fronts defined on open subsets $U_i(\subset \boldsymbol{R}^m)$ containing p_i, respectively, satisfying the above two conditions* (1), (2). *Suppose that f_2 is U_2-proper at p_2 and $f_2^{-1}(f_2(p_2)) = \{p_2\}$. If there exists a local diffeomorphism Θ on $\boldsymbol{R}^{m+1}$ such that $f_1(p_1) = \Theta \circ f_2(p_2)$ and*

$$\Theta^{-1}(f_1(U_1)) \cap B^{m+1}(f_2(p_2), r) \subset f_2(U_2) \cap B^{m+1}(f_2(p_2), r), \qquad \text{(E.3)}$$

then there exist neighborhoods $V_i(\subset U_i)$ of p_i $(i = 1, 2)$ and a diffeomorphism $\varphi \colon (V_1, p_1) \to (V_2, p_2)$ such that $\Theta \circ f_1 = f_2 \circ \varphi$ holds on V_1. In particular, f_1 and f_2 are right–left equivalent.

Remark. We cannot drop the condition that f_2 is U_2-proper. In fact, the condition $f_2^{-1}(f_2(p_2)) = \{p_2\}$ implies only the existence of a neighborhood $V(\subset U_2)$ of p_2 so that f_2 is V-proper (see Theorem E.1.11), but V may not coincide with U_2, in general.

We take the Legendrian lift (see Chapter 10)

$$L_{f_i} := (f_i, \nu_i) \colon U_i \longrightarrow \boldsymbol{R}^{m+1} \times S^m$$

of f_i $(i = 1, 2)$. Since f_i $(i = 1, 2)$ is a front, L_{f_i} is an immersion. By replacing f_1 by $\Theta \circ f_1$, we may assume that Θ is the identity map and we set $\mathrm{P} := f_1(p_1) = f_2(p_2)$. Then (E.3) is reduced to

$$f_1(U_1) \cap B^{m+1}(\mathrm{P}, r) \subset f_2(U_2) \cap B^{m+1}(\mathrm{P}, r). \qquad \text{(E.4)}$$

Since L_{f_2} is an immersion, there exists a connected relatively compact neighborhood V_2 such that $\overline{V_2} \subset U_2$ and $L_{f_2}|_{V_2}$ is an embedding. By Proposition E.1.15, there exists a connected relatively compact neighborhood $V_1(\subset U_1)$ of p_1 such that

$$f_1(V_1) \subset f_2(V_2), \qquad \overline{V_1} \subset U_1. \qquad \text{(E.5)}$$

Since L_{f_1} is an immersion, we may assume that $L_{f_1}|_{\overline{V_1}}$ is an embedding. Here, we remark that an *embedding* is an injective immersion that gives a diffeomorphism between the domain and the image (see [22, Proposition 2.10] for details).

We set $g_i := f_i|_{\overline{V_i}}$ $(i = 1, 2)$. We also set $L'_{f_2} := (f_2, -\nu_2)\colon U_2 \longrightarrow$ $\boldsymbol{R}^{m+1} \times S^m$, and

$$A_+ := \{p \in \overline{V_1}\,;\, L_{f_1}(p) \in L_{f_2}(\overline{V_2})\},$$
$$A_- := \{p \in \overline{V_1}\,;\, L_{f_1}(p) \in L'_{f_2}(\overline{V_2})\}.$$

Since $\overline{V_2}$ is compact, so are $L_{f_2}(\overline{V_2})$ and $L'_{f_2}(\overline{V_2})$. Moreover, we can rewrite

$$A_+ = (L_{f_1}|_{\overline{V_1}})^{-1}(L_{f_2}(\overline{V_2})), \quad A_- = (L_{f_1}|_{\overline{V_1}})^{-1}(L'_{f_2}(\overline{V_2})), \qquad \text{(E.6)}$$

and so $A_\pm$ are closed subsets of $\overline{V_1}$.

Since $(f_2(p_2), \nu_2(p_2)) \neq (f_2(p_2), -\nu_2(p_2))$, taking sufficiently small V_2, we may assume that

$$L_{f_2}(\overline{V_2}) \cap L'_{f_2}(\overline{V_2}) = \emptyset. \qquad \text{(E.7)}$$

We set

$$L_i := L_{f_i}|_{\overline{V_i}} \quad (i = 1, 2), \qquad L'_2 := L'_{f_2}|_{\overline{V_2}}$$

and

$$\mathcal{R} := \mathcal{R}_1 \cap \mathcal{R}_2,$$

where $\mathcal{R}_i$ $(i = 1, 2)$ is the set of regular values of the map g_i. We also set

$$O_i := g_i^{-1}(\mathcal{R}), \quad S_i := \overline{V_i} \setminus O_i \ (i = 1, 2).$$

The following assertion holds.

Proposition E.2.2. O_1 *is a dense subset of* $\overline{V_1}$.

Proof. We suppose that O_1 is not dense in $\overline{V_1}$. Then S_1 has an interior point q. Since ∂V_1 has no interior point and g_1 gives an immersion on an open dense set of V_1 (cf. the initial assumption (2)), we may assume that $q \in V_1$. Then there exists an open neighborhood $W(\subset V_1)$ of q such that $W \subset S_1$ and $g_1|_W$ is an immersion. Let Π be the tangent hyperplane of $g_1(W)$ at $g_1(q)$, and let $\pi : \boldsymbol{R}^{m+1} \to \Pi$ be the canonical orthogonal projection. Since $\Pi \circ g_1$ is an immersion at q, $g_1(q) \in \pi \circ g_1(W)$ is an interior point of $\pi \circ g_1(W)$. In particular, $\pi \circ g_1(W)$ is not a measure zero set in Π, where we identify Π with $\boldsymbol{R}^m$. On the other hand, since $W \subset S_1$, we have

$$g_1(W) \subset g_1(S_1) \subset g_1(\overline{V_1}) \setminus \mathcal{R}$$
$$\subset g_2(\overline{V_2}) \setminus \mathcal{R} = g_2(\overline{V_2}) \setminus g_2(O_2)$$
$$\subset g_2(\overline{V_2} \setminus O_2) = g_2(S_2).$$

Thus, $\pi \circ g_1(W) \subset \pi \circ g_2(S_2)$. Since S_2 is the singular set of $\pi \circ g_2$, the Sard theorem implies that $\pi \circ g_1(W)$ is a measure zero set in Π, a contradiction. $\qquad \square$

Proposition E.2.3. *We have*

$$A_+ \cup A_- = \overline{V_1}. \tag{E.8}$$

Proof. We fix an arbitrary $a \in \mathcal{R}$, and show $g_i^{-1}(a)$ $(i = 1, 2)$ are finite sets. It is enough to show this for $i = 1$, namely, showing it for g_1. We assume that the inverse image $g_1^{-1}(a)$ is an infinite set. Since $\overline{V_1}$ is compact, taking a sequence $\{q_j\}_{j=1}^{\infty} \subset g_1^{-1}(a)$ consisting of distinct points, it has an accumulation point $q \in \overline{V_1}$. Taking a subsequence of $\{q_j\}_{j=1}^{\infty}$ if necessary, we may assume that it converges to q. Since $g_1(q_j) = a$, by the continuity of g_1, it holds that $g_1(q) = a$. By the definition of $\mathcal{R}$ and because $a \in \mathcal{R}$, q is a regular point of g_1. Thus, there exists a neighborhood W of q such that $g_1|_W$ is an embedding. Since $\{q_j\}_{j=1}^{\infty}$ converges to q, we have

$$g_1(q_j) = a = g_1(q),$$

contradicting the fact that $g_1|_W$ is injective.

Hence there exist positive integers k, k' such that

$$g_1^{-1}(a) = \{x_1, \ldots, x_k\}, \qquad g_2^{-1}(a) = \{y_1, \ldots, y_{k'}\}.$$

Since L_i are embeddings on $\overline{V_i}$ $(i = 1, 2)$, $\nu_1(x_j) \in S^m$ $(j = 1, \ldots, k)$ are mutually distinct, and $\nu_2(y_l) \in S^m$ $(l = 1, \ldots, k')$ are also mutually distinct. Thus, the images of g_i $(i = 1, 2)$ near a are finitely many hypersurfaces in $\mathbf{R}^{m+1}$ that intersect transversally to each other. In particular, since $f_1(\overline{V_1}) \subset f_2(\overline{V_2})$, $k \le k'$ holds. Thus, changing the order appropriately, we may assume

$$L_1(x_j) = L_2(y_j) \quad \text{or} \quad L_2'(y_j) \quad (j = 1, \ldots, k). \tag{E.9}$$

Namely,

$$g_1^{-1}(a) \subset A_+ \cup A_-$$

holds. In particular, we have

$$O_1 = \bigcup_{a \in \mathcal{R}} g_1^{-1}(a) \subset A_+ \cup A_-. \tag{E.10}$$

Since A_1 and A_2 are closed subsets of $\overline{V_1}$, by taking the closure of (E.10), Lemma E.2.2 yields (E.8). $\qquad\square$

Under the assumptions of Theorem E.2.1, the following lemma holds.

Lemma E.2.4. $L_1(p_1)$ *coincides with* $L_2(p_2)$ *or* $L_2'(p_2)$.

Proof. Take a sequence $\{q_k\}_{k=1}^{\infty}$ in O_1 which converges to p_1. We set $Q_k := f_1(q_k)$. Noting that f_1 is an immersion at q_k, let T_k be the tangent hyperplane of g_1 at $g_1(q_k)$. Since $g_2^{-1}(Q_k)$ is a finite set, there exists a point $q'_k \in g_2^{-1}(Q_k)(\subset O_2)$ such that the tangent hyperplane of g_2 at $g_2(q'_k)$ coincides with T_k. Then $L_1(q_k) = L_2(q'_k)$ or $L_1(q_k) = L'_2(q'_k)$ holds. Since $O_2 \subset \overline{V}_2$ and $\{q'_k\}_{k=1}^{\infty}$ is a sequence of $\overline{V}_2$, there is an accumulation point $q' \in \overline{V}_2$. Then $L_1(p_1) = L_2(q')$ or $L_1(p_1) = L'_2(q')$ holds. In particular, $g_1(p_1) = g_2(q')$ holds. By (E.6), we have $q' = p_2$. So we get the conclusion. $\qquad\square$

Proof of Theorem E.2.1. Replacing ν_2 by $-\nu_2$ if necessary, we may assume that $L_1(p_1) = L_2(p_2)$. By Proposition E.2.2, we have (E.8). By (E.7), $A_+ \cap A_- = \emptyset$. Since $\overline{V}_1$ is connected, either $\overline{V}_1 = A_+$ or $\overline{V}_1 = A_-$ holds. Since $L_1(p_1) = L_2(p_2)$, A_+ is non-empty. Thus $\overline{V}_1 = A_+$ holds and so $L_1(\overline{V}_1) \subset L_2(\overline{V}_2)$. Since L_2 is an embedding, we can define a continuous map $\varphi : \overline{V}_1 \to \overline{V}_2$ by

$$\varphi := L_2^{-1} \circ L_1 : \overline{V}_1 \to \overline{V}_2.$$

By definition $L_2 \circ \varphi = L_1$ holds on $\overline{V}_1$, which implies

$$f_2 \circ \varphi = f_1.$$

Since φ is an injective continuous map from the compact space $\overline{V}_1$ to the Hausdorff space $\overline{V}_2$, it gives a homeomorphism between $\overline{V}_1$ and $\varphi(\overline{V}_1) \subset \mathbf{R}^m$. By the invariance of domain [22], $V'_2 := \varphi(V_1)$ is a connected open subset of $\mathbf{R}^m$. Thus, we have $L_1(V_1) = L_2(V'_2)$, and φ gives a homeomorphism between V_1 and V'_2. Moreover, since f_i $(i = 1, 2)$ are wave fronts, L_1 and L_2 are embeddings and then $\varphi : V_1 \to V'_2$ is a diffeomorphism. In particular $f(V_1) = f_2(V'_2)$ holds. Replacing V_2 by V'_2, we obtain the assertion. $\qquad\square$

Appendix F

Geometry of Cross Caps

F.1. Normal Forms for Cross Caps

Cross caps are known as the most generic singular points appearing on smooth maps from a two-dimensional manifold into $\boldsymbol{R}^3$ (cf. [90]). In Section 3.3 of Chapter 3, we proved a criterion for cross caps. Using this, we give here a normal form for cross caps, as in the formula given in Proposition 5.5.8 of Chapter 5, as follows.

We consider a C^∞-map $f\colon U \to \boldsymbol{R}^3$ from a domain $U \subset \boldsymbol{R}^2$ into $\boldsymbol{R}^3$ having a cross cap at $p \in U$. We set $p = (0,0)$. Then there exists a local coordinate system (u,v) centered at $p \in U$ satisfying $f_v(0,0) = \boldsymbol{0}$ which is compatible with the orientation of $\boldsymbol{R}^2$. Such a coordinate system is said to be *admissible*. Since the rank of the Jacobi matrix of f is one, we have $f_u(0,0) \neq \boldsymbol{0}$. Moreover, by Theorem 2.5.6 in Chapter 2 (criterion for a cross cap),

$$\{f_u(0,0), f_{uv}(0,0), f_{vv}(0,0)\}$$

gives a basis of $\boldsymbol{R}^3$ at $f(p)$. Like as in the case of cuspidal edges, swallowtails and cuspidal cross caps (cf. Section 2.7 in Chapter 2), we introduce several lines and planes at cross cap singular points.

Definition F.1.1. We fix a C^∞-map $f : (U; u, v) \to \boldsymbol{R}^3$ having a cross cap singularity at $(u,v) = (0,0)$, where $(U; u, v)$ is an admissible coordinate neighborhood of $(0,0)$. We call the line

$$\{f(0,0) + t f_u(0,0); t \in \boldsymbol{R}\}$$

the *tangent line* at the cross cap. The plane in $\boldsymbol{R}^3$ passing through $f(0,0)$ spanned by $f_u(0,0)$ and $f_{vv}(0,0)$ is called the *co-normal plane*. (In [28], this plane is called a *principal plane*, which plays an important role in analyzing the shape of cross caps.) On the other hand, the plane passing

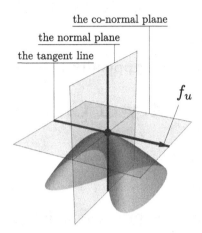

the co-normal plane

the normal plane

the tangent line

f_u

Fig. F.1.   The tangent line, normal plane and co-normal plane of a cross cap.

through $f(0,0)$ perpendicular to the tangent line is called the *normal plane* (see Fig. F.1).

We note that the tangent line, co-normal plane and normal plane do not depend on the choice of admissible coordinate system (cf. (3.10) and (3.11) in Chapter 3). To study the geometry of cross caps, the following "normal form" of an admissible coordinate system plays an important role:

Theorem F.1.2 (cf. [7, 87, 19]). *Let $f\colon U \to \boldsymbol{R}^3$ be a C^∞-map having a cross cap at $p \in U$. Then there exist a_{20}, $a_{11} \in \boldsymbol{R}$, $a_{02} \in (0,\infty)$ and an admissible coordinate system centered at p such that*

$$f(u,v) = \left(u, uv + O(u,v)^3, \frac{1}{2}\left(a_{20}u^2 + 2a_{11}uv + a_{02}v^2\right) + O(u,v)^3 \right),$$

$$\tag{F.1}$$

after appropriately rotating and translating in $\boldsymbol{R}^3$, where $O(u,v)^3$ stands for the terms whose degrees with respect to (u,v) are greater than or equal to three. In this expression, $e_1 := (1,0,0)$ points in the direction of the tangent line, $e_3 := (0,0,1)$ together with e_1 spans the co-normal plane, and $e_2 := (0,1,0)$ together with e_3 spans the normal plane. Moreover, a_{20}, a_{11} and a_{02} can be considered as geometric invariants of f at p.

Proof. By a suitable translation in $\boldsymbol{R}^3$, we may assume that $p = (0,0)$ and $f(p) = \boldsymbol{0}$ ($\boldsymbol{0} := (0,0,0)$). We write $f(u,v) = \big(x(u,v), y(u,v), z(u,v)\big)$. By taking an admissible coordinate system (u,v) on U, $f_v(0,0) = \boldsymbol{0}$ holds. Since the rank of the differential df of f at $(0,0)$ is one, we have $f_u(0,0) \neq \boldsymbol{0}$.

Thus, applying a rotation of $\boldsymbol{R}^3$ if necessary, we may assume that $f_u(0,0) = (\alpha, 0, 0) \neq \boldsymbol{0}$ and $\alpha > 0$. Then

$$x(0,0) = y(0,0) = z(0,0) = 0,$$
$$x_v(0,0) = y_u(0,0) = y_v(0,0) = z_u(0,0) = z_v(0,0) = 0,$$
$$x_u(0,0) = \alpha > 0$$

hold. Setting $(\tilde{u}, \tilde{v}) := \big(x(u,v), v\big)$, we have

$$(J :=) \begin{pmatrix} \tilde{u}_u & \tilde{u}_v \\ \tilde{v}_u & \tilde{v}_v \end{pmatrix} = \begin{pmatrix} x_u & x_v \\ 0 & 1 \end{pmatrix}.$$

Thus, it holds that $\det(J) = \alpha(> 0)$ at $(0,0)$. In particular, $(\tilde{u}, \tilde{v})$ gives an admissible coordinate system on a neighborhood of $(0,0)$. We rewrite this coordinate system as (u, v). Then $x = u$. By the criterion for cross caps, $f_{vv}(0,0) \neq \boldsymbol{0}$. Since

$$f_{vv}(0,0) = (0, y_{vv}(0,0), z_{vv}(0,0))$$

is perpendicular to $(1,0,0)$, we may assume $f_{vv}(0,0) = (0,0,\beta)$ and $\beta > 0$ by taking a suitable rotation of $\boldsymbol{R}^3$ which fixes the x-axis. Then f is expressed as

$$f(u,v) = \Big(u, \frac{1}{2}(b_{20}u^2 + 2b_{11}uv) + O(u,v)^3,$$

$$\frac{1}{2}(c_{20}u^2 + 2c_{11}uv + \beta v^2) + O(u,v)^3 \Big),$$

where b_{20}, b_{11}, c_{20} and c_{11} are constants. By the criterion, $f_{uv}(0,0)$ and $f_{vv}(0,0)$ are linearly independent, so it holds that $b_{11} \neq 0$. We set

$$(\hat{u}, \hat{v}) := \Big(u, \varepsilon(b_{11}v + \frac{1}{2}b_{20}u) \Big) \qquad \big(\varepsilon := \operatorname{sgn}(b_{11}) \big).$$

Then $(u,v) \mapsto (\hat{u}, \hat{v})$ is an orientation preserving coordinate transformation. Moreover, since

$$\boldsymbol{0} = f_v(0,0) = f_{\hat{u}}(0,0)\hat{u}_v(0,0) + f_{\hat{v}}(0,0)\hat{v}_v(0,0) = f_{\hat{v}}(0,0)\varepsilon b_{11},$$

$(\hat{u}, \hat{v})$ is an admissible coordinate system. Since $y(\hat{u}, \hat{v}) = \varepsilon \hat{u}\hat{v} + O(\hat{u}, \hat{v})^3$ and

$$0 < \beta = z_{vv}(0,0) = (\varepsilon b_{11})^2 z_{\hat{v}\hat{v}}(0,0),$$

rewriting $(\hat{u}, \hat{v})$ as (u,v), f is then expressed as

$$f(u,v) = \Big(u, \varepsilon uv + O(u,v)^3, \frac{1}{2}(a_{20}u^2 + 2a_{11}uv + a_{02}v^2) + O(u,v)^3 \Big),$$

where $a_{02} > 0$. If $\varepsilon = 1$, then we have the assertion. If $\varepsilon = -1$, considering the parameter change $(u, v) \mapsto (-u, -v)$ and the rotation $(x, y, z) \mapsto (-x, -y, z)$ in $\boldsymbol{R}^3$, we obtain (F.1).

Finally, we show the invariance of a_{20}, a_{11}, a_{02}. We assume that $f(u, v)$ is expressed by (F.1), and let $(\tilde{u}, \tilde{v})$ be another such admissible coordinate system. It holds that

$$\boldsymbol{0} = f_{\tilde{v}}(0, 0) = u_{\tilde{v}}(0, 0) f_u(0, 0) + v_{\tilde{v}}(0, 0) f_v(0, 0) = \tilde{u}_v f_u(0, 0).$$

Since $f_u(0, 0) \neq \boldsymbol{0}$, we have

$$\tilde{u}_v(0, 0) = 0. \tag{F.2}$$

Under the coordinate system $(\tilde{u}, \tilde{v})$, we assume that there exist $T \in \mathrm{SO}(3)$ and constants $\tilde{a}_{20}, \tilde{a}_{11}, \tilde{a}_{02}(> 0)$ such that

$$f(u, v) = T \circ \tilde{f}(\tilde{u}, \tilde{v}),$$

$$\tilde{f}(\tilde{u}, \tilde{v}) = \left(\tilde{u}, \tilde{u}\tilde{v} + O(\tilde{u}, \tilde{v})^3, \frac{1}{2} \left(\tilde{a}_{20}\tilde{u}^2 + 2\tilde{a}_{11}\tilde{u}\tilde{v} + \tilde{a}_{02}\tilde{v}^2 \right) + O(\tilde{u}, \tilde{v})^3 \right).$$

$$\tag{F.3}$$

Since $f_u(0, 0) = T \circ \tilde{f}_{\tilde{u}}(0, 0)$, the isometry T of $\boldsymbol{R}^3$ fixes the tangential direction $(1, 0, 0)$ at $\boldsymbol{0}$. Moreover, T also preserves the normal plane and co-normal plane In particular, T can be written in the following form:

$$T = \begin{pmatrix} \varepsilon & 0 & 0 \\ 0 & \varepsilon_1 & 0 \\ 0 & 0 & \varepsilon_2 \end{pmatrix}. \tag{F.4}$$

Since $T \in \mathrm{SO}(3)$, we have

$$\varepsilon \varepsilon_1 \varepsilon_2 = 1. \tag{F.5}$$

Comparing the first and second components of $T\tilde{f}(\tilde{u}, \tilde{v})$ and $f(u, v)$, we can write

$$u = \varepsilon \tilde{u}, \quad uv - \varepsilon_1 \varepsilon u \tilde{v} = \varphi(u, v),$$

where $\varphi(u, v)$ is a function belonging to $O(u, v)^3$. Since $\varphi(0, v) = 0$, by the division lemma, there exists a smooth function $\psi(u, v)$ such that $\varphi(u, v) = u\psi(u, v)$ (cf. Proposition A.1 in Appendix A), and so we have

$$u = \varepsilon \tilde{u}, \quad v = \varepsilon_1 \varepsilon \tilde{v} + \psi(u, v), \quad \psi \in O(u, v)^2.$$

Since (u, v) and $(\tilde{u}, \tilde{v})$ are compatible with the orientation, we have $\varepsilon_1 = 1$, that is,

$$\tilde{u} = \varepsilon u, \quad \tilde{v} = \varepsilon v + O(u, v)^2, \tag{F.6}$$

$$\varepsilon_2 = \varepsilon. \tag{F.7}$$

Substituting these into the right-hand side of $f(u,v) = T\tilde{f}(\tilde{u},\tilde{v})$, we have

$$a_{20}u^2 + 2a_{11}uv + a_{02}v^2 = \varepsilon(\tilde{a}_{20}u^2 + 2\tilde{a}_{11}uv + \tilde{a}_{02}v^2) + O(u,v)^3.$$

Since a_{02} and $\tilde{a}_{02}$ are positive, comparing the left- and right-hand sides of this equality, we have $\varepsilon = 1$ and $\tilde{a}_{ij} = a_{ij}$ $(i+j=2)$. $\square$

We call the coordinate system (u,v) in Theorem F.1.2 a *normal coordinate system* (of *second degree*), and Eq. (F.1) is called the *Bruce–West normal form* (of *second degree*) of a cross cap (cf. [7]).

Remark F.1.3. In [7, 87], it is shown that for any cross cap f, there exists a coordinate system (u,v) so that, by appropriately rotating and translating in $\boldsymbol{R}^3$, $f(u,v)$ is expressed as

$$f(u,v) = (u, uv + b(v), a(u,v)),$$
$$b(0) = b'(0) = b''(0) = 0, \quad a(0,0) = a_u(0,0) = a_v(0,0) = 0. \quad \text{(F.8)}$$

These two functions $a(u,v)$ and $b(v)$ are uniquely determined and can be considered as invariants of cross cap singular points (see [31]).

F.2. Fundamental Properties of Cross Caps

In this section, using the Bruce–West normal form (F.1), we discuss differential geometric properties of cross caps. The following assertion gives formulas for the three invariants a_{20}, a_{11} and a_{02}.

Theorem F.2.1 (cf. [29]). *Let* $f\colon U \to \boldsymbol{R}^3$ *be a cross cap at* $p \in U$, *and let* (u,v) *be an admissible coordinate system centered at* p. *Then the coefficients* a_{20}, a_{11} *and* a_{02} *as in* (F.1) *are expressed as follows:*

$$a_{20} = \frac{|f_u \times f_{vv}|}{4|f_u|^3 \det(f_u, f_{uv}, f_{vv})^2}\Big(\det(f_u, f_{uu}, f_{vv})^2$$
$$+ 4\det(f_u, f_{uv}, f_{vv})\det(f_u, f_{uv}, f_{uu})\Big), \qquad \text{(F.9)}$$

$$a_{11} = \frac{1}{2|f_u| \det(f_u, f_{uv}, f_{vv})^2}$$

$$\times \Big(2\det(f_u, f_{uv}, f_{vv})\, (f_u \times f_{uv}) \cdot (f_u \times f_{vv})$$

$$- |f_u \times f_{vv}|^2 \det(f_u, f_{uu}, f_{vv}) \Big), \tag{F.10}$$

$$a_{02} = \frac{|f_u|\, |f_u \times f_{vv}|^3}{\det(f_u, f_{uv}, f_{vv})^2}. \tag{F.11}$$

Proof. Let $(\tilde{u}, \tilde{v})$ be another admissible coordinate system. We set

$$P := f_u \times f_{vv}, \quad \tilde{P} := f_{\tilde{u}} \times f_{\tilde{v}\tilde{v}}, \quad Q := f_u \times f_{uv}, \quad \tilde{Q} := f_{\tilde{u}} \times f_{\tilde{u}\tilde{v}},$$

$$A := \det(f_u, f_{uv}, f_{vv}), \qquad \tilde{A} := \det(f_{\tilde{u}}, f_{\tilde{u}\tilde{v}}, f_{\tilde{v}\tilde{v}}),$$

$$B := \det(f_u, f_{uu}, f_{vv}), \qquad \tilde{B} := \det(f_{\tilde{u}}, f_{\tilde{u}\tilde{u}}, f_{\tilde{v}\tilde{v}}),$$

$$C := \det(f_u, f_{uv}, f_{uu}), \qquad \tilde{C} := \det(f_{\tilde{u}}, f_{\tilde{u}\tilde{v}}, f_{\tilde{u}\tilde{u}}).$$

One can easily check that

$$P = \tilde{u}_u (\tilde{v}_v)^2 \tilde{P}, \quad Q = (\tilde{u}_u)^2 \tilde{v}_v \tilde{Q} + \tilde{u}_u \tilde{v}_u \tilde{v}_v \tilde{P},$$

$$A = (\tilde{u}_u)^2 (\tilde{v}_v)^3 \tilde{A}, \quad B = (\tilde{u}_u)^2 (\tilde{v}_v)^2 (\tilde{u}_u \tilde{B} + 2\tilde{v}_u \tilde{A}),$$

$$C = (\tilde{u}_u)^2 \tilde{v}_v \big((\tilde{u}_u)^2 \tilde{C} - \tilde{u}_u \tilde{v}_u \tilde{B} - (\tilde{v}_u)^2 \tilde{A} \big).$$

Using these formulas, it can be checked that the right-hand sides of (F.9), (F.10) and (F.11) are independent of the choice of admissible coordinate system. Finally, using the Bruce–West normal form as in Theorem F.1.2, these three equalities can be checked. □

These three invariants a_{20}, a_{11}, a_{02} are *intrinsic*, that is, they are determined by the first fundamental form of f.

Corollary F.2.2. *Let (u, v) be an admissible local coordinate at a cross cap at $p \in U$ of $f : U \to \mathbf{R}^3$, and let $ds^2 = E\, du^2 + 2\, F\, du\, dv + G\, dv^2$ be the first fundamental form of f. Then the right-hand sides of (F.9), (F.10) and (F.11) can be written in terms of E, F, G and their first- and second-order derivatives.*

Proof. We first prove that a_{02} is intrinsic: Since $f_v(0,0) = \mathbf{0}$, the formulas

$$f_u \cdot f_u = E, \quad f_u \cdot f_{uv} = F_u, \quad f_u \cdot f_{vv} = F_v, \tag{F.12}$$

$$f_{uv} \cdot f_{uv} = \frac{G_{uu}}{2}, \quad f_{uv} \cdot f_{vv} = \frac{G_{uv}}{2}, \quad f_{vv} \cdot f_{vv} = \frac{G_{vv}}{2} \tag{F.13}$$

hold at $(0,0)$. Using these, we can prove

$$\det(f_u, f_{uv}, f_{vv})^2 = \det\left(\begin{pmatrix} f_u \\ f_{uv} \\ f_{vv} \end{pmatrix}(f_u, f_{uv}, f_{vv})\right) = \det\begin{pmatrix} E & F_u & F_v \\ F_u & G_{uu}/2 & G_{uv}/2 \\ F_v & G_{uv}/2 & G_{vv}/2 \end{pmatrix}$$
$$(\text{F.14})$$

and

$$|f_u|^2 = E, \quad |f_u \times f_{vv}|^2 = (f_u \cdot f_u)(f_{vv} \cdot f_{vv}) - (f_u \cdot f_{vv})^2 = \frac{EG_{vv}}{2} - (F_v)^2$$

at $(0,0)$. Thus, a_{02} is intrinsic.

We next show that a_{20} and a_{11} are intrinsic. In fact, at $(0,0)$, we have that

$$\det(f_u, f_{uu}, f_{vv}) = \frac{1}{\det(f_u, f_{uv}, f_{vv})} \det\left(\begin{pmatrix} f_u \\ f_{uv} \\ f_{vv} \end{pmatrix}(f_u, f_{uu}, f_{vv})\right)$$

$$= \frac{1}{\det(f_u, f_{uv}, f_{vv})} \det\begin{pmatrix} f_u \cdot f_u & f_u \cdot f_{uu} & f_u \cdot f_{vv} \\ f_{uv} \cdot f_u & f_{uv} \cdot f_{uu} & f_{uv} \cdot f_{vv} \\ f_{vv} \cdot f_u & f_{vv} \cdot f_{uu} & f_{vv} \cdot f_{vv} \end{pmatrix},$$

and

$$\det(f_u, f_{uv}, f_{uu}) = \frac{1}{\det(f_u, f_{uv}, f_{vv})} \det\left(\begin{pmatrix} f_u \\ f_{uv} \\ f_{vv} \end{pmatrix}(f_u, f_{uv}, f_{uu})\right)$$

$$= \frac{1}{\det(f_u, f_{uv}, f_{vv})} \det\begin{pmatrix} f_u \cdot f_u & f_u \cdot f_{uv} & f_u \cdot f_{uu} \\ f_{uv} \cdot f_u & f_{uv} \cdot f_{uv} & f_{uv} \cdot f_{uu} \\ f_{vv} \cdot f_u & f_{vv} \cdot f_{uv} & f_{vv} \cdot f_{uu} \end{pmatrix}.$$

Then one can easily verify that the right-hand side is intrinsic, using (F.12), (F.13) and

$$f_u(0,0) \cdot f_{uu}(0,0) = \frac{E_u(0,0)}{2},$$

$$f_{uv}(0,0) \cdot f_{uu}(0,0) = F_{uu}(0,0) - \frac{E_{uv}(0,0)}{2},$$

$$f_{vv}(0,0) \cdot f_{uu}(0,0) = F_{uv}(0,0) - \frac{E_{vv}(0,0)}{2}.$$

Moreover, using (F.13) and (F.14), we have

$$(f_u \times f_{uv}) \cdot (f_u \times f_{vv}) = \det\begin{pmatrix} f_u \cdot f_u & f_u \cdot f_{uv} \\ f_{vv} \cdot f_u & f_{vv} \cdot f_{uv} \end{pmatrix} = \det\begin{pmatrix} E & F_u \\ F_v & G_{uv}/2 \end{pmatrix},$$

at $(0,0)$. So a_{20} and a_{11} are intrinsic. $\qquad\square$

Behavior of the Gaussian curvature and a_{20}. In this subsection, we investigate the behavior of the Gaussian curvature around a cross cap singularity. We fix a cross cap map-germ $f : U \to \mathbf{R}^3$ such that (u, v) is a normal form of f (cf. (F.1)) at the cross cap $(0, 0)$. We then set

$$u = r \cos \theta, \quad v = r \sin \theta. \tag{F.15}$$

Then (r, θ) gives a polar coordinate associated with (u, v): Although the unit normal vector field ν is not well-defined at $(u, v) = (0, 0)$, the unit normal vector ν can be expressed as a smooth function of r, θ as follows:

Lemma F.2.3. *The unit normal vector field ν of the cross cap map-germ f given in (F.1) has an expression*

$$\nu = \frac{1}{A_\theta}(0, -a_{11} \cos \theta - a_{02} \sin \theta, \cos \theta) + O_1,$$

where O_i $(i = 1, 2)$ is a function such that O_i/r^i is bounded near $(0, 0)$, and

$$A_\theta := \sqrt{\cos^2 \theta + (a_{11} \cos \theta + a_{02} \sin \theta)^2} > 0. \tag{F.16}$$

Proof. By (F.1), we have

$$f_u = (1, v, a_{20}u + a_{11}v) + O(u, v)^2, \quad f_v = (0, u, a_{11}u + a_{02}v) + O(u, v)^2.$$

Since $O(u, v)^2 = O_2$, we have

$$f_u \times f_v = r((0, -a_{11} \cos \theta - a_{02} \sin \theta, \cos \theta) + O_1),$$

$$|f_u \times f_v| = r\sqrt{A_\theta + O_1}.$$

Substituting these into $\nu = (f_u \times f_v)/|f_u \times f_v|$, we obtain the conclusion.   □

Using this lemma, one can easily prove the following assertion:

Proposition F.2.4 ([19]). *We let*

$$ds^2 = E\,du^2 + 2\,F\,du\,dv + G\,dv^2, \quad II = L\,du^2 + 2\,M\,du\,dv + N\,dv^2$$

be the first and second fundamental forms of f, respectively. Then

$$E = 1 + O_2, \quad F = O_2, \quad G = r^2(A_\theta^2 + O_1),$$

$$L = \frac{a_{20} \cos \theta}{A_\theta} + O_1, \quad M = -\frac{a_{02} \sin \theta}{A_\theta} + O_1, \quad N = \frac{a_{02} \cos \theta}{A_\theta} + O_1.$$

Moreover, the Gaussian curvature K and the mean curvature H can be expressed as

$$H = \frac{1}{r^2}\left(\frac{a_{02} \cos \theta}{2A_\theta^3} + O_1\right), \tag{F.17}$$

$$K = \frac{a_{02}}{r^2 A_\theta^4}\left(a_{20} \cos^2 \theta - a_{02} \sin^2 \theta + O_1\right). \tag{F.18}$$

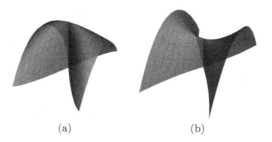

(a)   (b)

Fig. F.2.   An elliptic cross cap (a) and a hyperbolic cross cap (b).

As a consequence, we get the following:

Corollary F.2.5. *If $a_{20} < 0$, then the Gaussian curvature is negative, and* $\lim_{r \to 0} K = -\infty$.

Remark F.2.6. Bruce and West defined ellipticity and hyperbolicity of cross caps. If $a_{20} > 0$, it is called *elliptic*, and if $a_{20} < 0$, then it is called *hyperbolic* (see Fig. F.2 and also Tari [79]).

Proposition F.2.7. *One of the principal curvatures of f diverges at a cross cap, and the other is bounded on $D_\varepsilon := \{(r, \theta) \,;\, \cos\theta \geq \varepsilon\}$ for each $0 < \varepsilon < 1$.*

Proof. By (F.17) and (F.18), the principal curvatures λ_i ($i = 1, 2$) as solutions of the quadratic equation $t^2 - 2Ht + K = 0$ are given by

$$\lambda_1 = \frac{B + \sqrt{B^2 - 4r^2 C^2}}{4 A_\theta^3 r^2}, \quad \lambda_2 = \frac{a_{02}(a_{20}\cos^2\theta - a_{02}\sin^2\theta + O_1)}{A_\theta(B + \sqrt{B^2 - 4r^2 C^2})},$$

where

$$B = a_{02}\cos\theta + O_1, \quad C := a_{20}a_{02}\cos^2\theta - (a_{02})^2 \sin^2\theta + O_2.$$

Thus, λ_1 diverges as $r \to 0$, and λ_2 is bounded on D_ε. $\square$

The behavior of the caustic of f at the cross cap $(0,0)$ is investigated by Fukui–Hasegawa [19], where they find an important concept called the "focal conics" at cross caps. The following is mentioned in [11, 17, 79] and is a consequence of Proposition F.2.7.

Corollary F.2.8. *Umbilic points do not accumulate at a cross cap.*

Proof. If not, there exists a sequence $\{(r_n, \theta_n)\}_{n=1}^{\infty}$ satisfying $r_n \to 0$, $0 \le \theta_n \le 2\pi$ and

$$\lambda_1(r_n, \theta_n) = \lambda_2(r_n, \theta_n) \qquad (n = 1, 2, \dots).$$

Taking a subsequence if necessary, we may assume that $\lim_{n \to \infty} \theta_n = \theta_\infty \in [0, 2\pi]$. By Proposition F.2.7, we may assume that $\cos \theta_\infty = 0$. By (F.18), we have

$$\lim_{n \to \infty} r_n^2 \lambda_1(r_n, \theta_n) \lambda_2(r_n, \theta_n) = \lim_{n \to \infty} r_n^2 K(r_n, \theta_n) = \frac{-a_{02}^2}{A_\theta^2} < 0,$$

contradicting the fact that $\lambda_1(r_n, \theta_n) = \lambda_2(r_n, \theta_n)$. □

We remark that the principal curvature flow at a cross cap is of index one, see [11].

As an aside, suppose that $f : \mathcal{M}^2 \to \mathbb{R}^3$ is a C^∞-map defined on a compact orientable 2-manifold $\mathcal{M}^2$ without boundary. If f admits only cross caps, then

$$\frac{1}{2\pi} \int_{\mathcal{M}^2} K \, dA = \chi(\mathcal{M}^2)$$

holds, that is, the well-known Gauss–Bonnet formula holds without any changes, see [28] for details. The possible symmetries of cross caps are discussed in [31].

The Fundamental Theorem of Frontals as Hypersurfaces

G.1. Properties of Coherent Tangent Bundles of Frontals

In this appendix, we shall introduce the fundamental theorem for frontals as hypersurfaces in the space forms according to the paper [72].

Let $(\mathcal{N}^{m+1}, g)$ be an oriented $(m+1)$-dimensional Riemannian manifold. A C^∞-map $f : \mathcal{M}^m \to \mathcal{N}^{m+1}$ defined on an m-manifold $\mathcal{M}^m$ is called a *frontal* if for each $p \in \mathcal{M}^m$, there exists a neighborhood U of p and a unit vector field ν along f defined on U such that $g\big(df(X), \nu\big) = 0$ holds for any vector field X on U that is, $\nu : U \to S(T\mathcal{N}^{m+1})$ is a smooth unit normal vector field, where $S(T\mathcal{N}^{m+1})$ is the unit tangent bundle of $\mathcal{N}^{m+1}$.

Moreover, if ν can be taken to be an immersion for each $p \in \mathcal{M}^m$, f is called a *front* or a *wave front*.

The subbundle $\mathcal{E}^f$ which consists of the vectors in the pull-back bundle $f^*T\mathcal{N}^{m+1}$ perpendicular to ν induces a bundle homomorphism (cf. (9.15))

$$\varphi_f : T\mathcal{M}^m \ni \boldsymbol{v} \longmapsto df(\boldsymbol{v}) \in \mathcal{E}^f.$$

Let ∇ be the Levi-Civita connection on $\mathcal{N}^{m+1}$. Then by taking the tangential part of ∇, it induces a connection D^f on $\mathcal{E}^f$. Let $\langle\, ,\, \rangle$ be a metric on $\mathcal{E}^f$ induced from the Riemannian metric g on $\mathcal{N}^{m+1}$. Then D^f is a metric connection on $\mathcal{E}^f$. Moreover, it holds that

$$D^f_X \varphi_f(Y) - D^f_Y \varphi_f(X) - \varphi_f([X,Y]) = 0 \quad (X, Y \in \Gamma(T\mathcal{M}^m)), \quad \text{(G.1)}$$

that is, $(\mathcal{M}^m, \mathcal{E}^f, \langle\, ,\, \rangle, D^f, \varphi_f)$ is a coherent tangent bundle (cf. Theorem 9.2.1 in Chapter 9), where $\Gamma(T\mathcal{M}^m)$ denotes the set of smooth vector fields on $\mathcal{M}^m$. A frontal f is called *co-orientable* if there is a unit normal vector field ν globally defined on $\mathcal{M}^m$. By definition, φ_f is co-orientable if and only if f is also. From now on, we assume that f is co-orientable. (If $\mathcal{M}^m$ is simply connected, f is always co-orientable, see Chapter 10.)

We denote by $(\mathcal{N}^{m+1}(c), g)$ the complete simply connected Riemannian manifold of constant sectional curvature c. We set

$$\mathcal{N}^{m+1} := \mathcal{N}^{m+1}(c),$$

and define the second bundle homomorphism (cf. Chapter 9)

$$\psi_f : T\mathcal{M}^m \ni \boldsymbol{v} \longmapsto \nabla_{\boldsymbol{v}} \nu \in \mathcal{E}^f.$$

As shown in Theorem 9.2.2 in Chapter 9, ψ_f gives the structure of a coherent tangent bundle, that is, it satisfies

$$D_X^f \psi_f(Y) - D_Y^f \psi_f(X) - \psi_f([X,Y]) = 0 \quad (X, Y \in \Gamma(T\mathcal{M}^m)). \quad (\text{G.2})$$

We let ds^2 be the first fundamental form of f, that is, it satisfies

$$ds^2(X,Y) = g(\varphi(X), \varphi(Y)) = g(df(X), df(Y)) \quad (X, Y \in \Gamma(T\mathcal{M}^m)),$$

where $\varphi := \varphi_f$. We can prove the following:

Proposition G.1.1 ([72, (2.1)]). *If* $f : \mathcal{M}^m \to \mathcal{N}^{m+1}(c)$ *is a frontal, then*

$$g(\psi(X), \varphi(Y)) = g(\varphi(X), \psi(Y)) \quad (X, Y \in \Gamma(T\mathcal{M}^m)) \quad (\text{G.3})$$

holds, where $\varphi := \varphi_f$ *and* $\psi := \psi_f$.

Proof. Since ν is perpendicular to $df(Z)$ for $Z \in T\mathcal{M}^m$, we have

$$g(\psi_f(X), \varphi_f(Y)) = g(\nabla_X \nu, df(Y)) = Xg(\nu, df(Y)) - g(\nu, \nabla_X df(Y))$$

$$= -g(\nu, \nabla_X df(Y)) = -g(\nu, \nabla_Y df(X) - df([X,Y]))$$

$$= -g(\nu, \nabla_Y df(X)),$$

which implies the assertion. □

Moreover, the following assertion holds:

Proposition G.1.2 ([72, Example 2.2]). *A frontal* $f : \mathcal{M}^m \to \mathcal{N}^{m+1}(c)$ *is a wave front if and only if*

$$\mathrm{Ker}(\varphi_p) \cap \mathrm{Ker}(\psi_p) = \{\mathbf{0}\} \quad (p \in \mathcal{M}^m) \quad (\text{G.4})$$

holds, where $\varphi := \varphi_f$ *and* $\psi = \psi_f$.

Proof. For the sake of simplicity,[1] we prove this for $c = 0$, that is $\mathcal{N}^{m+1}(0) = \boldsymbol{R}^{m+1}$ is the Euclidean space. Suppose that $v \in T_p\mathcal{M}^m$ belongs to $\mathrm{Ker}(\varphi_p) \cap \mathrm{Ker}(\psi_p)$. Then $df(v) = \boldsymbol{0}$ and $\nabla_v\nu = \boldsymbol{0}$ hold, where ∇ is the Levi-Civita connection of $\boldsymbol{R}^{m+1}$. Since $d\nu(v) = \nabla_v\nu$ holds, v gives the kernel of the map $d\nu$ at p, that is, v belongs to the kernel of the differential of the map $L := (f, \nu) : \mathcal{M}^m \to \boldsymbol{R}^{m+1} \times \boldsymbol{R}^{m+1}$. Since f is a wave front if and only if L is an immersion, we obtain the assertion. $\square$

Also, one can show the following:

Fact G.1.3 ([72, Proposition 2.4]). Let $f : \mathcal{M}^m \to \mathcal{N}^{m+1}(c)$ be a co-orientable frontal, and ν a unit normal vector field. Then the following identity (i.e., the Gauss equation) holds:

$$\langle R^D(X,Y)\xi, \zeta \rangle$$
$$= c \det \begin{pmatrix} \langle \varphi(Y), \xi \rangle & \langle \varphi(Y), \zeta \rangle \\ \langle \varphi(X), \xi \rangle & \langle \varphi(X), \zeta \rangle \end{pmatrix} + \det \begin{pmatrix} \langle \psi(Y), \xi \rangle & \langle \psi(Y), \zeta \rangle \\ \langle \psi(X), \xi \rangle & \langle \psi(X), \zeta \rangle \end{pmatrix}, \quad (G.5)$$

where $\varphi = \varphi_f$ and $\psi = \psi_f$, X and Y are vector fields on $\mathcal{M}^m$, ξ and ζ are sections of $\mathcal{E}^f$, and R^D is the curvature tensor of the connection $D := D^f$:

$$R^D(X,Y)\xi := D_X D_Y \xi - D_Y D_X \xi - D_{[X,Y]}\xi.$$

G.2. Construction of Wave Fronts via φ and ψ

Conversely, we can prove the following:

Theorem G.2.1. Let $(\mathcal{M}^m, \mathcal{E}, \langle \, , \, \rangle, D, \varphi)$ be a coherent tangent bundle over a simply connected manifold $\mathcal{M}^m$ and $\psi : T\mathcal{M}^m \to \mathcal{E}$ a bundle homomorphism such that $(\mathcal{M}^m, \mathcal{E}, \langle \, , \, \rangle, D, \psi)$ is also a coherent tangent bundle. Let p be a point of $\mathcal{M}^m$ and $v_1, \ldots, v_m$ an orthonormal basis of $\mathcal{E}_p$. Moreover, let $w_1, \ldots, w_{m+1}$ be a positively oriented orthonormal basis of $T_\mathrm{P}\mathcal{N}^{m+1}(c)$ at a fixed point P on $\mathcal{N}^{m+1}(c)$. Suppose that φ and ψ satisfy (G.3) and (G.5). Then there exist

---

[1]For the case for $c \neq 0$, f can be considered as a map into $\boldsymbol{R}^{m+2}$, since the sphere and the hyperbolic space are hyperquadrics of $\boldsymbol{R}^{m+2}$ (when $c < 0$, $\boldsymbol{R}^{m+2}$ should be regarded as the Lorentz–Minkowski space). So we can use the derivative d on $\boldsymbol{R}^{m+2}$ and ∇ can be considered as its tangential part for the hyperquadric. So we can modify the proof for $c = 0$.

- *a unique frontal $f \colon \mathcal{M}^m \to \mathcal{N}^{m+1}(c)$ such that $f(p) = \mathrm{P}$,*
- *a unique section[2] $\nu \in \Gamma(\mathcal{M}^m, f^*S(T\mathcal{N}^{m+1}(c)))$, and*
- *a unique injective bundle isomorphism $\iota \colon \mathcal{E} \to \mathcal{E}^f$,*

satisfying the following properties:

(a) *ν gives a unit normal vector field of f,*

(b) *$\iota \circ \varphi = \varphi_f$, in particular $df(X) = \varphi_f(X)$ holds for $X \in \Gamma(T\mathcal{M}^m)$, where $\Gamma(T\mathcal{M}^m)$ denotes the set of vector fields on $\mathcal{M}^m$,*

(c) *$\iota \circ \psi = \psi_f$, in particular $d\nu(X) = \nabla_X \nu = \psi_f(X)$ holds for $X \in \Gamma(T\mathcal{M}^m)$,*

(d) *the connection D can be identified with D^f via ι,*

(e) *$\iota(\boldsymbol{v}_j)$ coincides with $\boldsymbol{w}_j$ for $j = 1, \dots, m$, and*

(f) *the unit normal vector ν_p at the base point p coincides with $\boldsymbol{w}_{m+1}$.*

Moreover, f is a wave front if and only if φ and ψ satisfy (G.4).

Proof. For the sake of simplicity, we only prove this assertion for $c = 0$, see [72] for the case of $c \neq 0$.

We let U be a connected open subset of $\mathcal{M}^m$ containing p such that $\mathcal{E}|_U$ is a trivial vector bundle. Then we can take a family of sections

$$V_j \in \Gamma(\mathcal{E}|_U) \quad (j = 1, 2, \dots, m)$$

defined on U such that

- $\{V_1, \dots, V_m\}$ gives an orthonormal basis of $\mathcal{E}_q$ at each point q on U,
- $V_j(p)$ $(j = 1, \dots, m)$ coincides with $\boldsymbol{v}_j$.

Suppose that there exist such a pair (f, ν) and ι satisfying (a)–(f). We set

$$F := (W_1, \dots, W_m, W_{m+1}),$$

where $W_j := \iota(V_j)$ $(j = 1, \dots, m)$ and $W_{m+1} := \nu$. Then F is a map from $U \subset \mathcal{M}^m$ to the special orthogonal group $SO(m+1)$.

Moreover, as an $\boldsymbol{R}^{m+1}$-valued 1-form, df can be written as

$$df := \sum_{i=1}^{m} \xi^i W_i + \eta\nu,$$

---

[2]$\Gamma(\mathcal{M}^m, f^*(S(T\mathcal{N}^{m+1}(c))))$ denotes the set of sections of the pull-back $f^*(S(T\mathcal{N}^{m+1}(c)))$ of the unit sphere bundle $S(T\mathcal{N}^{m+1}(c))$.

where $\xi^1, \ldots, \xi^m$ and η are 1-forms defined on U. Since

$$0 = df(X) \cdot \nu = \eta(X)$$

holds for $X \in \Gamma(TU)$, we have $\eta = 0$ on U. So we can write

$$df := \sum_{i=1}^{m} \xi^i W_i. \tag{G.6}$$

By (b), we have

$$\varphi = \sum_{i=1}^{m} \xi^i V_i. \tag{G.7}$$

On the other hand, we can write

$$D_X^f W_j = \sum_{i=1}^{m} \omega_j^i(X) W_i, \quad \nabla_X \nu = -\sum_{i=1}^{m} h^i(X) W_i,$$

where ω_j^i and h^k are 1-forms on U. By (d), we have

$$D_X V_i = \sum_{j=1}^{m} \omega_i^j(X) V_j. \tag{G.8}$$

In addition, (c) implies

$$\psi(X) = -\sum_{i=1}^{m} h^i(X) V_i. \tag{G.9}$$

Since $dW_{m+1}(X) = d\nu(X) = \nabla_X \nu$ and

$$dW_j(X) \cdot W_i = \omega_j^i(X),$$
$$dW_j(X) \cdot \nu = (\nabla_X W_j) \cdot \nu = X(W_j \cdot \nu) - W_j \cdot \nabla_X \nu$$
$$= -W_j \cdot \nabla_X \nu = -W_j \cdot D_X^f \nu$$
$$= W_j \cdot \left(\sum_{k=1}^{m} h^k(X) W_k \right) = h^j(X),$$

setting $\Omega := (\omega_j^i)$, we have that

$$dF = (dW_1, \ldots, dW_m, dW_{m+1}) = (dW_1, \ldots, dW_m, d\nu)$$
$$= (W_1, \ldots, W_m, \nu) \begin{pmatrix} \Omega & -\boldsymbol{h} \\ \boldsymbol{h}^T & 0 \end{pmatrix} = F \begin{pmatrix} \Omega & -\boldsymbol{h} \\ \boldsymbol{h}^T & 0 \end{pmatrix},$$

where ^{T} denotes transposition, and $\boldsymbol{h}^T := (h^1, \ldots, h^m)$.

Under the observations above, we shall prove the theorem. We regard the identities

$$df = \sum_{i=1}^{m} \xi^j W_j, \tag{G.10}$$

$$dF = F \begin{pmatrix} \Omega & -\boldsymbol{h} \\ \boldsymbol{h}^T & 0 \end{pmatrix} \tag{G.11}$$

as a system of differential equations for unknown functions $F : U \to \mathrm{SO}(m+1)$ and $f : U \to \boldsymbol{R}^{m+1}$, where ξ^j, h^j $(j = 1, \ldots, m)$ and $\Omega = (\omega_j^i)$ are defined by (G.7), (G.9) and (G.8), respectively.

The condition $D_X \varphi(Y) - D_Y \varphi(X) - \varphi([X, Y]) = 0$ for φ is equivalent to

$$d\xi^j + \sum_{l=1}^{m} \omega_l^j \wedge \xi^l = 0 \quad (j = 1, \ldots, m). \tag{G.12}$$

Similarly, since $D_X \psi(Y) - D_Y \psi(X) - \psi([X, Y]) = 0$, it holds that

$$dh^j + \sum_{l=1}^{m} \omega_l^j \wedge h^l = 0 \quad (j = 1, \ldots, m). \tag{G.13}$$

Using matrix notations, this is equivalent to

$$d\boldsymbol{h} + \Omega \wedge \boldsymbol{h} = \boldsymbol{0}. \tag{G.14}$$

Since D is a metric connection, it holds that

$$\omega_i^j = -\omega_j^i, \quad \text{that is, } \Omega^T = -\Omega. \tag{G.15}$$

In addition, the condition (G.3) implies

$$0 = g(\varphi(X), \psi(Y)) - g(\varphi(Y), \psi(X))$$

$$= \sum_{i=1}^{m} \xi^i(X)h^i(Y) - \sum_{i=1}^{m} \xi^i(Y)h^i(X) = \sum_{i=1}^{m} \xi^i \wedge h^i(X, Y),$$

so we have

$$\sum_{i=1}^{m} \xi^i \wedge h^i = 0. \tag{G.16}$$

Using these, we now check the integrability of (G.11) and (G.10). We let

$$\hat{\Omega} := \begin{pmatrix} \Omega & -\boldsymbol{h} \\ \boldsymbol{h}^T & 0 \end{pmatrix}.$$

Then the integrability of (G.11) is equivalent to

$$O = d\hat{\Omega} + \hat{\Omega} \wedge \hat{\Omega} = \begin{pmatrix} d\Omega + \Omega \wedge \Omega - \boldsymbol{h} \wedge \boldsymbol{h}^T & -d\boldsymbol{h} - \Omega \wedge \boldsymbol{h} \\ d\boldsymbol{h}^T + \boldsymbol{h}^T \wedge \Omega & -\boldsymbol{h}^T \wedge \boldsymbol{h} \end{pmatrix}. \qquad (G.17)$$

The top-left component of the identity (G.17) is equivalent to

$$d\omega_i^j + \left(\sum_{k=1}^m \omega_k^j \wedge \omega_i^k \right) - h^j \wedge h^i = 0,$$

which is equivalent to the condition (G.5). On the other hand, the bottom-right of the right-hand side is

$$-\boldsymbol{h}^T \wedge \boldsymbol{h} = -\sum_{i=1}^m h^i \wedge h^i = 0,$$

since $h^i \wedge h^i = 0$ for each $i = 1, \ldots, m$. The top-right and the bottom-left components of the right-hand side of (G.17) vanish because of (G.14) and (G.15). Thus, the integrability of (G.11) has been shown. On the other hand, the integrability of (G.10) corresponds to the fact that the right-hand side of (G.10) is closed. By (G.11), it holds that

$$dW_i = \left(\sum_{j=1}^m \omega_i^j W_j \right) + h^i W_{m+1},$$

where W_j $(j = 1, \ldots, m)$ are the first m columns of F satisfying (G.11): Hence

$$d\left(\sum_{i=1}^m \xi^i W_i \right) = \sum_{i=1}^m \left(d\xi^i W_i - \xi^i \wedge dW_i \right)$$

$$= \sum_{i=1}^m \left(d\xi^i - \sum_{j=1}^m \xi^j \wedge \omega_j^i \right) W_i + \left(\sum_{i=1}^m \xi^i \wedge h^i \right) W_{m+1} = \boldsymbol{0}$$

because of (G.12) and (G.16).

Thus, there exist a unique smooth map $F : U \to SO(m+1)$ and $f : U \to \boldsymbol{R}^{m+1}$ satisfying (G.11) and (G.10) with initial conditions

$$F(p) = (\boldsymbol{w}_1, \ldots, \boldsymbol{w}_{m+1}) \quad \text{and} \quad f(p) = \text{P}. \qquad (G.18)$$

By (G.10), we have

$$df(V_i) = \sum_{j=1}^m \xi^j(V_i) W_j \quad (i = 1, \ldots, m),$$

which are perpendicular to W_{m+1}. So $\nu := W_{m+1}$ is the unit normal vector field of f. For each $q \in U$, we define a linear map $\iota_q : \mathcal{E}_q \to \mathcal{E}_{f(q)}^f$ by

$$\iota_q(V_j(q)) = W_j(q) \quad (j = 1, 2, \ldots, m). \qquad (G.19)$$

Then ι gives a bundle isomorphism from $\mathcal{E}|_U$ to $\mathcal{E}^f|_U$. By definition, ι satisfies (b), (c), (d) on U. Moreover, (e) and (f) hold because of the choice of the initial conditions.

Next we show that f and $\nu(= W_{m+1})$ do not depend on the choice of a frame field $V_1, \ldots, V_m$. In fact, let $\tilde{V}_1, \ldots, \tilde{V}_m$ be another orthonormal frame on U such that $\tilde{V}_i(p) = v_i$ for $i = 1, \ldots, m$. Then we can write

$$(\tilde{V}_1, \ldots, \tilde{V}_m) = (V_1, \ldots, V_m)G, \tag{G.20}$$

where $G : U \to \mathrm{SO}(m)$ is a smooth map such that $G(p) = \mathrm{id}$. If we set

$$\tilde{\Omega} := G^{-1}\Omega G + G^{-1}dG, \quad \tilde{h} := G^{-1}h$$

then the equations (G.10) and (G.11) with respect to $(\tilde{V}_1, \ldots, \tilde{V}_m)$ just turn out to be

$$d\tilde{f} = \sum_{i=1} \tilde{\xi}^j \tilde{W}_j, \quad d\tilde{F} = \tilde{F}\begin{pmatrix} \tilde{\Omega} & -\tilde{h} \\ \tilde{h}^T & 0 \end{pmatrix}, \tag{G.21}$$

where

$$\begin{pmatrix} \tilde{\xi}^1 \\ \vdots \\ \tilde{\xi}^m \end{pmatrix} = G^{-1}\begin{pmatrix} \xi^1 \\ \vdots \\ \xi^m \end{pmatrix}.$$

Hence the solutions $\tilde{F}$ and $\tilde{f}$ of (G.21) with initial condition (G.18) are given by

$$(\tilde{W}_1, \ldots, \tilde{W}_m, \tilde{W}_{m+1}) := \tilde{F} = F\begin{pmatrix} G & 0 \\ 0 & 1 \end{pmatrix}, \quad \tilde{f} = f.$$

In particular, we have

$$(\tilde{W}_1, \ldots, \tilde{W}_m) = (W_1, \ldots, W_m)G, \quad \tilde{W}_{m+1} = W_{m+1},$$

and the smooth map f and the unit normal vector field $\tilde{\nu} := \tilde{W}_{m+1}$ with respect to $(\tilde{V}_1, \ldots, \tilde{V}_m)$ coincide to those with respect to $(V_1, \ldots, V_m)$. Moreover, if we set $\tilde{\iota} \colon \mathcal{E}|_U \to \mathcal{E}^f|_U$ by $\tilde{\iota}(\tilde{V}_i) = \tilde{W}_i$, we have

$$(\tilde{\iota}(\tilde{V}_1), \ldots, \tilde{\iota}(\tilde{V}_m)) = (\tilde{W}_1, \ldots, \tilde{W}_m) = (W_1, \ldots, W_m)G$$
$$= (\iota(V_1), \ldots, \iota(V_m))G.$$

By (G.20), this implies $\tilde{\iota}(\tilde{V}_i) = \iota(\tilde{V}_i)$ for $i = 1, \ldots, m$, that is, the bundle isomorphism ι does not depend on the choice of $(V_1, \ldots, V_m)$.

Finally, we show that f and ν are defined globally on $\mathcal{M}^m$. For each $q \in \mathcal{M}^m$, there exists a continuous curve $\gamma : [0, 1] \to \mathcal{M}^m$ such that $\gamma(0) = p$ and $\gamma(1) = q$. Since $\gamma([0, 1])$ is compact, this can be covered by a finite family $\mathcal{U} := \{U_i\}_{i=1}^k$ of coordinate neighborhoods of $\mathcal{M}^m$ such that each restriction $\mathcal{E}|_{U_j}$ is a trivial bundle and $U_i \cap U_{i+1} \neq \emptyset$ for $i = 1, \ldots, k-1$. So we can apply the above argument for $U := U_i$ for $i = 1, 2, \ldots, k$ inductively. Consequently, f and ι are defined on $U_1 \cup \cdots \cup U_k$. In particular, $f(q)$ and $\iota : \mathcal{E}_q \to (\mathcal{E}^f)_q$ are determined independently of the choice of $\mathcal{U}$. Moreover, since $\mathcal{M}^m$ is simply connected, they are determined independently of the choice of γ. So f and ι can be canonically extended on $\mathcal{E}$ and $\mathcal{M}^m$, respectively. $\square$

Appendix H

The Half-space Theorem for Swallowtails

In this appendix, we show the following assertion:

Theorem H.1. *Let* $f : U \to \mathbf{R}^3$ *be a* C^∞-*map defined on a domain* $U(\subset \mathbf{R}^2)$ *such that* $p \in U$ *is a swallowtail singular point of* f. *Then the section of the image of* f *by the normal plane of* f *at* $f(p)$ *lies on one side of limiting tangent plane* (*see Fig. 2.11 in Chapter 2*).

Without loss of generality, we may assume that $f(p) = \mathbf{0}(= (0, 0, 0))$. We can take a local coordinate system (u, v) centered at p so that the u-axis is the singular curve. Since $p = (0, 0)$ is a swallowtail, the null vector field $\eta(u)$ along the u-axis can be taken as

$$\eta(u) := \partial_u + \varepsilon(u)\partial_v \quad (\varepsilon(0) = 0, \ \varepsilon'(0) \neq 0),$$

where

$$\partial_u := \partial/\partial u, \quad \partial_v := \partial/\partial v.$$

Setting $\tilde{u} = \varepsilon(u)$ and $\tilde{v} = v$, we see that $(\tilde{u}, \tilde{v})$ is a coordinate system since $\varepsilon'(0) \neq 0$. The coordinate system $(\tilde{u}, \tilde{v})$ satisfies that the $\tilde{u}$-axis is the singular curve, and $\eta(\tilde{u}) := \partial_{\tilde{u}} + \tilde{u}\partial_{\tilde{v}}$ is a null-vector field of f along the $\tilde{u}$-axis. We rewrite $(\tilde{u}, \tilde{v})$ as (u, v). Since $f_u(0, 0) = \mathbf{0}$, we have $f_v(0, 0) \neq \mathbf{0}$. So, the section of the image of f by the normal plane of f at $f(p)$ is given by

$$S := \{f(u, v) \, ; \, f(u, v) \cdot f_v(0, 0) = 0\}.$$

If we set $h(u, v) := f(u, v) \cdot f_v(0, 0)$, then we have

$$h(0, 0) = 0, \quad h_v(0, 0) = f_v(0, 0) \cdot f_v(0, 0) \neq 0.$$

By the implicit function theorem, exists a function $v(u)$ satisfying $v(0) = 0$ and $h(u, v(u)) = 0$. In particular, if we set

$$\hat{c}(u) := f(c(u)), \quad c(u) := (u, v(u)),$$

then $\hat{c}(u)$ gives a parametrization of the set S. We set

$$g(u) := \hat{c}(u) \cdot \nu(0,0) = f(u, v(u)) \cdot \nu(0,0),$$

where ν denotes the unit normal vector field of f. To show Theorem H.1, it is sufficient to show that

$$g'(0) = g''(0) = g'''(0) = 0, \quad g^{(4)}(0) \neq 0. \tag{H.1}$$

To prove this, we prepare the following two lemmas:

Lemma H.2. *We have* $f_{uu} = -f_v$, $f_{uuu} = -2f_{uv}$ *and* $f_{uuuu} = -3f_{uuv}$ *at* $p = (0,0)$.

Proof. Since $\eta = \partial_u + u\partial_v$ points in the null direction of f, we have

$$(f_\eta =)f_u + uf_v = \mathbf{0}$$

on the u-axis. So we have

$$(f_\eta)_u = f_v + uf_{uv} + f_{uu} = \mathbf{0}, \quad (f_\eta)_{uu} = 2f_{uv} + uf_{uuv} + f_{uuu} = \mathbf{0},$$
$$(f_\eta)_{uuu} = 3f_{uuv} + uf_{uuuv} + f_{uuuu} = \mathbf{0}$$

on the u-axis. Substituting $u = 0$, we obtain the conclusion. □

Lemma H.3. *We have* $v'(0) = 0$ *and* $v''(0) = 1$.

Proof. Differentiating $h(u, v(u)) = f(u, v(u)) \cdot f_v(0,0) = 0$, we have

$$(f_u(u, v(u)) + v'(u)f_v(u, v(u))) \cdot f_v(0,0) = 0, \tag{H.2}$$
$$(f_{uu}(u, v(u)) + 2v'(u)f_{uv}(u, v(u))$$
$$+ v'(u)^2 f_{vv}(u, v(u)) + v''(u)f_v(u, v(u))) \cdot f_v(0,0) = 0. \tag{H.3}$$

Since $f_u(0,0) = \mathbf{0}$, $f_v(0,0) \neq \mathbf{0}$ and $v(0) = 0$, substituting $u = 0$ into (H.2), we have $v'(0) = 0$. Then (H.3) reduces to

$$(f_{uu}(0,0) + v''(0)f_v(0,0)) \cdot f_v(0,0) = 0.$$

By Lemma H.2, we have $f_{uu}(0,0) = -f_v(0,0)$ and so $v''(0) = 1$. □

Proof of Theorem H.1 Setting $\nu_0 := \nu(0,0)$, we have

$$g'(u) = (f_u(c(u)) + v'(u)f_v(c(u))) \cdot \nu_0,$$

$$g''(u) = (f_{uu}(c(u)) + 2v'(u)f_{uv}(c(u)) + v'(u)^2 f_{vv}(c(u))$$
$$+ v''(u)f_v(c(u))) \cdot \nu_0,$$

$$g'''(u) = (f_{uuu}(c(u)) + 3v'f_{uuv}(c(u))) + 3v'v''f_{vv}(c(u)) + 3v'^2 f_{uvv}(c(u))$$
$$+ 3v''f_{uv}(c(u)) + v'^3 f_{vvv}(c(u)) + v'''f_v(c(u))) \cdot \nu_0,$$

$$g^{(4)}(u) = (f_{uuuu}(c(u)) + 4v'f_{uuuv}(c(u)) + 6v'^2 f_{uuvv}(c(u)) + 6v''f_{uuv}(c(u))$$
$$+ 12v''v'f_{uvv}(c(u)) + 6v''v'^2 f_{vvv}(c(u)) + 4v'^3 f_{uvvv}(c(u))$$
$$+ 4v'''f_{uv}(c(u)) + v'^4 f_{vvvv}(c(u)) + 4v'v'''f_{vv}(c(u))$$
$$+ 3v''^2 f_{vv}(c(u)) + v^{(4)}f_v(c(u))) \cdot \nu_0.$$

Since ∂_u points in the null direction at $(0,0)$, we have $f_u(0,0) = \mathbf{0}$. By Lemma H.3 and the fact $f_v(0,0) \cdot \nu(0,0) = 0$, we have

$$g'(0) = 0, \qquad g''(0) = f_{uu}(0,0) \cdot \nu_0, \tag{H.4}$$

$$g'''(u) = \Big(f_{uuu}(c(u)) + 3f_{uv}(c(u))\Big) \cdot \nu_0, \tag{H.5}$$

$$g^{(4)}(u) = \Big(f_{uuuu}(c(u)) + 6f_{uuv}(c(u))$$
$$+ 4v'''f_{uv}(c(u)) + 3f_{vv}(c(u))\Big) \cdot \nu_0. \tag{H.6}$$

By (H.4) and Lemma H.2, we have $g''(0) = 0$. By Lemma H.2, we have $f_{uuu} = -2f_{uv}$ at $(0,0)$. Since $f_{uv} \cdot \nu = -f_u \cdot \nu_v$ and $f_u(0,0) = \mathbf{0}$, (H.5) yields that $g'''(0) = 0$ and $f_{uv}(c(0)) \cdot \nu_0 = 0$. By Lemma H.2, we have $f_{uuuu}(0,0) = -3f_{uuv}(0,0)$. So we obtain

$$g^{(4)}(0) = 3(f_{uuv}(0,0) + f_{vv}(0,0)) \cdot \nu_0.$$

On the other hand, since $f_u + uf_v = \mathbf{0}$ along the u-axis, there exists a smooth $\mathbf{R}^3$-valued function $\varphi(u,v)$ such that

$$f_u(u,v) + uf_v(u,v) = v\varphi(u,v). \tag{H.7}$$

Then we have $v\varphi(u,v) \cdot \nu(u,v) = 0$. By the continuity of φ, we have

$$\varphi(u,v) \cdot \nu(u,v) = 0. \tag{H.8}$$

Differentiating (H.7) by u and v, we have

$$f_{uuv} + f_{vv} + uf_{uvv} = \varphi_u + v\varphi_{uv}.$$

Substituting $u = v = 0$, $f_{uuv} + f_{vv} = \varphi_u$ holds at $(0,0)$. Regarding the identity (H.8), we have

$$g^{(4)}(0) = 3\varphi_u(0,0) \cdot \nu(0,0) = -3\varphi(0,0) \cdot \nu_u(0,0). \tag{H.9}$$

Since swallowtails are non-degenerate singular points, we have

$$0 \neq \lambda_v = \det(f_u, f_v, \nu)_v = \det(f_{uv}, f_v, \nu) = \det(\varphi, f_v, \nu)$$

at $(u,v) = (0,0)$. In particular, $\{\varphi, f_v, \nu\}$ is a frame near the origin. Since swallowtails are fronts, $f_u(0,0) = \mathbf{0}$ implies $\nu_u(0,0) \neq \mathbf{0}$. Since

$$\nu_u \cdot f_v = -\nu \cdot f_{vu} = \nu_v \cdot f_u,$$

we have $\nu_u \cdot f_v = 0$ at $(0,0)$. This with the fact $\nu_u \cdot \nu = 0$ yields $\varphi \cdot \nu_u \neq 0$ at $(0,0)$, because $\{\varphi, f_v, \nu\}$ is a linearly independent set. Hence (H.9) implies $g^{(4)}(0) \neq 0$.

Bibliography

[1] V. I. Arnol'd, S. M. Gusein-Zade and A. N. Varchenko, *Singularities of Differentiable Maps. Vol. I.* Monog. Math., Vol. 82, Birkhauser Inc., 1985.

[2] V. I. Arnol'd, *Topological Invariants of Plane Curves and Caustics*, AMS, Univ Lectures **5**, Providence, 1994.

[3] V. I. Arnol'd, The geometry of spherical curves and the algebra of quaternions, *Russian Math. Surv.* **59** (1995), 1–68.

[4] D. Bleecker and L. Wilson, Stability of Gauss maps, *Illinois J. Math.* **22** (1978), 279–289.

[5] E. Brieskorn and H. Knörrer, *Plane Algebraic Curves*, Birkhäuser, 1986.

[6] J. W. Bruce, Geometry of singular sets, *Math. Proc. Cambridge Philos. Soc.* **106** (1989), 495–509.

[7] J. W. Bruce and J. M. West, Functions on a crosscap, *Math. Proc. Cambridge Philos. Soc.* **123** (1998), 19–39.

[8] J. W. Bruce and T. C. Wilkinson, *Folding Maps and Focal Sets*, Singularity Theory and its Applications, Part I (Coventry, 1988/1989), Lecture Notes in Math., Vol. 1462, Springer, Berlin, 1991, pp. 63–72.

[9] N. Dutertre and T. Fukui, On the topology of stable maps, *J. Math. Soc. Japan* **66** (2014), 161–203.

[10] W. Domitrz and M. Zwierzyński, The Gauss–Bonnet theorem for coherent tangent bundles over surfaces with boundary and its applications, *J. Geom. Anal.* **30** (2020), 3243–3274 (cf. https://doi.org/10.1007/s12220 -019-00197-0).

[11] R. Garcia, C. Gutierrez and J. Sotomayor, Lines of principal curvature around umbilics and Whitney umbrellas, *Tohoku Math. J.* (2) **52** (2000), 163–172.

[12] J. J. Koenderink, *Solid Shape*, MIT Press Series in Artificial Intelligence, MIT Press, Cambridge, MA, 1990.

[13] Y. Chen and M. Kossowski, Global differential geometry of 1-resolvable C^∞ curves in the plane, *Ann. Global Anal. Geom.* **16** (1998), 173–188.

[14] J. Damon, P. Giblin and G. Haslinger, *Local Features in Natural Images via Singularity Theory*, Lecture Notes in Mathematics, Vol. 2165, Springer, [Cham], 2016.

[15] A. C. C da Silva, *Lectures on Symplectic Geometry*, Lecture Notes in Mathematics, Springer Verlag, 2008.

[16] T. Fukui, Local differential geometry of cuspidal edge and swallowtail, 2017, preprint. (www.rimath.saitama-u.ac.jp/lab.jp/Fukui/preprint/CE_ST.pdf).

[17] T. Fukui and J. J. Nuno-Ballesteros, Isolated roundings and flattenings of submanifolds in Euclidean spaces, *Tohoku Math. J.* **57** (2005), 469–503.

[18] S. Fujimori, K. Saji, M. Umehara and K. Yamada, Singularities of maximal surfaces, *Math. Z.* **259** (2008), 827–848.

[19] T. Fukui and M. Hasegawa, Fronts of Whitney umbrella – Differential geometric approach via blowing up, *J. Singul.* **4** (2012), 35–67.

[20] D. Fuchs and S. Tabachnikov, *Thirty Lectures on Classic Mathematics*, American Mathematical Society, 2007.

[21] T. Fukunaga and M. Takahashi, Framed surfaces in the Euclidean space, *Bull. Braz. Math. Soc.* **50** (2019), 37–65.

[22] M. Golubitsky and V. Guillemin, *Graduate Texts in Mathematics: Stable Mappings and Their Singularities*, Vol. 14, Springer-Verlag, 1973.

[23] M. J. Greenberg and J. R. Harper, *Algebraic Topology: A First Course*, Mathematics Lecture Note Series, CRC Press, 1981.

[24] H. Gounai and M. Umehara, Caustics of convex curves, *J. Knot Theory Ramifications* **23** (2014), 1–28.

[25] R. Hartshorne, *Algebraic Geometry*, Springer-Verlag, New York, 1997 (cf.ht tps://www.springer.com/jp/book/9780387902449).

[26] M. W. Hirsch, *Graduate Texts in Mathematics: Differential Topology*, Vol. 33, Springer-Verlag, New York-Heidelberg, 1976.

[27] F. Hirzeburuch, Topological Method in Algebraic Geometry, Springer-Verlag, 1956, 1962 and 1966.

[28] M. Hasegawa, A. Honda, K. Naokawa, K. Saji, M. Umehara and K. Yamada, Intrinsic properties of surfaces with singularities, *Int. J. Math.* **26** (2015), doi:10.1142/S012697X150008X.

[29] M. Hasegawa, A. Honda, K. Naowaka, M. Umehara and K. Yamada, Intrinsic invariants of cross caps, *Selecta Math. New Ser.* **20** (2014), 769–785.

[30] A. Honda, K. Naokawa, K. Saji, M. Umehara and K. Yamada, Duality on generalized cuspidal edges preserving singular set images and first fundamental forms, *J. Singularities* **22** (2020), 59–91.

[31] A. Honda, K. Naokawa, K. Saji, M. Umehara and K. Yamada, Symmetries of cross caps, preprint, arXiv:2105.01967.

[32] A. Honda, K. Naokawa, K. Saji, M. Umehara and K. Yamada, A generalization of Zakalyukin's Lemma, and symmetries of surface singularities, to appear in J. Singul.

[33] A. Honda, K. Naokawa, M. Umehara and K. Yamada, Isometric deformations of wave fronts at non-degenerate singular points, *Hiroshima Math. J.* **50** (2020), 269–312.

[34] P. Hartman and L. Nirenberg, On spherical image maps whose Jacobians do not change sign, *Amer. J. Math.* **81** (1959), 901–920.

[35] G. Ishikawa, Singularities of Frontals, Singularities in Generic Geometry, Adv. Stud. Pure Mathematics, Vol. 78, Mathematical Society of Japan, Tokyo, 2018, pp. 55–106.

[36] S. Izumiya, Legendrian dualities and spacelike hypersurfaces in the lightcone, *Mosc. Math. J.* **9** (2009), 325–357.

[37] S. Izumiya, M. C. Romero Fuster, M. A. S. Ruas and F. Tari, *Differential Geometry from a Singularity Theory Viewpoint*, World Scientific Publishing Co. Pte. Ltd., Hackensack, NJ, 2016.

[38] S. Izumiya, K. Saji and N. Takeuchi, Flat surfaces along cuspidal edges, *J. Singul.* **16** (2017), 73–100.

[39] S. Izumiya and K. Saji, The mandala of Legendrian dualities for pseudospheres in Lorentz-Minkowski space and "flat" spacelike surfaces, *J. Singul.* **2** (2010), 92–127.

[40] S. Izumiya, K. Saji and M. Takahashi, Horospherical flat surfaces in hyperbolic 3-space, *J. Math. Soc. Japan* **62** (2010), 789–849.

[41] S. Kobayashi and K. Nomizu, *Foundations of Differential Geometry*. Vol. I, Reprint of the 1963 original. Wiley Classics Library. A Wiley-Interscience Publication. John Wiley & Sons, Inc., New York, 1996.

[42] S. Kobayashi and K. Nomizu, *Foundations of Differential Geometry*. Vol. II, Reprint of the 1969 original. Wiley Classics Library. A Wiley-Interscience Publication. John Wiley & Sons, Inc., New York, 1996.

[43] J. J. Koenderink, What does the occluding contour tell us about solid shape?, *Perception* **13** (1984), 321–330.

[44] M. Kokubu, W. Rossman, K. Saji, M. Umehara and K. Yamada, Singularities of flat fronts in hyperbolic 3-space, *Pacific J. Math.* **221** (2005), 303–351.

[45] M. Kokubu, W. Rossman, K. Saji, M. Umehara and K. Yamada, Addendum: Singularities of flat fronts in hyperbolic 3-space, *Pacific J. Math.* **294** (2018), 505–509.

[46] D. Korteweg, Sur les points de plissement, *Arch. Néeerl*, **24** (1891), 57–98.

[47] M. Kossowski, The Boy–Gauss–Bonnet theorems for C^∞-singular surfaces with limiting tangent bundle, *Ann. Global Anal. Geom.* **21** (2002), 19–29.

[48] M. Kossowski, Realizing a singular first fundamental form as a nonimmersed surface in Euclidean 3-space, *J. Geom.* **81** (2004), 101–113.

[49] R. Langevin, G. Levitt and H. Rosenberg, Classes d'homotopie de surfaces avec rebroussements et queues d'aronde dans $\boldsymbol{R}^3$, *Canad. J. Math.* **47** (1995), 544–572.

[50] H. Levine, Mappings of manifolds into the plane, *Amer. J. Math.* **88** (1966), 357–365.

[51] B. Malgrange, Ideals of Differentiable Functions, Tata Institute of Fundamental Research Studies in Mathematics, Vol. 3, Tata Institute of Fundamental Research, Oxford University Press, 1967.

[52] J. Martinet, *Singularities of Smooth Functions and Maps*, London Math. Soc. Lecture Note Series, Vol. 58. Cambridge University Press, 1982.

[53] L. F. Martins and K. Saji, Geometric invariants of cuspidal edges, *Canad. J. Math.* **68** (2016), 445–462.

[54] L. F. Martins, K. Saji, M. Umehara and K. Yamada, Behavior of Gaussian curvature and mean curvature near non-degenerate singular points on wave fronts, *Springer Proc. Math. & Stat.* **154** (2016), 247–282.

[55] W. S. Massey, *Graduate Texts in Mathematics: A Basic Course in Algebraic Topology*, Vol. 127, Springer-Verlag, 1991.

[56] M. Melko and I. Sterling, Application of Soliton theory to the construction of pseudospherical surfaces in $\mathbf{R}^3$, *Ann. Global Anal. Geom.* **11** (1993), 65–107.

[57] J. W. Milnor, *Topology from the Differentiable Viewpoint*, Based on notes by David W. Weaver, The University Press of Virginia, Charlottesville, Va. 1965.

[58] J. W. Milnor and D. Stasheff, *Characteristic Classes*, Annals of Mathematics Studies, Princeton University Press, 1974.

[59] S. Murata and M. Umehara, Flat surfaces with singularities in Euclidean 3-space, *J. Differential Geom.* **82** (2009), 279–316.

[60] J. R. Munkres, *Elementary Differential Topology*, rev. ed., Princeton University, Press, Princeton, NJ, 1966.

[61] K. Naokawa, Singularities of the asymptotic completion of developable Mobius strips, *Osaka J. Math.* **50** (2013), 425–437.

[62] S. Ohno, T. Ozawa and M. Umehara, Closed planar curves without inflections, *Proc. Amer. Math. Soc.* **141** (2013), 651–665.

[63] R. Oset Sinha and F. Tari, On the flat geometry of the cuspidal edge, *Osaka J. Math.* **55** (2018), 393–421.

[64] I. R. Porteous, The normal singularities of a submanifold, *J. Differ. Geom.* **5** (1971), 543–564.

[65] I. R. Porteous, *Geometric Differentiation: For the Intelligence of Curves and Surfaces*, 2nd ed., Cambridge University Press, 2001.

[66] J. R. Quine, A global theorem for singularities of maps between oriented 2-manifolds, *Trans. Amer. Math. Soc.* **236** (1978), 307–314.

[67] K. Saji, Criteria for D_4 singularities of wave fronts, *Tohoku Math. J.* (2) **63** (2011), 137–147.

[68] K. Saji, M. Umehara and K. Yamada, Behavior of corank one singular points on wave fronts, *Kyushu J. Math.* **62** (2008), 259–280.

[69] K. Saji, M. Umehara and K. Yamada, The geometry of fronts, *Ann. Math.* **169** (2009), 491–529.

[70] K. Saji, M. Umehara and K. Yamada, A_k singularities of wave fronts, *Math. Proc. Cambridge Phil. Soc.* **146** (2009), 731–746.

[71] K. Saji, M. Umehara and K. Yamada, The duality between singular points and inflection points on wave fronts, *Osaka. J. Math.* **47** (2010), 591–607.

[72] K. Saji, M. Umehara and K. Yamada, A_2-singularities of hypersurfaces with non-negative sectional curvature in Euclidean space, *Kodai Math. J.* **34** (2011), 390–409.

[73] K. Saji, M. Umehara and K. Yamada, Coherent tangent bundles and Gauss-Bonnet theorems for wave fronts, *J. Geom. Anal.* **22**, (2012), 383–409.

[74] K. Saji, M. Umehara and K. Yamada, An index formula for a bundle homomorphism of the tangent bundle into a vector bundle of the same rank, and its applications, *J. Math. Soc. Japan* **69** (2017), 417–457.

[75] S. Shiba and M. Umehara, The behavior of curvature functions at cusps and inflection points, *Differential Geom. Appl.* **30** (2012), 285–299.

[76] I. M. Singer and J. A. Thorpe, *Lecture Notes on Elementary Topology and Geometry*, Scott, Foresman and Co., Glenview, Ill. 1967.

[77] M. Spivak, *A Comprehensive Introduction to Differential Geometry*. Vol. I, 2nd ed., Publish or Perish, Inc., Wilmington, Del., 1979.

[78] M. Spivak, *Calculus*, Cambridge University Press, 2006.

[79] F. Tari, On pairs of geometric foliations on a cross-cap, *Tohoku Math. J.* **59** (2007), 233–258.

[80] C. H. Taubes, *Differential Geometry. Bundles, Connections, Metrics and Curvature*, Oxford Graduate Texts in Mathematics, Vol. 23, Oxford University Press, Oxford, 2011.

[81] M. Umehara, *Differential Geometry of Surfaces with Singular Points*, (Japanese) Joy of Math. Eds. K. Ueno, T. Sunada and H. Arai, Nippon Hyoron Sha Co., Ltd. Publishers, 2005, pp. 50–64.

[82] M. Umehara, *Tokuiten wo motsu kyokusen to kyokumen no kikagaku*. (Japanese) [Geometry of curves and surfaces with singular points], Seminar on Mathematical Sciences, Vol. 38, Keio University, Department of Mathematics, Yokohama, 2009.

[83] M. Umehara and K. Yamada, Applications of a completeness lemma in minimal surface theory to various classes of surfaces, *Bull. Lond. Math. Soc.* **43** (2011), 191–199.

[84] M. Umehara and K. Yamada, *Differential Geometry of Curves and Surfaces* [Translated from the second (2015) English edition by Wayne Rossman], World Scientific Publishing Co. Pte. Ltd., Hackensack, NJ, 2017.

[85] F. W. Warner, *Foundations of Differentiable Manifolds and Lie Groups*, Scott, Foresman and Co., Glenview, Ill.-London, 1971.

[86] F. W. Wells, *Differential Analysis on Complex Manifolds*, Springer-Verlag, New York Inc, 1980.

[87] J. West, The differential geometry of the cross-cap, PhD. thesis, University of Liverpool, 1995.

[88] H. Whitney, Differentiable even functions, *Duke Math. J.* **10** (1943), 159–160.

[89] H. Whitney, The singularities of a smooth n-manifold in $(2n-1)$-space, *Ann. Math.* **45** (1944), 247–293.

[90] H. Whitney, The general type of singularity of a set of $2n - 1$ smooth functions of n variables, *Duke Math. J.* **10** (1943), 161–172.

[91] H. Whitney, On singularities of mappings of Euclidean spaces. I. Mappings of the plane into the plane, *Ann. Math.* **62** (1955), 374–410.

[92] V. M. Zakalyukin, Reconstructions of fronts and caustics depending on a parameter and versality of mappings, *J. Soviet Math.* **27** (1984), 2713–2735.

Index

Printed in the United States
by Baker & Taylor Publisher Services